AF797316

EVOLUTION REVOLUTION

Evolution is true

Darwin is wrong

This changes everything

Alan Bennett

LEXEM PUBLISHING
Published by Lexem Publishing.
Los Angeles, CA.
First published in 2014 by Lexem Publishing.
Copyright © Alan Bennett, 2014.
All rights reserved.
Unless otherwise noted, all illustrations are copyright© Alan Bennett, 2013 and 2014.
ISBN 978-0-9891891-0-1
No part of this publication may be reproduced, stored in any type of retrieval system, or transmitted by any means or in any form (including but not limited to mechanical, print photocopy, electronic, recorded) without the prior written permission of both the copyright holder and the publisher.

The photographing, scanning, uploading, downloading or distribution of this book via the Internet or any other means, without the express written permission of both the copyright holder and publisher is prohibited and illegal.

While the author has made every effort to ensure the accuracy of telephone numbers, internet addresses, mailing addresses, university positions and titles, citation and reference information neither the author nor the publisher assume and responsibility for errors or changes that occur after publication. Neither the author or publisher has control over, and does assume any responsibility for the content of third party internet sites.

For Judy, life's great gift to me.

A Snapshot of the Revolution In Ten Quotes

The New History of Evolution

1. "Darwin was less important than we thought."

Jim Secord, historian, Cambridge University

2. "Darwin ended his life more or less in a sense of confusion."

Simon Conway Morris, Cambridge University

3. Many of today's stories about Darwin and evolution are "nonsense." "Nonetheless...variations on them are made regularly throughout the media – even by those who really ought to know better."

Jim Endersby, historian, University of Sussex

The Modern Rejection of Darwinism

4. "In the past few years every element of [the Darwinian] paradigm has been attacked... Development, ecology, genetics and paleontology all provide new perspectives on how evolution operates."

Douglas Erwin, Smithsonian Natural History Museum

5. "Surprising discoveries have radically altered our views of evolution, contrary to the previously accepted neo-Darwinian view."

Edward Ziff, New York University biologist

6. Neo-Darwinism "is totally wrong...Every major tenet of it is wrong."

"It will look ridiculous in retrospect, because it is ridiculous."

Lynn Margulis, National Medal of Science Winner

7. "Evolution by natural selection...functions as an anti-theory... with exactly the opposite effect of real theories...[It]stops thinking [and] covers up embarrassing shortcomings."

Robert Laughlin, Nobel prize winning physicist

The Emerging Revolution

8. "We are witnessing a major paradigm shift in the life sciences,"

James Shapiro, geneticist, University of Chicago

9. "We are witnessing a transition from the science of the past to

the science of the future...with its hope for new discoveries, new concepts, and new wisdom."

David Pines and Robert Laughlin, physicists

10. "Can we harness the exquisite complexity of nature to transform the future? Nature tells us it can be done and provides us with models."

A group of leading scientists in *Science for the 21st Century*

Scientists say that with these new discoveries we can grasp how nature builds complex structures, repairs damaged hearts, and creates helpful adaptations. We can learn nature's design rules and mimic green plants to capture solar energy, make genetic changes that cure crippling diseases, develop more protein-rich foods, new medicines and futuristic industrial materials. "These concepts may seem fanciful," these scientists write, "but with the revolution now taking place, the possibilities seem endless. Nature has done it. Why can't we?"

The answer is, we can, using science's new theory of evolution.

Contents

INTRODUCTION

In 2008, Claudia Castillo, a young mother of two, was struggling for her life. Tuberculosis was destroying the windpipe in her throat. In the past, this was virtually untreatable. But that year scientists created the world's first "tailor-made" body part for her. In a laboratory, they reset the growth signals in a few cells from her own body which then grew into a new wind pipe, which they surgically placed in her throat. Astonishingly, the wind pipe continued to grow, connecting to the throat as though it were a natural part of Castillo's body. She made a complete recovery.[1]

In 2010, Ciaran Finn-Lynch, a 10-year-old boy, faced a similar problem. This time, doctors reset the growth signals of cells inside his throat, and grew the new windpipe there. In 2011, Dr. Paolo Macchiarini, the key developer of this procedure, revealed that his team could now grow a new windpipe inside the body "within two days or one week."[2]

Doctors have discovered how to tell the body to heal itself. This is medicine from Star Trek – except it is here today. While not as fast as the ten-second cures of Dr. "Bones" McCoy, the idea is the same: activate the body's natural ability to repair itself.

These new procedures are vastly more effective than current treatments because all the complex, difficult work is done naturally by the body. The head of Regenerative Medicine at University College London, Dr. Martin Birchall, says that "from here we can now go on and build all sorts of other things"– repairing damaged hearts, diseased livers, and perhaps even injured brains.[3]

The Revolution in Medicine

This breakthrough is part of a broad revolution in medicine built around a single idea: stimulating the body's own "self-assembly" system to replace damaged tissue. Just months ago, at Purdue University's Center for Paralysis Research, scientists decoded the signals that tell damaged nerve cells how to repair themselves, restoring movement to paraplegic animals.[4]

We are all intuitively familiar with the "self-assembly" process because we see it every day. This is how our body naturally heals a cut finger, mends a broken bone and replaces billions of different cells every day, knowing precisely which ones to make and where to place them.

Self-assembly is how a salamander can grow a new tail, and why a tree sprouts new branches at precisely the spot where a limb was cut off. In each case, chemical signals are saying "a part is missing here." The self-assembly process then kicks in, replacing the missing part. This is what happened inside Ciaran Finn-Lynch's throat, and with the damaged nerves at the Paralysis Research Center. This natural self-assembly process is what makes the field of regenerative medicine possible, and is transforming every aspect of health care. "We need to think about how to make regenerative medicine a key part of our health care," says Birchall. "The idea is to use the reactions of the body" to heal itself.[5]

Will the body's natural self-assembly and self-repair process allow us one day to halt cancers, reverse Alzheimer's disease, slow the aging process, cure crippling diseases and rebuild the scarred faces of those children in the "Operation Smile" commercials? The answer is yes, yes, yes, yes and yes.

The Revolution in Science: Discovering How Nature Makes Things

These developments are part of a larger revolution in science that is transforming every area of our lives. The key discovery is that everything in nature – from stars and planets, to algae, antelopes and the human brain – makes itself using the self-assembly process. Scientists say it is like building with Legos®, where a few simple parts can combine to form anything from a simple row boat to a complex ocean liner. Nature's parts come with forces which naturally attract each other, much like magnets. These parts literally pull themselves together – they "self-assemble"– to create more complex forms.

Everything complex in the universe emerged from this simple process, thus scientists dubbed it "Complexity Theory." Nature's smallest elements combine in increasingly complex ways to form atoms, then stars and planets. Each new level creates a form which guides development of the next level, pushing nature up the ladder of complexity.

This simple process organizes billions of atoms into colossal stars and millions of stars into vast galaxies. It turns atoms into plants, animals and people, and organizes life at every level, from sea creatures to ant colonies and entire ecosystems. It is nature's universal process for making and organizing everything.

Research into Complexity Theory and the self-assembly process has grown explosively over the last decade and is now "big science." It applies to areas far beyond medicine – agriculture, industry, information systems, technology – and is altering our basic ideas about everything from creating eco-friendly new materials, to how social networks form and cities grow. It is so important that many scientists believe it will define the scientific agenda for the 21st century.[6]

The science editors at the BBC called Complexity Theory one of the three great conceptual breakthroughs in 20th century science, along with Einstein's Relativity Theory and the Quantum Revolution.[7] The Kyoto Prize committee said these ideas "profoundly influenced a wide range of basic sciences and brought about one of the most dramatic changes in mankind's view of nature since Sir Isaac Newton."[8]

THE REVOLUTION IN EVOLUTION: DISCOVERING NATURE'S HIDDEN EVOLUTION PROCESS

Self-assembly also reveals a powerful, previously unknown evolution process. The rules for making things are also the rules for changing them. Evolution is simply change over time. Thus, self-assembly reveals how things are made, how they change and how they evolve. This makes self-assembly nature's most basic mechanism of evolution, with two important consequences.

First, it unites the entire natural world in *a single evolutionary process.* This starts with stars, galaxies and the Universe, which are all on a 14 billion-year evolutionary journey guided by the rules of self-assembly.

These same rules also guide the evolution of planets, like the Earth. Today, geologists link volcanoes, mountain ranges, earthquakes, tsunamis and a host of other events in a single "self-managing" system with these rules. They can track the Earth's history from its beginning as a giant gas ball through every stage of its evolution right up to the present.

These same processes also made the first life forms possible, which then began their own evolution journey, leading to flowers, fish and, eventually, to us.

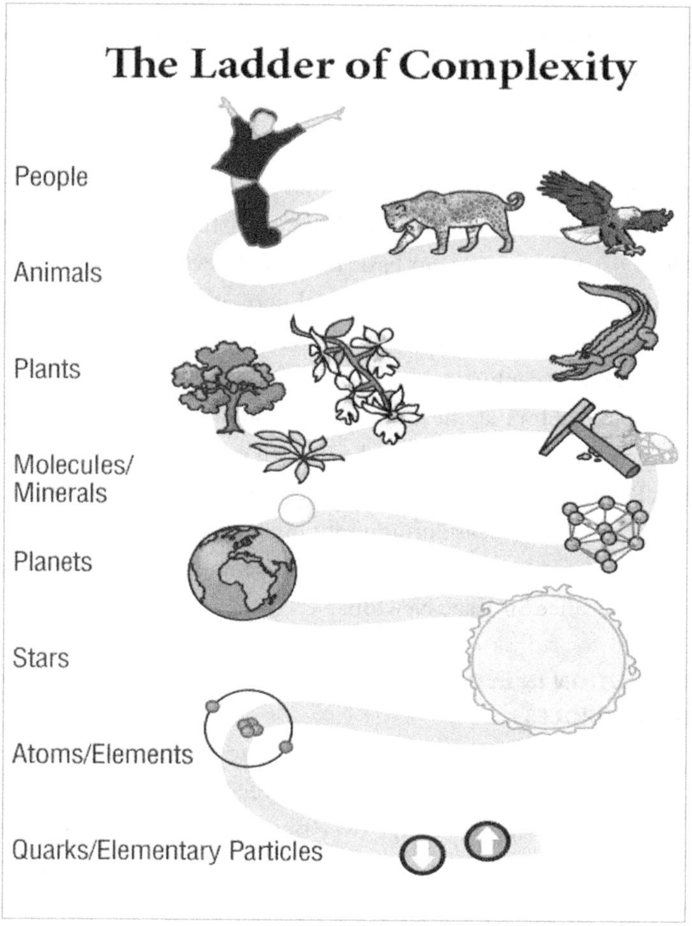

Second, self-assembly explains how the evolution process works *at every level of nature* – how, quarks can evolve to become stars, planets and eventually snowflakes, sunflowers and sea otters. It resolves many mysteries about evolution – why we can't have both arms and wings (even though both are pretty handy), why sex evolved (it's really not necessary for reproduction), and why turtles have a shell. And it explains things once thought too complex to ever unravel, like how the human brain evolved.

Self-assembly Contradicts Darwin's Version of Evolution

There is, however, a problem with all this. Nature's self-assembly process flatly contradicts Darwin's version of evolution. Distilled to its essence,

Darwinism makes six claims about how evolution works (see chapter 16). Self-assembly makes six exactly opposite claims. Each of these claims has now been tested. Darwinism fails on every claim, while self-assembly proves correct for each one.

Nothing in these discoveries means evolution is untrue – just Darwin's version. This is unsurprising to thousands of scientists. Despite popular perception, they have been waving their hands for decades with the message that Darwin's theory doesn't match the evidence in their laboratories and field research. Even those who say Darwin has some things right believe his "natural selection" is, at best, only part of the evolution story.

Those in the vanguard of this revolution are all staunch evolutionists and respected leaders in their fields – from internationally renowned researchers to the new generation of scientists today transforming our concepts of disease, childhood development, brain processes and dozens of other fields. Their revolutionary claim is: Evolution is true. Darwin is wrong. And this changes everything.

In books, professional journals and at science conferences, they say that nature never worked as Darwin claimed, and perpetuating his theory misdirects research and impedes discovery. In 2010 this argument spilled into the pages of *The New York Times* in an article entitled "Darwinism Must Die So That Evolution May Live."[9]

The divide between the Darwinists and other scientists has progressed from conflicts to bitter battles (dubbed the "Darwin wars"). Today, leaders in every field say Darwinism is not just an old theory that needs updating, but it is fundamentally wrong. This is a comprehensive rejection of Darwin from within science. The higher we go up the ladder of scientific success (bigger discoveries, more awards), the more dismissive become the assessments of Darwinism. Those at the top are scathing. "It is totally wrong. Every major tenet of it is wrong," said National Medal of Science winner and renowned geneticist Lynn Margulis.

National Medal of Science winner Carl Woese, whose discoveries revolutionized biology, said science needs to be "rescued" from Darwinism.[10] Research in many areas of genetics, biology and medicine had hit an impasse from what he called "dogmatic Darwinism." Like many others, Woese campaigned for a new, modern "21st century theory of evolution." "We can't expect to explain evolution," he said, "if we stay locked in the classical Darwinian mode of thinking."[11]

Despite popular perception, Darwin's theory has precious little supporting evidence. Frequently invited to speak to other scientists, Lynn Margulis often began her talks by asking if anyone in the audience knew of a proven example of Darwin's natural selection. After a bit of silence, someone would throw out an idea. But another scientist would quickly chime in with "Oh, that's been disproven." After a few more tries, the audience would go silent. No one had a documented example of Darwin's evolution in action. Margulis then began her talk, providing a dozen well-proven examples of evolution through another process, which these scientists could verify in their own labs the next day.

Not only is there little supporting evidence for Darwinism, but it never resolved any of evolution's important questions. In 2009, *The New York Times* reported that one of evolution's most fundamental issues, the origin of new species – supposedly settled by Darwin in *The Origin of Species* – remains "not only one of the field's most active areas of study, but also one of its most contentious."[12] After 150 years, Darwinism could not even explain that.

Scientists of every stripe are also deeply troubled by the fact Darwin's theory doesn't integrate with the other physical sciences, which all mesh together in a seamless whole. The sole exception is Darwin's theory of evolution, which is not only sits outside this well integrated system, but is in constant conflict with the other sciences.

For example, Darwinism claims evolution is unpredictable and directionless. The other sciences, however, say the entire natural world is governed by laws which predict how things change. And predictability, *by definition*, gives direction to evolution. This is how cosmologists can predict the evolutionary future of stars and galaxies, and geologists can predict the future movement of continents and oceans.

The daunting challenge for Darwinists is this: if evolutionary change in every science – physics, chemistry, cosmology, geology and all the others – is predictable, why would evolution in biology be unpredictable? If the evolution of atoms, molecules, minerals, stars, planets, continents, mountains and oceans are predictable, why would the evolution of plants and animals suddenly be unpredictable? Would nature really have two types of laws, one that makes biological evolution directionless, and another that makes the evolution of everything else in the universe predictable?

For 150 years, Darwinists have insisted the answer is "yes." But today many leading scientists call this absurd, unscientific, and are saying

"Enough!" Nobel prize winning physicist Robert Laughlin says "[Darwin's] evolution by natural selection...functions as an anti-theory...with exactly the opposite effect of real theories...[it] stops thinking [and is] called upon to cover up embarrassing shortcomings."[13]

Celebrated Harvard paleontologist Stephen Jay Gould went further, calling Darwinism a "cult," which denied established facts and offered unproven explanations for evolution.

A PARADIGM SHIFT: THE NEW EVOLUTION

What is going on here?

Why is Darwinism so at odds with the other sciences? Why did it never resolve important issues in evolution, like the origin of species? And why, after 150 years, do we have (at best) only a handful of examples of natural selection?

"Where does the faith in natural selection come from?" asks legendary Harvard geneticist Richard Lewontin. "Certainly not from inside science." Darwinism's persistence was based on ideology. "Born in ideological struggle," says Lewontin, "Darwinism has never escaped."[14]

Lynn Margulis said the Darwinists "hijacked the field of evolutionary biology" to promote an ideological agenda.[15] Other leading scientists agree. "At present," say historian Thomas Hall, "Darwinism remains... not because of its scientific value, but because... of its ideological importance."[16]

What were these ideological issues? How do they relate to evolution? And how did they manage to work their way so deeply into science?

Evolution Revolution answers those questions.

It also reveals that today, entirely out of the public eye, a sweeping change is underway. "Major changes in our understanding of evolution have gone virtually unnoticed," writes New York University biologist Edward Ziff.[17] "Surprising discoveries have radically altered our views of evolution, contrary to the previously accepted neo-Darwinian view."

"We are witnessing a major paradigm shift in the life sciences," says University of Chicago geneticist James Shapiro.[18]

Darwin's ideas pervade our concepts of biology. Replacing them is revolutionizing our approach to health and medicine. For example, in a process Darwinists said was impossible, science has discovered that our personal experiences change our genes (what we eat, drink, breathe, experience mentally and emotionally), and these changes are *directly inherited by our children* through a process called "epigenetics." This has profound implications. Scientists have linked these directly inherited changes to both good effects (the ability to resist infections and cancer), and bad effects (birth defects, ADD and other medical and mental problems – see chapter 22).

Darwin's claim that evolution is directionless is also untrue. The same laws work throughout all of nature, meaning that evolution at every level of nature is predictable. Scientists say this means evolution has direction, it is headed someplace. This includes the evolution of life.

Since scientists can predict the evolutionary future of stars and planets, does this mean they can also predict the evolutionary future of life? If so, where is it going? And what about human evolution, does it also have direction? And if it does, can we see where it is headed?

Evolution Revolution answers these questions.

Darwin's ideas also permeate how we see ourselves, painting a nihilistic picture of life, devoid of the hope and meaning which have always animated the human spirit. He created an image of Man as a violent, sex-driven primate, and society as a brutish, "survival of the fittest" struggle. If Darwin had evolution wrong how does this alter our ideas about human nature, society and ourselves? How does this change the way we raise our children and attempt to solve social problems?

And what about the ceaseless social war over evolution and education? The real issue in this bitter battle is the question "What does it mean to be human?" Both sides see this as a fight to the death. The New Evolution is reversing all the old Darwinists claims about evolution. Will these discoveries reveal an evolutionary science everyone can embrace? Will it show we have been fighting an unnecessary social war?

Evolution Revolution also resolves some of evolution's deepest mysteries. For example, all living things are made from the same material – 96% water, with a bit of calcium and minerals. Arranged one way this mineral water becomes a tulip, arranged another way it becomes a tarantula or a tiger. All three are made from the same matter. The only difference between them is how that mineral water is organized.

How can rearranging water and minerals create different creatures and traits? How can reorganizing mineral water *create new behavior*?

These questions are part of a larger mystery about life and existence itself. At every level of nature, there is no new matter, only old matter organized in new ways. The universe began with tiny particles called quarks, which possess a few simple properties, but nothing more. Everything that followed came from organizing quarks into different patterns.

How can a "pattern" turn quarks into radiant stars, rocky planets, rain forests, orchids, and elephants? How can an organization pattern turn lifeless molecules into breathing, feeling creatures – zebras, flamingos, pandas and people? Why does one pattern lead to an orchid, and another to an elephant?

In a series of remarkable discoveries, Complexity Theory can now answer these questions, explaining how something non-physical – like a pattern – can create something physical, like a tulip or tiger, and how it can create new behaviors and abilities, like sight, smell, hearing and even human consciousness.

It also reveals that nature can become more complex and adaptive only through cooperation, making evolution an inherently cooperative process. (Memo from Nature to Man: Cooperation is the key to life and building a better future. Corny? Perhaps, but overwhelmingly true.)

Complexity Theory also offers a hopeful view of the future – a message of optimism and empowerment which validates the human spirit – grounded in solid science. A pioneer in this new science, physicist David Pines says we are witnessing a major transition, "a shift from the science of the past to the science of the future, with its hope for new discoveries, new concepts, and new wisdom."[19]

In *Complex Systems: Science for the 21st Century*, 100 of America's leading scientists explained the remarkable potential: "Can we harness the exquisite complexity of nature? Nature tells us it can be done and provides us with models."[20] These scientists say that with self-assembly, we can grasp how nature builds complex structures and invents new forms, how she repairs damage, solves problems and creates helpful adaptations. We can learn nature's design rules and mimic green plants which capture solar energy and make genetic changes which cure crippling diseases.

These discoveries are transforming agriculture, industry and our handling of the environment. They help explain the life-cycles of oceans, forests and entire eco-systems, and offer insights into how to keep them healthy. They reveal how to develop more protein-rich foods and futuristic eco-friendly new industrial materials. In short, these advances give humanity unprecedented tools for changing the world and our own lives by working through natural processes, rather than against them.

"These concepts may seem fanciful," these scientists write, "but with the revolution now taking place, the possibilities seem endless. Nature has done it. Why can't we?"[21]

The answer is, we can. These advances are all part of the Evolution Revolution.

PART 1

HOW WE DECIDE WHAT IS TRUE

How do we know if something is true?

For most of human history, people looked at the world around them and organized what they saw into a coherent story we call culture. This story explained both *how* the world worked, and *why* it worked that way. Explanations for new events which fit this narrative were considered true. Those that did not were viewed as suspect or false.

In the century before Darwin, Western society developed a profoundly new way to determine what is true. Understanding this new idea, and how it changes our beliefs about nature, is central to understanding why scientists today say evolution is true, but Darwinism is wrong.

Chapter One

From Prophecy to Prediction

A NEW CONCEPT OF TRUTH

Nature and Nature's laws lay hid in night;
God said 'Let Newton be' and all was light.
Alexander Pope

How do we know what is true?

For thousands of years people explained nature by observing the world around them and developing the best stories they could imagine for what they saw.

In Western Europe a new approach to determining the truth began to emerge in the 1500s. Francis Bacon summarized this in 1620 based on two principles. First, simply observing nature was replaced with systematic testing. This gave a more accurately picture of how nature behaves. Second, testing isolated the precise cause and effect relationship between events, called "natural laws," which explained why nature acts as it does.

Systematic testing revealed how nature works, and natural laws explained why it works that way. This new way to determine truth was called "science."

Science revolutionized how people determined truth. Its assertions could be tested and proven true or false, trumping all other explanations for events – logic, opinion, ideology or religious dogma. Its explanations for nature were also universal, true at all times and in all places overriding the different beliefs of individual societies and cultures.

Careful testing improved people's knowledge about nature. But it was natural laws, with their ability to reliably predict the future based on cause and effect connections, that made science unique and gave it power. In 1660, when England's leading naturalists founded The Royal Society to advance knowledge, they invoked this principle as their guiding precept. Natural laws, with predictable outcomes, became the new criteria for determining what was true about nature.

Discovering The Hidden Laws of Nature

A quarter-century later, this revolutionary idea exploded across Europe when Isaac Newton published his famous laws that explained the movement of stars and planets.

Beginning deep in the primeval past, humans could see that nature had patterns and tried to decode the hidden rules which guided them. Understanding the sun, moon and stars were particularly important. Their cycles were intimately connected to human existence, from the life-giving fertility connected to the moon, to the life-sustaining rays of the sun and the change of seasons heralded by the stars. Their power was undeniable. People sought to understand what moves them, and predict their behavior. Our first great monuments – Stonehenge, Buena Vista in Peru, Nabta in Africa – were built to answer these questions.[1] Life depended on it.

Over the millennia, facts accumulated but no one could explain how or why the planets moved as they did. Then, in 1687, using the new tools of science, Isaac Newton stunned the world by explaining both. He discovered in nature a hidden set of laws which answered the ancient quest to precisely predict the future course of the planets.

The effect on society was electrifying. With natural laws it was possible to *know the future*. As this idea raced through Western society, it transformed people's thinking in ways barely imaginable today. The practical implications were immense. Using Newton's laws, *anyone* could predict the future movement of a planet. One could, with mathematical precision, also determine the energy needed to pull a plow, hoist stone blocks up a cathedral wall or sail from London to the West Indies. The same formulas also predicted the path of cannonballs, flying arrow, colliding billiard balls and everything else that moved.

A Revolution in How Society Decides What is True

As this revolution swept through the West, it also transformed the mental landscape. The idea that the future obeys laws – that *the future can be predicted* – altered how people thought about the natural world, a change still rippling through the mind of Man.

In particular, the predictive power of laws changed the way people thought about their personal lives. Since ancient times, people had known that much of their life was determined by powerful, unseen forces in nature, which were arbitrary and often cruel. This was "fate." One's life was governed by this fate, which could not be changed. Most cultures advised accepting your fate, rather than fighting it.

The discovery of natural laws radically reversed this belief about life. "Knowledge is power," said Bacon. Knowing the future means you can alter it. Changing the future is the most extraordinary power imaginable, so potent that it was previously attributed only to gods. Suddenly, man had this power in his grasp. In a collective gasp, people in the West realized that with this knowledge they could *change the future*. They could *alter their own personal destiny* – the idea was breathtaking.

Looking back from the 21st century, with our awareness of science's limits, it is hard for us to grasp how comprehensively this idea changed society. The predictive power of laws represented an exhilarating transfer of control from nature to Man. People could throw off the chains of fate, alter their destiny and control nature as it had previously controlled them. It took society 200 years to get over the dizzying sensation.

Natural laws also transformed society's ideas about truth. Prior to Newton, explanations for events were developed *after* the fact; you started with an event, then reasoned back to determine the cause. Sometimes the cause was easy to see, but most often events were complex and explanations were mere conjecture. It was impossible to say with certainty which factor was responsible for the event, leading to endless debates.

The predictive power of laws decisively changed this. Questions could be tested and proven true or false. Humankind now had a method to determine truth that was beyond dispute. Science became the new judge of reality. Predictability became the new arbiter of what is true.

A New Reality: The Mechanical Universe

Around the same time, French mathematician René Descartes began analyzing nature by breaking it into increasingly smaller pieces. Unlike Aristotle, who believed the whole is something greater than its parts, Descartes believed that the whole is precisely the sum of its parts, no more and no less. He famously described this idea by comparing the universe to a clock – a huge collection of mechanical cogs pushing each other around. This clock metaphor proved persuasive and by the mid-1600s the idea that nature was like a clock had been widely adopted.

Descartes understood that there were important differences between his clock metaphor and real nature, but these were swept away in the euphoria over Newton's discoveries. His precise plotting of movement turned metaphor into reality: the universe is not *like* a clock, it *is* a clock. However complex it may be, nature is at heart a collection of mechanical cogs.

The search for predictability quickly became a quest to find nature's hidden mechanical processes. This was the mindset animating Victorian scientists. Good science, they believed, would be mechanical, like Newton's cosmos. Nature could be reduced to parts and the forces pushing them around. Attempts to understand the natural world in any other way were seen as unscientific and simply irrational.

Three Consequences of a Mechanical World

This mechanical idea of nature led to three ideas, which science has now abandoned, but which were central to Victorian thought, and to our understanding of how Darwin got evolution wrong.

First is **reductionism**, the idea that a thing is merely the sum of its parts. The behavior of everything in nature – stars, planets, earthworms, eagles – could be reduced to the behavior of atoms. "Newton's success in mathematically describing planetary movements were powerful influences toward the easy belief that reductionism would provide all the necessary answers" to nature, says University of Edinburgh biologist Anthony Trewavas.[2]

The second idea is **determinism**, the belief that everything in the future is determined solely by past events. In 1814, Pierre Simon Laplace confidently claimed that if one knew the activity of all the atoms in the universe, they could predict all future events. "Present events are connected with preceding ones," he said, "a thing cannot occur without a cause which produces it." If we knew the prior events, said Laplace, the future would "appear before our eyes."[3]

Third is the idea that **cause** of changes comes from **outside forces**. Newton's laws said that nothing changed unless an external force pushed it in a new direction. In biology, this reduced life to a series of "stimulus-response" effects, like Pavlov's dogs, conditioned to salivate at the sound of a bell before being fed.

The mechanical model of nature implied that everything in nature was based on simple clockwork principles. Man was no exception. Since ancient times, most cultures described Man as one part animal instincts and one part a uniquely human mind. In the 1800s, a desire to make Man part of the mechanical universe led some to define psychology as a sub-discipline of biology, which in turn could be reduced to the activity of atoms. Human behavior was simply the sum of the behavior of our atoms. Man was no longer unique, merely more complex.

This idea went hand-in-hand with the belief that most people are irrational, and with many superstitious beliefs as easily seen in the jumble of conflicting religious ideas they held. "If we go back to the beginning," wrote Baron D'Holbach in 1770, "we shall find that ignorance and fear created the gods...that weakness worships them; that credulity preserves them and that custom, respect and tyranny support them."[4]

A MECHANICAL VIEW OF NATURE

Initially the mechanical idea of nature proved helpful.

Scientific progress in the 1800s was impressive. James Clerk Maxwell unlocked the secrets of electricity. Antoine Lavoisier's discoveries led to a 100-year-long revolution in chemistry. Biologists Jakob Schleiden and Theodor Schwann discovered cells, the foundation of all modern biology. Louis Pasteur discovered germs, created vaccines against fatal diseases and developed the pasteurization process that protected food from bacteria.

The mechanical model proved so successful that Victorian scientists were lulled into thinking they knew more than they did. Many believed nature held few remaining secrets, and these would soon be explained.

In 1900, the world's leading physicist, William Thompson (Lord Kelvin), began advising young people not to go into physics because the field had little left to discover. Well known physicist Henri Poincaré declared that only a few loose ends remained in physics, which he outlined in a short list, predicting each would soon be resolved.

OVERTURNING THE MECHANICAL VIEW OF NATURE

But physics was on the verge of a convulsive upheaval. The tidy mechanical universe which had reigned supreme for 200 years would not survive the first few years of the new century.

Poincaré's "loose ends" involved some puzzling facts that did not fit the mechanical model of nature, such as the properties of light, and the motion of particles in a glass of water. In 1905, a young Albert Einstein, trying to win a degree from Zürich University professors skeptical of his scientific abilities, set out to solve three of these "loose ends." He succeeded.

The new physics which emerged from these and other discoveries, called the Quantum Revolution, revealed how nature works at the smallest level

– atoms, electrons and particles. It paved the way for the modern nuclear energy industry, the electronics, computer, information and Internet revolutions, and to life-saving medical tools, such as MRIs and CT scans, now used to diagnose disease, saving countless lives.

It also plunged science into previously unknown, utterly strange dimensions of nature, and generated a broad range of mind-bending ideas. In this new world, a particle could be in two places at the same time. Matter, long thought to be made of physical "stuff," was found to be more complex. Forces, "fields," and other newly discovered parts of nature were "material," but had no physical dimensions.

Nor was "cause" a simple mechanical process, one part pushing another, as in Newton's universe. Some changes occur without any external influence. And while future events are predictable, they are not mechanically determined like a clock, but instead occur within a range of possibilities that scientists called "probabilities."

This spelled the end of the mechanical concept of nature.

More than that, it transformed our most basic concepts of reality. As one scientist said, nature is not just stranger than we imagine, it is stranger than we *can* imagine. Science needed a new vocabulary – an entirely new mindset – to understand and explain the natural world.

One after another, each area of science tumbled into the same new reality discovered by physics. A broad revolution in thinking cascaded through the sciences, providing a radical new picture of how nature works. Each science underwent a dramatic transformation, creating new ideas, languages and entirely new theories to explain the new reality they had found.

DISCOVERING THE UNIVERSE IS ALIVE – AND EVOLVING

Cosmology's revolution was particularly dramatic, leading to a revision of ideas about the origin of everything that exists.

Astronomers had long believed the universe was largely frozen in place, the stars circling eternally. Questions of how this vast stellar world came to exist were met with a blank stare and shrug of the shoulders.

In 1924, using the powerful Mount Wilson telescope, Edwin Hubble discovered the universe holds millions of other galaxies, all billions of miles away from our Milky Way. The universe was staggeringly larger than anyone ever imagined.

EDWIN HUBBLE

In 1928, Hubble startled the world again by discovering these distant galaxies are in motion, racing away from the Earth at almost unimaginable speeds – millions of miles per hour. Working back to the start of this expansion, Belgian physicist Georges Lemaitre discovered that the universe had a beginning. Fourteen billion years ago, a microscopic, but unimaginable dense ball of matter experienced a massive explosion (called the "Big Bang), and began to expand outward, creating the galaxies, stars, planets and the cosmic world we see today.

This meant that the universe is alive, part of a 14 billion year evolution process that is still unfolding. Remarkably, scientists found that with this knowledge, they could predict the *evolutionary future* of stars, galaxies and the Universe itself.

These discoveries forced a wholesale change in cosmic thought and theory. The static model of the universe was dead. In its place was a dynamic, living universe, with a dramatic history of change and evolution.

Each of the sciences had similar experiences, and each discovered an evolution process at work. The origin of the Earth, for example, previously a mystery, was revealed to be part of the same ongoing evolutionary process seen in the heavens. Geologists linked volcanoes, mountain ranges, earthquakes, tsunamis and a host of other events under a single set of laws and found they could track the Earth's history from its beginning as a giant gas ball 5 billion years ago, through every evolutionary stage right up to the present. They could even predict *future evolutionary changes* of the Earth.

The End of Mechanical Man: Mind as More Than Biology

Even the social sciences, long mired in 19th-century ideas about human behavior, were caught up in this transformation. In the process old ideas about Man, society, human nature and human history were replaced.

The old belief that human behavior is nothing but mechanical responses, called "Behaviorism," treated psychology as a branch of biology. It eliminated all reference to the mind from psychology, reducing thoughts to nothing more than brain chemistry. (Freud's seething unconscious was considered an unfit topic for scientific psychology.)

Behaviorism explained some simple activities but proved woefully inadequate at explaining any of the important things about Man. The vast gap between this mechanical idea of the mind and the real world was exposed in 1959, when the leading Behaviorist of the day, BF Skinner, in *Verbal Behavior*, purported to explain that children learn language through the same processes Pavlov used to train his dogs.[5]

The book was an embarrassment and the assessment scathing. In a short article famed MIT linguist Noam Chomsky demolished the Behaviorist model and held the theory up to ridicule. Skinner's claims were "empty" and "pointless," wrote Chomsky. Behaviorism couldn't explain complex mental activity. Worse, it wasn't even real science, said Chomsky, offering only the "illusion of a rigorous scientific theory."

Within a decade, Behaviorism collapsed as an explanation for human behavior.

Casting off Behaviorism and its 19th-century ideas revolutionized psychology, refocusing the field on the unique mental abilities of humans.

This refocus was part of a broad movement in the social sciences which became known as the "cognitive revolution:" the idea that human mental processes are *not reducible to biology* and should be studied on their own. By the 1980s, leading cognitive psychologists said that "from the early 20th century...psychology was dominated by Behaviorism [but these] assumptions proved quite false [and] cognitive psychology has dominated... since the 1960s."[6]

Predictability Determines Truth in Science

As we go forward, keep two ideas in mind. First, in science, truth is determined by laws that make predictions about the future -- forecasts which can be tested and proven true or false. (I know that a "future prediction" is redundant, but stick with me on this, there's a reason for it.)

Second, why was evolution the only science *not* to experience a revolution in ideas and discoveries during the 20th century? Why was it the only science still based on 19th century concepts?

Chapter Two

From Prediction to Discovery

HOW SCIENCE REDEFINES THE TRUTH

The 20th-century revolution in science was not just an expansion of our knowledge. It overturned our most basic concepts of nature. This startling transformation raises two questions crucial for understanding evolution and, more broadly, the role of science in society.

First, if scientific facts reflect Truth, in the big sense, shouldn't they remain constant – otherwise they're not really true?

Second, when science presents us with a new Truth, how can we tell if it is correct? How can we tell if the new Truth is, well, true?

Science is an Approximation of Reality

The problem is that while science gives us an increasingly accurate picture of nature, this picture constantly changes as our knowledge grows. Thus, science never produces "truth" in any absolute sense, but only an ever shifting *approximation of reality*.

Part of the problem is that we don't really know how much we don't know (like physicists before Einstein). The other problem is that often what scientists believe is absolute truth turns out to simply wrong. In his PBS series on the history of science, "The Ascent of Man," Jacob Bronowski showed that at any point in time, 80% of what science calls "fact" is so incomplete as to mislead us, or is just plain wrong.

For example, in 1900, science confidently told us the universe is like a clock. In 1905, Einstein revealed it is nothing like a clock. In the 1920s, it was discovered the universe is not a static set of circling stars, as long thought, but was filled with billions of galaxies we never knew existed, all flying away from each other at millions of miles an hour.

In 2006, cosmic reality changed again when the universe was discovered to contain a lot more "stuff" than previously thought. Called "dark matter" and "dark energy" (because no one has a clue what it's made of), this new stuff makes up a whopping 95% of the universe. What we once thought of as "everything" is a mere 5% of what exists. Physicist Edward Witten says this "greatly changed how we think about the laws of nature."[1] Yes it does.

It also changes how we think about scientific "truth."

New Evidence Forces Science to Change Its Theories

If science gives us only *an approximation of reality*, how does science "change its mind" about what is true when it learns more about nature? And crucial, what does this mean for our understanding of evolution, and human nature?

In his famous 1962 book on scientific change, Thomas Kuhn gave the idealized view.[2] Each science starts with a theory which integrates current knowledge into a coherent story. Over time, new discoveries emerge which can't be explained by the old story. As the discrepancy between fact and theory grows, scientists search for a new explanation. Once found, the old story is relegated to the history books and the new theory becomes scientific "truth."

Kuhn called this abrupt change in thinking a "paradigm shift," and said it is central to scientific progress. New discoveries lead to a new theory, which in turn leads to a wave of new insight and understanding about nature, and points the way to yet more new discoveries.

The real scientific world is more complicated.

Although we think of science as a fact-driven enterprise – new discoveries enthusiastically embraced, old ideas easily jettisoned – history tells a different story. As the great Max Planck said, "an important scientific innovation rarely makes its way by gradually winning over and converting its opponents. It rarely happens that Saul becomes Paul. What does happen is that the opponents gradually die out, and that the new generation is familiarized with the ideas from the beginning."[3]

Unsurprisingly, scientists are human first and subject to the same emotions as the rest of us. Those who have built their career on a flawed theory are, understandably, reluctant to see their life's work relegated to a footnote in the history of science under "Discarded Ideas." Who would not choke at saying "everything I've done for the last 40 years is worthless."

Then there are the economics of the situation. The jobs of the old guard depend on their defending a set of old ideas. And with the old guard in control of who gets hired and fired, can a Young Turk oppose them? As Upton Sinclair noted, "it's difficult to get a man to understand something when his salary depends on him not understanding it."

Another problem is that it is hard to build a scientific career by saying "I don't know." Scientists succeed by explaining how the world works. Like the rest of us, when scientists are confronted with baffling evidence they invent the best explanation they can imagine, often using ideas in general circulation, and call it a "scientific" theory. Over the last century, explanations of how the brain works have changed in lockstep with advances in technology. The brain was compared to an electrical grid (1920), a telephone switchboard (1940s), a computer processor (1970s) and a software program (1990s). Today, neuroscientists say that none of these examples bears much resemblance at all to how the brain actually works.

And there is the problem of ideology. As any scientist can tell you, science is filled with professional rivalries and ideological beliefs. A surprising number of scientists are willing to twist evidence or even invent facts to further their personal and political agendas.

Emotions, economics and ideology – all are powerful obstacles to getting at the truth. Over the last century, the pattern of change in each of the sciences has looked very much the same. New discoveries led to bitter disputes that regularly degenerated into pitched battles as the old guard actively undermined conflicting research and attempted to suppress new ideas. There were real causalities – crushed careers, brilliant scientists driven into obscurity, and important discoveries delayed for decades or generations.

Just how difficult it can be to dislodge a failed theory is seen in the revolution which transformed our ideas about the continents, earthquakes and volcanoes. How geology abandoned a set of silly 19th century ideas (like "land bridges") and became a modern science is similar to what is happening in evolution today. And it's helpful to see a snapshot of how it works.

Is the Earth a Jigsaw Puzzle?

> "America is getting further away from Europe. This is not a political statement but a geological fact. Just as the Pacific is getting smaller, the Red Sea bigger, the Himalayas are still going up and one day the Horn of Africa will be a large

> island. This is the theory of Plate Tectonics, a revolutionary idea in 20th century geology that saw the continents of Earth to be dancing to the music of deep time. A dance of incredible slowness, yet powerful enough to throw up the mountains and pour away the oceans. Plate Tectonics was a genuine scientific revolution. It made geologists profoundly re-think what the Earth was, how it worked and how it related to all the things in it."
>
> "Plate Tectonics," *In Our Time*, BBC [4]

Looking at a map of the Earth, people are often struck by how South America and Africa seem to fit together like pieces of a jigsaw puzzle. Were they once united deep in the ancient past? This strange possibility was first proposed by 16th-century Flemish mapmaker Abraham Ortelius. South America was torn away from Africa "by earthquakes and floods," wrote Ortelius. "The vestiges of the rupture reveal themselves, if someone... considers carefully the coasts."

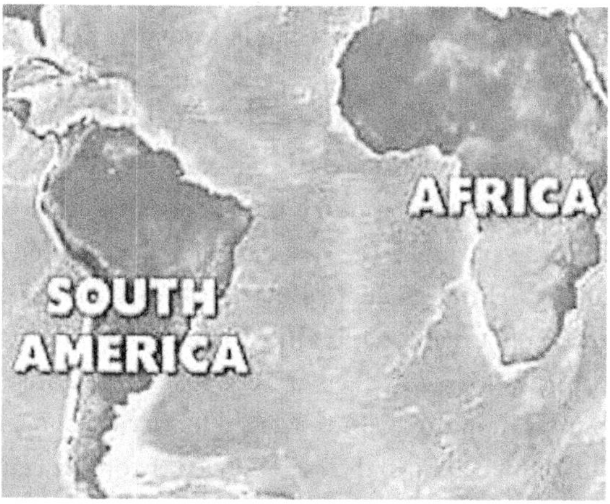

The idea that the continents were once joined captured the imagination of many, including Benjamin Franklin, and a growing body of evidence supported the idea. In the 19th century, geologists found rock layers of the same age and type in matching parts of South America, Africa and India. The same plant and animal fossils were found in identical rock layers spread across four continents. Could these formations and fossils be so similar if the continents had not at some point been joined?

And then there were some truly bizarre discoveries. In mountain ranges all across the globe, for example, geologists found the fossils of plants and animals which live only in the deep oceans. What were these creatures doing sitting on top of the Alps? Geology was chock-full of such strange facts and puzzling questions. How could one make sense of it all?

Inventing the Best Story Possible

Victorian geologists explained the similarities between continents by inventing a mechanical idea called "land bridges" – large landmasses which would appear, join the continents, then disappear. How and why they moved no one could explain. "[They] went up and down like a series of random elevators," says geologist Richard Corfield, "then disappeared as completely as the lost continent of Atlantis."[4]

The strange discovery of deep ocean fossils on mountain tops was explained by the equally strange idea of a "wrinkled Earth."[5] It was claimed that over time, the Earth had wrinkled up like a drying prune, forcing some ocean floors to rise.

There was little evidence for these explanations and field geologists knew they made no sense. Nonetheless, they dominated academic thinking and became embedded in geology texts as established fact.

As younger geologists asked how, *precisely*, do these processes work, explanations like "land bridge" became untenable. But rather than abandon these ideas, geology's old guard tried to bolster them with ever stranger inventions. Lynne Frostick, Director of the Hull Environment Research Institute, remembers being taught "about things called...inclines which dropped down and then came back up again for no explained reason."[4] It was very much like a jigsaw which didn't fit together and which had no underlying principles," says Frostick. "It was very confusing."

Today, these old ideas – land bridges, inclines and all the rest – are gone from geology texts. What happened? How were geology's old guard and their 19th-century ideas replaced?

How this occurred is a microcosm of how science changes its theories. It explains how preposterous ideas like "land bridges" appear as "scientific" fact in textbooks, why leading scientists defend these ideas, and why the popular press promotes them. And it reveals why it is so difficult to get rid of old theories, no matter how little sense they make or how much evidence piles up showing they're wrong.

Wegener Discovers the Earth is Alive – and Evolving

In 1922, German geologist Alfred Wegener challenged the accepted theory in geology with an idea he called "Continental Drift." The continents, he said, are gradually sliding across the earth's surface, creating and destroying oceans as they move and forging new mountain ranges as they collide.

When fully fleshed out, Wegener's idea accomplished three important things: It replaced implausible mechanisms like land bridges with a single process that explained many previously confusing facts, like ocean fossils on mountain tops. It also integrated geology's different disciplines under a single explanation. And it integrated geology with discoveries in other sciences, such as climatology and biology, by showing how the movement of continents changed ocean currents which dramatically altered weather patterns and the distribution of plant life around the world.

Third, Continental Drift made predictions – the essence of real science. This theory could be put to the test and proven true or false.

ALFRED WEGENER

Continental Drift was a brilliant idea, but it contradicted traditional theory and the geological establishment would have none of it. Wegener met a wave of rejection. One geologist bellowed, "if continental drift is happening then everything we've done for the last 70 years is wrong."[4]

Wegener's presentations at Harvard and elsewhere were howled down. In 1928, a geological conference in the US declared moving continents are "impossible." After that, says geologist Joe Cann, "you couldn't teach continental drift at all in America."[4]

Wegener spent his life defending Continental Drift against a steady stream of ridicule and attack. He was particularly frustrated by geologists refusal to look at the facts as a whole. He realized the great mysteries of geology would be resolved only when all the facts were integrated in a single explanation:

> "Scientists still do not appear to understand that the truth of the matter can *only* be reached by combining all this evidence. It is only by combining the information furnished by all the earth sciences that we can hope to determine 'truth.'"[11]

It would take another 40 years for this to happen, and then only after a mass of new evidence overwhelmed the old ideas, and the old guard had, mercifully, retired.

The transformation began in the 1940s when mountain ranges were discovered running down the middle of all the oceans. These ranges connected with each other around the globe, like seams on a baseball. Each mountain range had a trench running down the middle from which molten lava spewed every day, pushing the ocean floor out in opposite directions. (It was to the deepest of these trenches that *Titanic* film director James Cameron descended in 2012.)

The idea that the ocean floors were split and pushing out became known as "sea floor spreading." The earth isn't expanding, so the mystery was "Where does the expanding sea floor go?" In 1962, Princeton geologist Harry Hess solved this mystery, discovering that when the expanding ocean floor reaches a continent it is forced underneath the land. The sea floor is like a conveyor belt, moving across the Earth's surface then folding underneath the continents where the rock and sediment is recycled.

The following year a young Cambridge University geologist, Fred Vine, and his supervisor Drummond Matthews, pointed out that the Earth's magnetic field has regularly flipped directions causing magnetic matter to change the direction in which it points. If the ocean floors are spreading, these changes should have created stripes in the sea beds, much like those on a candy cane.

Over the next few decades, tests were run and the magnetic stripes were found, just as predicted. The work of Hess, Vine, and others led to a new map of the earth which explained and integrated previously conflicting data.

Facts Trump Theory

In science, facts eventually triumph over theory, but the process can be long and painful. Geology's old guard, heavily invested in their "land bridges," resisted the new evidence and attacked Continental Drift for decades. In 1965, Joe Cann, a newly minted Cambridge University geologist, remembers discussing sea floor spreading during a job interview at Columbia University. "I was told afterwards 'I think you blew it there, Joe,'" says

Cann. "The next day the head of the Geology Department called me in and said 'Cann, very interesting, very interesting, but I can show you...that if the continents were moving the sediment would all be absolutely deformed, and it's not deformed, at all. So the continents cannot possibly be moving."[4] Cann eventually secured a position, but not at Columbia.

Lynne Frostick says the new ideas were resisted "because there was an enormous amount invested in the previous ideas. I don't think anyone was ready to give up on those ideas. It was only people who really thought beyond the accepted that would take this theory on."[4]

It was not until the late 1970s, half a century after Wegener first presented the compelling evidence for Continental Drift, that it became the accepted theory of geology.

'Continental Drift' Transformed Geology

Wegener's theory, now called Plate Tectonics, revolutionized geology by doing three important things. First, it explained the facts – all of them. It made sense of what is seen in the real world: the shape of continents, changes in the ocean floor, the appearance of sea fossils on mountain tops.

It explains that the continents were once joined in a single land mass called "Pangaea," but then began to slide across the earth, creating earthquakes and new oceans as they move, and new mountain ranges as they grind into each other and push some of the sea floor into the air.

Map of Pangaea

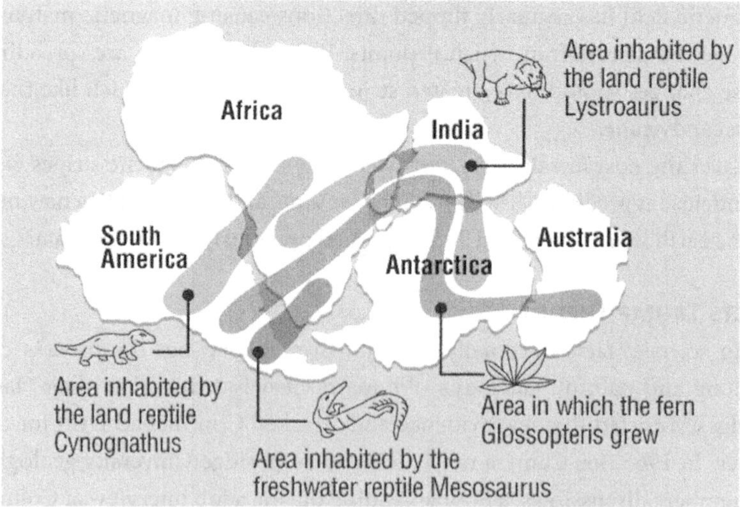

Once all the continents were joined together.

Second, it unified the facts into a single coherent system. "It changed the way we think because now we can fit everything together," says Lynne Frostick. "It was a eureka moment because before that we had lots of [facts] but nothing joined together. Before that geology was in compartments... and you learned it in compartments. What this did was put the whole thing together."[4]

Beginning in the 1980s, geology's previously isolated parts were integrated into a single system – the Earth Sciences. Seemingly unrelated events – Japanese earthquakes, the rise of the Himalayas, emergence of the Hawaiian Islands, the explosion of Mount Vesuvius that buried Pompeii – now fit together. "It's a huge idea because it joins everything up," says Frostick. "Before that we didn't have anything joined up. It makes people look at things holistically. It is the prelude to the Earth Sciences as this one big system that works together and the driving force in this is Plate Tectonics."[4]

Plate Tectonics also integrated geology with other sciences – oceanography, climatology, paleontology, biology – showing how the movement of continents powerfully alters climate patterns which then effect plants, animals and entire ecosystems. "The essence of Plate Tectonics is that it demonstrates the interconnectivity of things," says geologist Richard Corfield, "the rock cycle and the climate cycle and ultimately the cycle of life, which seems bound up with Plate Tectonics in a fundamental way."[4]

Third, by making geology predictable, and thus truly scientific, it led to a rush of new discoveries. The *future landscape* of the Earth is now known – where continents will move, new mountain ranges will form and oceans will appear and disappear. "These changes are all a result of the predictable processes of Plate Tectonics," says geologist Ann Egger.[7] "Before Wegener, few had conceived of such a world. His continental drift theory was the first step in the development of Plate Tectonic Theory, the foundation upon which modern geology is built."

The Three Traits That Distinguish Good Theories from Bad

In short, Wegener's theory explained, integrated and pointed the way to new discoveries. And this answers our second question: how can we tell good scientific ideas from bad? With hindsight, it's easy to see which ideas are right. But in the midst of an emotion-laden battle, with careers on the line and complex scientific facts tossed in all directions, how do we distinguish good theory from bad? And can ordinary people tell the difference, or only experts with specialized knowledge?

The answer is we can. All good theories do the same three things as Continental Drift.

They **explain**. They make sense of all the evidence (what is actually seen). Importantly, facts are never twisted to fit the theory. Theory changes to match the evidence – always.

They **integrate**. They unite the facts in a single coherent system. Seemingly unrelated events are connected, old mysteries clarified, and disputes resolved. When future discoveries arise, a good theory easily integrates them into the wider body of knowledge.

They **predict**. Good theories make predictions that can be tested and confirmed. This is where science becomes most exciting and helpful to society: these predictions point the way to new discoveries and new benefits. The germ theory of disease made predictions about infections that transformed medicine, saving millions of lives. The Periodic Table, when introduced in the 1800s, was powerful in part because it revealed the existence of undiscovered elements.

These three attributes – explain, integrate, predict – are how we tell good theory from bad. They transform information into understanding, facts into science. And you don't have to be a rocket scientist to tell if a theory is doing these three things. Understanding nature can require highly specialized knowledge. But common sense tells us if a theory helps clarify natural events, integrate evidence, and can be tested, or if it creates confusion, conflict and can not be tested.

Distinguishing good science from bad has important consequences. Scientific ideas determine many of our beliefs about life – how to stay healthy, raise our children, and solve problems. Scientific ideas determine which medical research gets funded, guide the development of new products, and frame our efforts to care for the Earth and give our children a better future.

Scientific theory plays a particularly important role in the social sciences and our concept of human nature and evolution. These ideas bear directly on how we view ourselves, manage our societies and attempt to solve pressing social problems. If our theories about human nature are wrong, then our ideas about self, society and human history are wrong, with profound personal and social consequences.

Bad Theory: Missed Chances to Solve Problems

Bad scientific theory leads to confusion and splits science into warring factions, making collaboration impossible and discovery more difficult. Worst of all, it sends scientific research in the wrong direction. Sometimes this is harmless. No one died because geology took 50 years to embrace Continental Drift. But bad theories can have tragic consequences, especially in biology, medicine and the social sciences.

First are the lost opportunities which can save lives or lift people out of poverty. Take the discovery of germs. The idea that invisible microorganisms (germs) cause infection and death emerged from the work of British physician Joseph Lister in the 19th century. Lister revolutionized surgery by using carbolic acid as an antiseptic to reduce infections (today we have "Listerine"). The French biologist Louis Pasteur showed these infections are due to microorganisms, and the German physician Robert Koch isolated the specific microorganisms that cause tuberculosis and cholera. Germs – bacteria and viruses – had been discovered.

Despite these impressive discoveries, the medical profession held on to their old ideas, unwilling to believe such small creatures could exist. Doctors continued to move from patient to patient without washing their hands. Surgeons continued to operate in their street clothes, spreading infection with every move.

It took a quarter of a century before hand washing became common practice in medicine. When the new "germ theory of disease" was finally accepted, it led to dramatic improvements in medicine and food production and the widespread immunization of children against disease. Sterilization was required in hospitals, pasteurization (boiling milk to kill bacteria) became common, and vaccination meant diseases like smallpox, chickenpox, mumps and measles, which previously killed children by the millions, now largely disappeared. Deaths from infectious disease plunged dramatically in the 20th century (from 30% of all deaths in 1900 to 5% in 2000). Had the old theories of disease been abandoned earlier, millions of lives would have been saved.

Bad Theory: Creating New Problems

The second effect is even more troubling. Bad theories always create *problems that would not otherwise exist*, often with tragic consequences. Take the death sentence imposed on many American women in the 1980s by assumptions about aging that proved wrong.

In 1900, few women lived past the age of 60. Improved nutrition and healthcare led to a dramatic increase in lifespan. By 1970, many women were living into their 70s and 80s, well past menopause, which began to emerge as a major medical problem.

When menopause begins, the body stops producing certain hormones, especially estrogen. Doctors reasoned that the problems caused by menopause could be eliminated by replacing the lost hormones. Estrogen became widely prescribed for post-menopausal women and the treatment soon appeared in medical textbooks as established scientific fact.

During this same period, heart attacks and stokes among women began to skyrocket. In the 1990s, doctors concerned about the increasing death rate from these events looked into the research underpinning estrogen therapy. In a surprise, they found very little. The estrogen treatment program was based largely on assumptions.

This led to a series of large studies and the results were a shock. "Hormone replacement therapy, already linked to increased risk of breast cancer, heart disease and stroke, nearly doubles a woman's risk of dying from lung cancer researchers report."[8] Later studies confirmed that hormone treatment "significantly increased the risk of a coronary event."[9] In 2009, Dr. Apar Ganti at the University of Nebraska Medical Center, said "these results, along with the findings showing no protection against coronary heart disease, seriously question whether hormone-replacement therapy has any role in medicine today."[10]

Over the last few years, estrogen has been discontinued for many women and death rates have fallen in lockstep. Today, the National Cancer Institute says they "advise women to use MHT [hormone replacement therapy] for the shortest time and at the lowest dose possible to control menopausal symptoms."[11]

Hormone replacement to treat menopausal problems, largely and untested theory, resulted in the unnecessary disability or death of perhaps one million women. The message is clear: bad science theory can kill people.

Why Didn't Evolution Have a Revolution?

What does all this mean for evolution? As we go forward there are two key ideas to remember. First, in science, facts always trump theory. Science is an ever changing approximation of nature. When facts and theory conflict, the *theory must always change to fit the facts*, never the other way around.

Second, a theory becomes scientific when it does those three key things: explain and make sense of facts, integrate previously conflicting facts which resolves old mysteries, and make predictions about future discoveries which opens the door to new insight, understanding and scientific progress.

Over the last century, every science underwent a revolution with these criteria (explain, integrate and predict) at their center. Except evolution. Alone among the sciences, evolution remained dominated by a19th-century idea: Darwinism. Supporters said this was because Darwin's natural selection is "the single best idea" in the annals of science. "If I were to give an award for the single best idea anyone ever had," wrote Daniel Dennett, "I'd give it to Darwin, ahead of even Newton or Einstein and everyone else."[12]

Detractors think Darwinism is just one more 19th-century theory full of explanations as silly as "land bridges." National Medal of Science winner Lynn Margulis said Darwinism "will look ridiculous in retrospect, because it is ridiculous."[13]

Is Darwin's theory the "single best idea anyone ever had," or just another "ridiculous" 19th century theory? The three traits of a good scientific theory – explain, integrate and predict – will guide our search for the answer.

PART 2

EVOLUTION IS TRUE, DARWIN IS WRONG

In the revolution that swept through the sciences in the 20th century, every field abandoned their Victorian era beliefs – except evolution. Among the sciences, only evolution is today still embroiled in battles over 19th century ideas.

These conflicts are bitter and never ending. In the other sciences new evidence resolves old questions and disputes. But not in evolution, where the Darwinists are perpetually at war with the new discoveries of genetics, biology and science in general.

To understand the New Evolution we must first clear away the dead underbrush of Darwinism, show why it is wrong but also why it persisted. We must understand the 19th century ideas that fuel today's fires in evolution, otherwise the old battles will simply roll on and on.

This understanding begins with a new generation of historians who have dramatically revised the history of evolution and Darwin's role in it.

First, they tell us what did *not* happen. Darwin did not discover evolution, which was a widely-accepted idea by the time he published.

Nor, as often claimed today, did his theory ignite a firestorm of conflict between science and religion. On the contrary, many Victorians enthusiastically embraced the idea of evolution as a scientific confirmation of their belief in God as a "Divine engineer" who set the laws of nature in motion.

More remarkable is what *did* happen. Surprisingly, the Victorian public saw Darwin's ideas as old and unoriginal, and he was largely ignored.

Scientists more loudly rejected Darwin's theory. By the 1850s, scientific ideas of evolution had become sophisticated (some surprisingly accurate by modern standards). When *The Origin of Species* was published in 1859 it encountered withering scientific criticism, not over evolution, but because Darwin's theory contradicted well-established scientific facts. From anatomy to zoology, nature simply doesn't work the way Darwin claimed. A leading biologist summed up the prevailing scientific view when he wrote of Darwin's ideas, "all that is true is old, all that is new is wrong."

Darwin's evolution work was widely seen as confused, full of unsupported speculation, with little hard evidence for the claims he made. In short, as *The Times* of London wrote in an 1871 review, Darwin's ideas were viewed as scientifically "reckless" and "irresponsible."

Chapter Three

DARWIN DIDN'T DISCOVER EVOLUTION

I have a confession. For years I have been Director of the Darwin Correspondence Project. But my name and reputation are built on a book showing Darwin was less important than we thought.[1]

Jim Secord, Cambridge University, 2009

Historian Jim Endersby says the story told today about Darwin and evolution goes roughly like this: "On November 24, 1859, Charles Darwin's *On the Origin of Species* appeared, and the world changed forever. An age of faith was plunged into profound religious doubt. Believers of every kind rose to pronounce anathema on Darwin's godless tract, sparking a fresh battle in the long-running war between science and religion. But while the reactionaries raged, the scientific community soon came to accept Darwin's theory of natural selection."[2]

The problem, says Endersby, is that "apart from the publication date, everything in the previous paragraph is nonsense." That story is a myth. "Nonetheless, statements like [that] summarize the popular view of Darwin," he says. "Variations on them are made regularly throughout the media – even by those who really ought to know better."[3]

The real story about Darwin and evolution is quite different, in fact the *precise opposite* of the myth that circulates today. Long before Darwin, Victorian society was awash with ideas about evolution. The big debate was not if evolution occurs, but how it works. Darwin added little to this discussion. His basic idea – natural selection – was 100 years old and already considered inadequate for explaining evolution. In a society captivated by the idea of evolution, Darwin's ideas were seen as outdated, unoriginal and, surprisingly, unscientific.

So what *did* happen?

And how did the current myths about Darwin and evolution arise?

Evolution: An Idea Older than Aristotle

The idea of evolution is almost as old as humanity.

The knowledge that traits are inherited goes back to ancient nomadic tribes, who were experts at breeding herds to produce stronger, sturdier animals. The Babylonians raised this practice to a high art with thoroughbred horses, tracking their breeding for generations on clay tablets.

In the Near East, 10,000 years ago, farmers selectively bred plants to produce larger grains. Chinese farmers did much the same to improve the quality of their rice. By 6,000 years ago, selective breeding had been perfected to the point where the Babylonians knew how to cross-pollinate date trees to produce sweeter or more exotic tasting fruit.

The first theories of evolution predate Aristotle. Over 2,500 years ago, Greek scholar Anaximander described an evolutionary process in which animals transform from one species to another, and he believed (correctly) that humans had evolved from a fish-like creature.[4]

Greek naturalist Empedocles later outlined a system in which animals naturally evolve by rearranging their body parts, and Theophrastus ("the father of botany") organized plants in a branching system that began with simple forms and showed increasingly complex plants emerging in a series of stages.

The founders of Christianity, 2,000 years ago, also believed a natural evolution process must exist. Church leaders like Augustine and Gregory of Nazianzus said some animals must have evolved from others, reasoning that Noah's Ark was not remotely large enough to hold all the different animals then known to exist.

The earliest Christians were also emphatic that the Genesis creation story was not a literal description of events, but a metaphor of what happened through natural processes. St. Paul said of Genesis, "all these things that happened to them were symbolic."

Augustine, the most influential thinker of Christianity's first 1,000 years, in *The Literal Meaning of Genesis*, declared "no Christian will dare say" Genesis should be read literally.[5] On the contrary, he said, creation occurred through natural processes, writing that "plant, fowl and animal life are not perfect, but created in a state of potentiality [which unfolds] over time" – as clear a statement of evolution as any modern scientist might make.[5]

There was an important reason for early Christianity's uncompromising position that Genesis should be read as a metaphor. They believed the natural world was God's creation, and no human interpretation of the Bible could contradict what God had created. Thus, said the early Church Fathers, scripture must *always* be interpreted in ways that are consistent with what is found in nature.

Medieval Christians made the same point. The most influential scholar of the era, Thomas Aquinas, defined Church teaching as "natural theology," the idea that reason and nature should guide the interpretation of religious doctrine. He argued that human interpretations of scripture should not determine how we view nature, rather nature, as God's creation, should determine how scripture is understood by humans.

This included Genesis.[6] Aquinas said that creation had emerged through natural processes, and held a remarkably modern idea of evolution (similar to Einstein's). Aquinas believed God supplied the raw material and rules of assembly, and after that nature did the rest:

> "Nature has the means by which it acts as it does... It is as if a shipbuilder were able to give to timbers [the ability] by which they would *move themselves* to form a ship."[7]

Aquinas' thinking became the standard of Church teaching from the middle ages through the modern era. (Aquinas' metaphor for evolution – timbers that "move themselves" to form a ship – also mirrors the modern concept of evolution through the process of "self-assembly." See chapter 14.)

The Western Idea of Progress

The concept of a "natural evolution" process really began to take hold in the 1700s, as the Christian belief that religious dogma should be consistent with nature became intertwined with two new ideas: the unique Western concept of "progress" and the discovery that civilizations change over time.

Most ancient societies viewed life as a circular affair. The ancient Chinese viewed history as a process of rise, decay and collapse, a cycle they saw repeated in one dynasty after another. In Mayan cosmology, time proceeds in a cycle of birth, growth, collapse and then a rebirth which restarts the cycle. The civilizations of India perfected this circular idea in the central Hindu belief that life is a series of reincarnations over thousands of years.

About 3,000 years ago, a radically different idea emerged among a nomadic tribe of herders in the Near East, the Hebrews. They said life has a purpose – to move closer to God. For them, history was not circular but linear, a journey forward toward God. Christianity, an offshoot of Judaism, absorbed the idea that history is linear and reframed it as progress forward, to achieve salvation.

During the Renaissance, Christians rediscovered the cultures of ancient

Greece and Rome and realized that progress to higher levels extended beyond individuals to entire societies and civilizations. During the Enlightenment, this insight crystallized into the unique Western concept of "progress," the idea that society, civilization and history all move forward.

VOYAGES OF DISCOVERY REVEAL HUMAN SOCIETY HAD EVOLVED

The idea of "historical progress" was later cemented in the Western mind by the "Voyages of Discovery." The Portuguese trips around Africa in the 15th century began the process, introducing Europeans to startlingly different cultures, like the huge Kingdom of Congo that sprawled across central Africa.

Sailing west in the 16th century, Spanish expeditions discovered the native societies of North America, the spectacular Aztec civilization of Mesoamerica, and the vast Inca Empire to the south. Heading east, the Dutch made similar discoveries in Southeast Asia.

Along with gold, spices and trading goods, these explorers returned with stories of exotic cultures and brought back art, clothing and sometimes the indigenous people themselves.

In the 18th century, the British began to systematically explore the globe. In 1768, Captain James Cook embarked on the first round-the-world voyage, arriving in Tahiti a year later to encounter a Polynesian culture unlike anything ever imagined in the West. Sailing on, Cook made contact with Australia's native aborigines, a society living in the most basic state of nature, with only a few primitive stone tools.

These encounters sent a shock through Europe. It was clear human societies had progressed from primitive tribes to more complex cultures and then on to advanced civilizations. Humankind had evolved!

LINNAEUS AND ORDER IN NATURE

To this swirl of ideas about evolution, Europeans added "transmutation." Medieval alchemists believed that the basic elements could be transformed through a process called "transmutation" (for example, turning iron into gold). The alchemists never achieved this, but their experiments showed that the elements can change their form, (creating the foundation of modern chemistry).

It was also widely believed that "transmutation" occurred in plants and animals, changing them from one form into something entirely different.

In the 1730s, Swedish botanist Carl Linnaeus began systematically classifying plants and animals based on their forms. His ideas, published as

Systems of Nature, spread rapidly throughout Europe and established three important ideas:

1. Many plants and animals share similar traits.
2. Based on these traits, all life can be integrated into a single system.
3. Nature progresses from simple to more complex forms.

The Hierarchy of Life

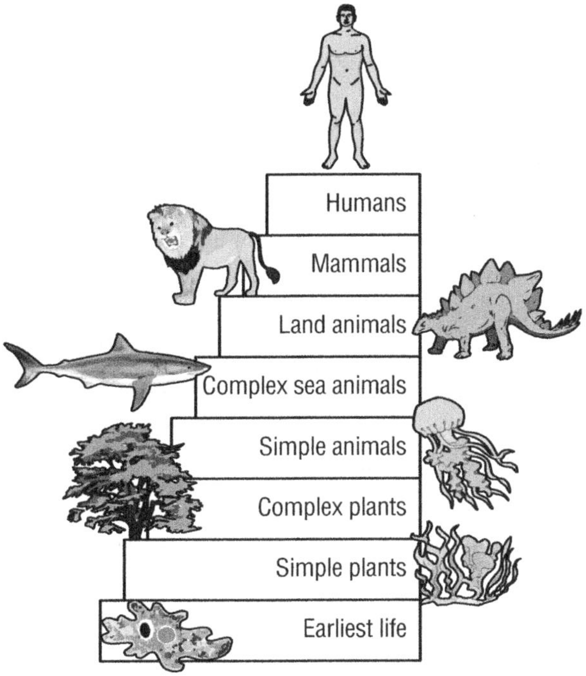

Life became more complex in stages.

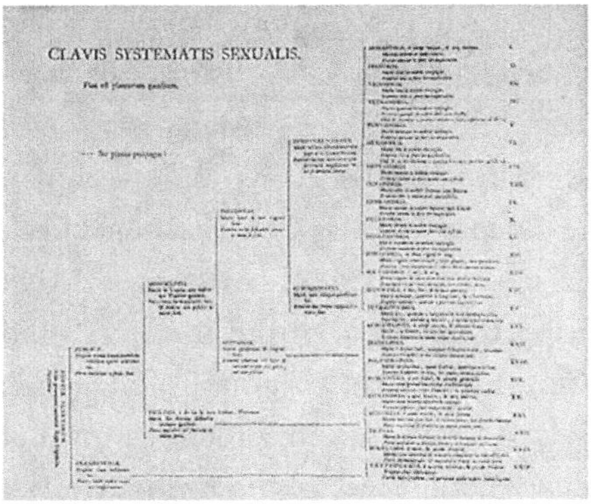

LINNAEUS'S BRANCHING SYSTEM THAT CONNECTED ALL OF LIFE.

Without specifying *how* these relationships arose, Linnaeus made it clear that nature is not a random collection of creatures. Everything in life is systematically connected. Nature has an underlying order.

1700: The First Attempts at a Theory of Evolution

The word "evolution" was first used in 1670 by the Royal Society to mean development ("the gradual and natural Evolution and Growth of the parts"), but was later expanded to include the change of species over time.[8]

The idea of evolution as change over time, coupled with the realization that plants and animals are unified in a single system that progressed from simple to complex, led to speculation in the 1700s about exactly how the evolution process might work.

Famed philosopher Gottfried Leibniz proposed a "law of continuity" in which "all the orders of natural beings form but a single chain, in which the various classes...are closely linked one to another."[9]

In 1745, Pierre Louis Maupertuis, director of the French Academy of Sciences, published a theory of evolution based on natural selection. Pointing to the artificial selection used by farmers to breed only the strongest animals, Maupertuis said a natural version of this "selection" process occurs when the differences in new born animals led some to survive, while others died:

> "Chance produced an innumerable multitude of individuals; a small number found themselves constructed in such a manner that the parts were able to satisfy its needs; in another infinitely greater number [of species], there was neither fitness nor order [and] all of these perished."[10]

In 1770, well-known Scottish naturalist James Burnett outlined a similar natural selection process to account for changes in plants and animals, even speculating humans had evolved from apes this way.[11] In the 1780s, Immanuel Kant proposed that the races of Man had evolved their different facial features and colors by adapting to their unique climates.[12]

A few years later French naturalist Georges-Louis Leclerc, the Comte de Buffon, knit these ideas into a single history of evolution in his *Epochs of Nature*. Starting with the origin of the stars, Buffon walked up the ladder of evolution through the origin of the Earth, plants, animals and then humans. Based on the cooling rate of iron, Buffon estimated the Earth was at least 75,000 years old, a dramatically older age than previous estimates.

In 1795, one of the era's most influential thinkers, the Marquis de Condorcet, said that even the *human mind* had evolved. In his *Historical Picture of the Progress of the Human Mind*, he said the mind had evolved through nine stages, becoming more complex and sophisticated at each new level.

1800: THE VICTORIANS EMBRACE EVOLUTION

Rather than being aghast at these ideas, Victorians embraced them. Evolution was nature's perfect match to the driving idea of the Victorian era – progress. "The general idea of evolution fitted the ideology of the age," says historian Peter Bowler.[13] Evolution was visible everywhere – in new machines, new clothing styles, new words, new social conventions. For Victorians, evolution was simply common sense.

In 1796, Charles Darwin's grandfather, Erasmus Darwin, in his book *Zoönomia*, described an evolution process where "all warm-blooded animals have arisen from one living filament." He said the first tiny creatures emerged from deep in the pond muck, and over eons had evolved into the many exotic forms we see today.

Erasmus Darwin became so well-known for this idea about evolution, that in the 1830s, *The Times* of London coined the term "Darwinize," meaning to believe life had evolved in a long, continuous chain.[14] Naturalists in Italy, France and elsewhere developed similar "chain of life" ideas.

By 1820, the pace of thinking about evolution became brisk, and the Victorian world was soon awash in new theories. At the University of Paris, French anatomist Étienne Geoffroy Saint-Hilaire pointed to the fossil record, and said life followed a specific evolutionary path from simple to complex forms. In 1826, at University College London, biologist Robert Grant proposed that all life "had evolved from a primitive model" to more complex forms based on natural laws.[15]

Especially common at the time were theories of evolution based on natural selection, the idea Pierre Maupertuis had introduced 80 years earlier. Artificial breeding was a Victorian passion, used to alter everything from garden plants to horses and lap dogs. It was especially popular with roses, creating an endless variety of new colors and shapes.

With artificial breeding, some creatures emerge hardier and better able to survive than others. It just made sense that the same thing occurred in the wild. Plants and animals with better traits would survive, while the weak would not. Hence the idea of "natural selection."

Many versions of the idea emerged. In 1819, for example, Sir William Lawrence, Sergeant Surgeon to the Queen (Victoria's personal physician), said evolution is a natural version of artificial breeding. In *The Natural History of Man*, he wrote that nature is dotted with the "occasional appearance of different characters." The new characters (or traits) that aid survival will spread, while those that hurt survival will die out, a process he called "selection."[16]

Lawrence, who later was president of the Royal College of Surgeons, was well-known, and his idea of evolution by natural "selection" circulated widely.

Evolution: A Best-Selling Idea

All this talk about evolution reached a crescendo in England in 1844, when Robert Chambers published *Vestiges of the Natural History of Creation*, creating a nationwide sensation.[17] Readers were spellbound by its vision of evolution. It began with the origin of the universe, progressed through the evolution of the stars, planets, oceans, plants and animals, and ended with the emergence of humans from apes. It was a comprehensive natural history of how everything in the universe had evolved, including Man.

And readers loved it.

Reaction was overwhelmingly positive, says Cambridge historian Jim Secord, while criticism was "feeble and limited."[18] *Vestiges* was favorably reviewed everywhere, from science journals like *The Lancet*, to religious publications like the Unitarian *Prospective* ("powerfully supportive"). Even the *Witness*, a newspaper for Scottish evangelical Christians, reviewed *Vestiges* approvingly.[19]

"It sparked one of the greatest sensations of the Victorian era," says Secord.[20] *Vestiges* became a national event, with thousands of people reading and reacting simultaneously, creating a huge, society-wide psychological effect. Discussed in kitchens, coffee houses, dinner parties and polite company everywhere, it even entered Buckingham Palace where every afternoon for a month, Prince Albert read it aloud to his young wife Victoria.[21] Reprinted widely in the United States and translated into German, Dutch and Italian, *Vestiges* ideas about evolution thoroughly permeated the Western psyche.

Belief in evolution became so pervasive that in 1857, Herbert Spencer (by far the most popular non-fiction author of the era) argued in *Progress: Its Law and Cause*, that evolution is woven throughout the *entire fabric of life*. Pointing to historical changes in language, poetry, music, government and society, Spencer said everything in the universe had gone through an evolution process, including the human mind.

Spencer's popularity was due in large part to this focus on evolution. Long before Darwin, some of evolution's most memorable phrases – "survival of the fittest" for example – flowed from Spencer's pen.

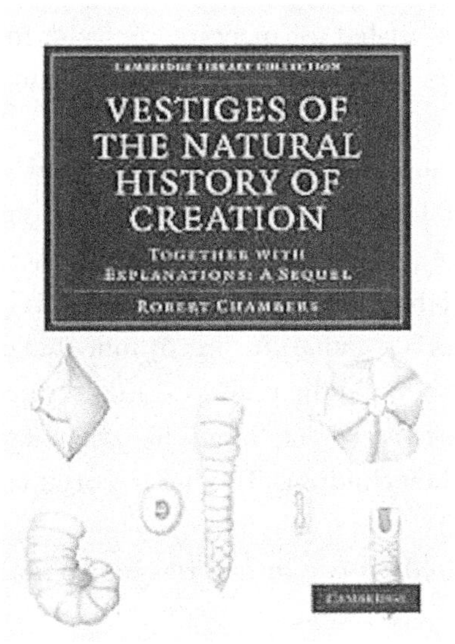

DISCOVERY 1: ADAPTIVE EVOLUTION: BEHAVIOR CHANGES BIOLOGY

Early in the 1800s, evolution had been distilled to two key questions: how do new types of life arise (the "origin of species" mystery), and how do they later adapt and change (the adaptive evolution problem)?

As the evidence for evolution grew, it led to two major discoveries about these questions. The first was how adaptive evolution works.

It was well-known that plants and animals adapt to new circumstances. The mystery was how these changes took place. In 1809, French biologist Jean-Baptiste Lamarck provided the answer.

It was common knowledge that children inherit their parent's traits, an idea widely embedded in folk wisdom – "like father, like son," "the apple never falls far from the tree."

It was also common knowledge that behavior changes the body. Muscles get bigger when used, and wither if unused. (Today we say "use it or lose it.")

In a stroke of insight, Lamarck realized that this "use it or lose it" principle applies to *all of biology*. The parts of the body that are used (or "stressed") grow stronger. Those that go unused get weaker. This is easily seen with muscles, but applies to every aspect of biology. For example, the immune system, when stressed by an invading virus, builds a stronger resistance to it. A stomach regularly stressed by strong spices develops a tolerance for them.

Even the brain rewires itself in response to mental stress or new behavior. Among London taxi drivers (who think a lot about where they're going), the brain regions related to "mapping" expand. In concert violinists, who practice six hours a day, the brain areas that control hand-muscles expand, while the "mapping" area may shrink.

Crucially, Lamarck discovered that the changes from "use and disuse" can be directly inherited by the next generation. A giraffe, constantly reaching for the high leaves on a tree, lengthens its neck and that increase is passed to its young. Lamarck called this the "inheritance of acquired characters"– changes in a parent's "characters" (or traits) are directly inherited by their children.

Today we know that while the process is more complex than Lamarck described, his basic idea is correct. Some changes in a parent's biology are directly inherited by their children. (This has important implications for life today. See chapter 22.)

Lamarck summarized his two discoveries in two laws, creating the first true scientific theory of evolution.[22]

First, he said, the use or disuse of a trait leads to adaptive change:

> "In every animal...sustained use of an organ strengthens that organ...while the constant disuse imperceptibly weakens it... and ends in its disappearance."

Second, the changes from "use and disuse" are inherited:

> "Everything nature has caused individuals to acquire or lose...it preserves by heredity and passes on to the new individuals descended from it."

In short, Lamarck said that behavior changes the body, and these changes are inherited by the next generation, creating adaptive evolution.

Lamarck's Discovery is a Scientific Success

Supported by a strong body of evidence, Lamarck's laws became the most widely accepted explanation for evolution throughout Darwin's lifetime and well into the 20th century.

Herbert Spencer's model of self-improvement, progress and evolution, for example, which became immensely popular in the late 19th century, relied on Lamarck's "inheritance of acquired characteristics" to explain how-improvements accumulated over time, leading to evolution and social progress.

Lamarck's theory proved successful because it excelled at the three things important to good science. It explained important aspects of evolution, such as the complex problem of proportional change. In the giraffe, for example, lengthening its neck, even slightly, requires thousands of adjustments, from microscopic blood vessels to large bones. These changes in turn require thousands of other adjustments throughout the body, from head to toe. All these changes always emerge in *perfect proportion* to each other. How this worked was a long-standing mystery which Lamarck's discovery resolved. Behavior, by its very nature, changes everything simultaneously. When you run, for example, everything in the leg changes together - muscles, tendons, bones, blood vessels, and all the rest.

Lamarck's theory also meshed well with what else was known about evolution. It was clear, even in the early 1800s, that evolution occurs in short, sharp bursts. Lamarck's process explained why. New behavior (or stress of any type) creates a cascade of changes throughout the body. A drop in temperature can simultaneously trigger the growth of new fur, a layer of fat for warmth, blood cells more resilient to the cold, and a host of other changes. This type of large scale change can reshape the body in just a few generations, explaining the rapid changes seen in evolution.

Lamarck also believed that, over time, these adaptive changes led to the emergence of new species, an idea later picked up by Darwin. This idea, however, proved untrue (for both Lamarck and Darwin). Nonetheless, Lamarck had answered the important question of how adaptive changes occur in nature.

THE DISCOVERY OF 'BASIC' FORMS

By the 1830s, scientists began to glimpse a second important truth about evolution. Despite the appearance of great diversity, nature uses only a few basic patterns. All plants - from algae, to orchids and giant redwoods - are built using just five basic patterns.

The Earth's one million types of insects - from a Fairyfly the size of a comma, to an Elephant beetle bigger than our palm, and the extinct Griffinfly with a two foot wingspan - all share the same body plan.

The 60,000 animals with four limbs (called vertebrates) - from fish to reptiles, rodents, dolphins, dogs, cats, birds and humans - all use the same basic body plan. Nature's wild variety of spiders all share a single body plan, as do all crustaceans (shrimp, crabs, lobsters and the like).

These nine patterns account for 95% of life. Six more patterns cover what's left. Nature's millions of wildly different creatures are all built from just 15 body plans.

Huge changes in size and shape don't alter the basic plan. Dogs, from a three-pound Chihuahua to a 300-pound Great Dane, are all built alike. All dinosaurs – from a 30-ounce Microraptor to a 30-ton Sauropod that stretched 100 feet – were built from the same body plan.

Nor do great differences in the environment alter the core biology of these plans. The first trees appeared 250 million years ago. In all that time, whether on the parched plains of Africa, the steamy Amazon jungles, or clinging to a frozen Alpine cliff, trees have remained resolutely "tree-like" and haven't evolve into something else.

Crocodiles, with jaws 10 times more powerful than a great white shark, have existed for 50 million years. They've endured millennia-long heat waves and many ice ages. Yet in all that time, despite all those extreme changes, they have remained crocodiles. (Only if crocodiles had changed their behavior would their bodies have changed. With that powerful jaw, it appears they didn't have to.)

In the early 1800s, well-known anatomists like Geoffroy Saint-Hilaire looked at these facts and saw powerful underlying laws at work. These laws limit the patterns that nature can form. These constraints explained why only a few basic body plans exist, and why creatures retain these forms for hundreds of millions of years.[23]

RICHARD OWEN

How natural laws determine the origin of new species was worked out in the 1840s by Richard Owen, the greatest anatomist of the Victorian era, with an intellect that bordered on genius.

As a young man Owen began a series of public lectures which astounded people and drew many leading scientific and social figures, including the Royal family. In one instance, he analyzed the bones of a huge mysterious creature, called *Megalosaurus*, and concluded it was not a lizard, as long thought, but an entirely new type of reptile which he named *Dinosauria* ("terrible reptile"). Owen had identified the first dinosaur!

Owen made other discoveries and his reputation grew rapidly. He was seen as the man who could solve mysteries and make sense of the past.

Owen's skills were broad. He became an expert to the government on a wide array of scientific issues, helped organize the great 1851 London Exhibition and was the driving force behind the creation of London's remarkable Natural History Museum. As the Museum's first director, he introduced the modern concept of the museum as a place to enrich the mind and fuel the imagination. "By making the Natural History Museum an institution for everyone," writes Bill Bryson. "Owen transformed our expectations of what museums are for."[24]

RICHARD OWEN

Owen even taught natural history to Victoria's and Albert's children, "astonishing the court with the news that tadpoles turn into frogs."[25]

In 1856, *The Times* of London declared "there is not a more distinguished man of science in the country." Owen was "the single most visible scientist of the British empire," says historian Nicolaas Rupke, his name mentioned alongside Isaac Newton.[26]

DISCOVERY 2: "LAWS OF STRUCTURE" DETERMINE NEW SPECIES

Owen's analytic skills and insights were remarkable – a century ahead of everyone else's. His insights allowed him to penetrate surface features and see the deep structural similarity between living things. And he sought to understand the underlying logic of these similarities.

Using work by German biologists as a foundation (Johann Goethe's idea of change through "metamorphosis," and George Bucher's 1836 idea that species are variations on a basic form which emerge over time through an internal "generative" process), by the1840s Owen succeeded in his quest.

Remarkably, Owen worked out the structural aspects of evolution so well that he was able to chart the entire history of human evolution from one species to the next, and realized that, over the course of many eons, we had evolved from fish. "We learn from past history," he wrote, that life "has advanced...from the first embodiment of the Vertebrate idea under its old Ichthyic [or fish] form until it became arrayed in the glorious garb of the Human form."[27] Owen summarized this evolutionary progression from fish to humans in a famous drawing.

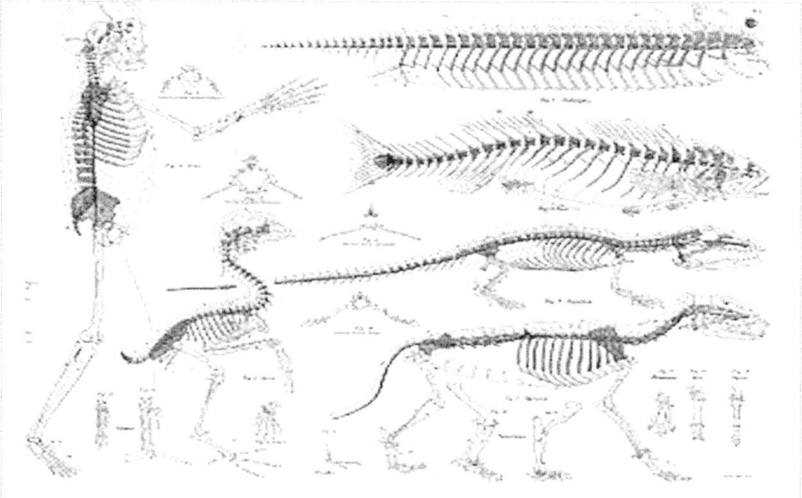

EVOLUTION FROM FISH TO HUMANS

Owen had brilliantly deciphered nature's underlying architectural logic, and grasped that *all* species arose this way. This led him to a remarkably modern concept of evolution: the change from one species to another is governed by a set of "structural" laws inherent in what he called "Nature."[28]

On a February evening in 1849, at one of his famous public lectures, Owen outlined his theory of evolution. He argued that change is governed by "organic laws" which determine the structure each new species can take. He then linked these structures into a single evolutionary progression.

"Owen revealed similarities in biological forms from species to species that suggested some underlying ideal plan," says biologist Michael Coates.[29]

"With example after example, Owen hammered home the point that structural [similarities] between species...cannot be explained simply by adaptation. The echoes of his words still reverberate."

Owen realized that he didn't know all the laws involved in evolution ("to what natural laws and secondary causes the orderly succession [occurs] we as yet are ignorant"), but it was clear to him that these laws of structure explained the origin of new species. "If we may conceive the existence of such [laws]," he wrote, "then we learn from past history" how life evolved from one species to the next.[30] (As we will see in chapter 18, modern science shows that on this he was correct.)

Owen said the number of ways that life's basic forms could be changed was "very far from being exhausted," thus new species would likely evolve in the future. [31]

Astonishingly, Owen even envisioned evolution on other planets! He wrote that if there is life on other planets (as some astronomers speculated) while some potential new species "may never be developed on this planet, it is quite conceivable that certain of them may be developed, [on] other planets of our system."[31] It is hard to imagine a more modern, less Biblically literal idea of evolution than this.

Owen with Fossil

Owen's Astonishing Analytic Abilities

At the 1854 British Association for the Advancement of Science meeting, Owen used his "organic laws" to demonstrate that it was impossible for apes, like the newly discovered gorilla, to have stood erect and evolved into humans. Man had evolved from an earlier species, said Owen, but it was not an ape.

Owen later showed there are other crucial differences between apes and humans. Darwin, predictably, opposed the idea writing, "I cannot swallow Man...distinct from a Chimpanzee."[32]

Today we know Owen was correct, and Darwin wrong. In 2009, results from the largest study on human origins ever undertaken were released. Led by Tim White at the University of California, this 15-year effort by a large team of international experts demonstrated that humans did not evolve from apes. "No modern ape is a realistic proxy" for humans, they concluded.[33]

"The idea that we evolved from something like a chimpanzee goes out the window," wrote Owen Lovejoy, a senior scientist on the project and one of the world's leading physical anthropologists.[34] "Human origins can no longer be sought by studying apes," said Lovejoy.[35] Owen, using purely his intellectual and analytic skills, discovered 150 years ago what modern science now proves.

(Anyone who loves science, and evolution, must stand in awe of Richard Owen. A man so far ahead of his time that he could discuss evolution on other planets, or state (correctly) that while humans had evolved from a primate, it was not from knuckle walking apes. And yet have the humility to understand, and say publicly, *how much he did not know.*)

Owen Anticipates Modern Ideas of Evolution

Owen's remarkable insight into human history was just one of many things he had right. Owen's approach to evolution anticipated the ideas central to modern evolutionary developmental biology ("Evo-Devo"), and he presaged modern Systems Biology, showing that physical processes (like "polarization" and "attraction") determine which forms nature can create, and thus which forms can appear in evolution.[36] "Many modern structuralists recognize their intellectual ancestors among the 19th-century... structuralists," like Richard Owen, says historian Ron Amundson.[37]

Owen's work anticipated other processes central to modern ideas about evolution. He argued that a set of "organizing forces" would lead nature to "mould specific forms."[28] Today, we recognize this as a forerunner to the "self-organization" process which underpins modern concepts of evolution.

As we saw, the idea that the natural world emerged through a process of "self-development," or "self-organization," had been gestating in Western thought for over 2,000 years, stretching from Owen and the "structuralists" to Aquinas, Augustine and on back to Aristotle and the ancient Greeks.

THE TWO MAJOR SCHOOLS OF EVOLUTION

The laws of structure and adaptive change formed the two big ideas about evolution in the Victorian era. Most scientists supported one or both as the solution to evolution's two big questions.

Owen's laws of structure explained a great deal about the origin of new species. The natural world is governed by laws. Clearly visible patterns in the history of life point to similar laws at work in the evolution of new plants and animals. Called the "Structure" school of evolution, it argued that natural laws determine both how new species arise, and the form they will take.

Lamarck's laws explained how species adaptively change after they emerge. The stress of new behavior alters the body, and some of these changes are directly inherited. Called the "Adaptation" school of evolution, it said stress (or lack of it) re-molds the body to meet new challenges.

Natural laws explain *what* can exist (the type of plants and animals that can evolve), and stress of new behavior explains *how* they change after that.

The Adaptionists saw adjustment as biology's central truth. The Structuralists believed biology's organizational logic reflected a deeper reality. Today, we know both represent major insights into the evolution process. Modern versions of these two ideas now form two pillars of the new evolution currently emerging from science.

Despite their differences, both schools agreed on one thing: evolution is governed by natural laws with predictable effects.

Chapter Four

DARWIN'S IDEAS WERE OLD

So if Darwin didn't discover evolution, what did he add to the discussion? Where did his ideas fit in this world bubbling with interest in evolution?

DARWIN (AGE 4) WITH SISTER CATHERINE

Born into a wealthy family, and with a substantial fortune in his future, Darwin was always more interested in riding, hiking, shooting birds and beetle hunting than a career.[1] In 1831, having finished university with little investment in his studies and no career plans, the 22-year-old Darwin paid for a cabin on the Royal Navy ship *Beagle,* then leaving on a five-year voyage to explore South America, the Pacific Islands, and Australia. An avid naturalist, Darwin used the trip to catalog thousands of plants and animals, and study the differences between species.

Darwin's Idea: Evolution by Death and Elimination

Later, while pondering how these differences arose, Darwin read Thomas Malthus' *Principles of Population*, one of the best-known books of the prior century. Malthus said that animals give birth to many young because most of them die quickly, at the hands of a predator or disease.[2] Without these natural killers, animal populations quickly outrun the food supply, leading to starvation. Whether through predators, disease or starvation, said Malthus, the weak die young, before they can reproduce. In each new generation the weak die quickly. Eventually the weak disappear altogether, leaving only the "most fit" as survivors.

Darwin thought that over long periods of time this could lead to new species. Take antelopes, which evolved from the cow. On the African plains, the slowest cows became lunch for lions and died before they could breed. Slightly faster cows survived and gave birth to young who, like themselves, were slightly faster.

Every year lions brought down the slowest cows in the group, leaving the next generation a bit quicker and slimmer. After many generations, only the very fast and sleek remained. A lumbering cow had evolved into a swift antelope.

That, in a nutshell, is Darwin's idea of evolution through natural selection: life is a competition in which the losers are eliminated before they can breed. The survivors have been "selected" to survive because of their superior biology. Darwin explained it like this:

> "As more individuals are produced than can possibly survive, there must in every case be a struggle for existence. Individuals having any advantage...would have the best chance of surviving... any variation *in the least degree injurious* would be *rigidly destroyed.* This preservation of favourable variations and rejection of injurious variations, I call Natural Selection."[3]

(If you think this sounds remarkably similar to Pierre Maupertuis' description of evolution through natural selection from a hundred years earlier, you are right. The idea, even the wording, is nearly identical.)

The Lack of Interest in Darwin

In an era fascinated by evolution, how did Darwin's theory fare? Despite the story we are told today, it hardly made a ripple. Darwin first presented his theory in an 1858 lecture at England's important Linnaean Society. The Society was founded to carry on Carl Linnaeus's effort to map the biological relationship between all plants and animals. If any group would be interested in Darwin's claim to have explained how *every* species had come to exist, it was the Linnaean Society.

Darwin's lecture, however, came and went with little notice. In his year-end report to the members, Linnaean president Thomas Bell didn't mention Darwin's theory. On the contrary, a disappointed Bell wrote:

> "The year which has passed...has not been marked by any of those striking discoveries which at once revolutionize... science; it is only at remote intervals that we can reasonably expect any sudden and brilliant innovation which shall produce a marked [advance in] knowledge...a Newton... a Davy [a pioneering chemist] or a Daguerre [inventor of photography], is an occasional phenomenon."[4]

Darwin's idea was certainly not up there with Newton, or even Daguerre. Bell, however, didn't even include Darwin's theory in the list of minor ideas worthy of note. In Bell's report, which covered everything of scientific interest in1858, Darwin's name appeared twice, as the person who recommended the Society publish an article by Alfred Wallace, and in a footnote for an insight Darwin had "deduced from the structure of coral reefs."[5] Apparently Bell thought Darwin's tidbit about sea life was more noteworthy than his theory of evolution.

Darwin received a single response to his lecture, from the eminent biologist Samuel Haughton, which Darwin summarized as saying "all that is new in the paper is false, and all that is true is old."[6] That succinctly expressed the scientific assessment of Darwin's theory: ideas that were either old or untrue.

Darwin as a Young Man

The Social Response: Indifference

The *Origin of Species*, published a year later, also drew little interest. This was especially disappointing for Darwin, write historians Janet Browne, Adrian Desmond and James Moore in their biography *Charles Darwin*. "Stripped of references," they say, *The Origin* "was aimed not at the specialists, but directly at the reading public."[7]

Reading was Victorian England's most popular form of education and entertainment. It was even a popular way to socialize. "The Victorians virtually invented mass literature," says historian Sally Mitchell. "High-speed presses, cheap wood-pulp paper...railways to send printed material quickly all over the country, and the steadily growing number of people who are literate enough to read for pleasure encouraged the publication of newspapers, magazines and novels at every price and for every taste."[8]

"Reading for pleasure was not only an individual and private occupation," says Mitchell, "but a communal activity shared with family and friends," who often read books aloud to each other in the evenings.

Victorians were particularly avid about books on nature and evolution. In 1859, the year Darwin published, *The Vestiges of Creation* was in its 10th edition. Charles Lyell's *Geological Evidence of the Antiquity of Man*, published around the same time, quickly sold out and went through three re-printings within months.[9]

Darwin was hoping for a similar response. Yet in 1859, as people snapped up 100,000 copies of *A Tale of Two Cities* and 40,000 copies of a new poetry book, *The Origin of Species* sold a mere 500 copies in book stores.[10] Even a new "ethics" book, *Essays and Reviews*, sold 22,000 copies that year, a number *The Origin* would not reach until the 20th century.[11]

Once, when asked what creates sales Darwin replied, "Getting talked about is what makes a book sell."[12] Clearly few people were talking about Darwin's book. "If you had asked any Victorian 'What is the big book on evolution' they would all have answered *Vestiges*," says historian Jim Secord, "even Darwin would have agreed."[13]

The Origin's sales were a profound statement of the public's indifference to Darwin's ideas. While the book was politely reviewed, it was seen as outdated, highly speculative for a supposed science book and, well, just plain tedious. While some praised its workman-like catalog of examples, others thought it read like a never ending specimen cabinet put into words.

Almost no one thought it contained any groundbreaking insights into evolution. For example, there was general agreement that animals adapt, and Lamarck had explained how this works. What did Darwin have to add to the topic? While *The Origin* held an eye-glazing list of animals adapting to their environment, it offered only speculation that these changes were the result of his natural selection process. (Huxley later wrote that "on the whole the supporters of Mr. Darwin's views in 1860 were numerically extremely insignificant."[14])

Darwin's theory failed to spawn much interest in the years that followed. Victorians had a passion for new discoveries, and public lectures were usually packed. A talk by Richard Owen on "Fossil Reptiles" (dinosaurs) would draw a capacity crowd of 800.[15] Even an esoteric topic like "Comparative Anatomy" could fill a lecture hall "to overflowing," says historian Nicolaas Rupke.[16] A breakthrough on the exciting topic of evolution would surely have packed a London lecture hall several nights running, but Darwin's ideas generated no such interest.

DARWIN'S IDEAS WERE OLD

Darwin's theory was ignored for several reasons. First, it was based on the 100-year-old idea of natural selection. This might have been interesting decades earlier, when little was known about evolution, but science had progressed by leaps and bounds since then.

Many thought the idea of natural selection had a "fatal flaw." If natural selection was at work, survival of the fittest should lead to the emergence of a single "most fit" creature in each locale. But this never happens. Exactly the opposite occurs. Nature teems with diversity.

Called the "diversity paradox," it asked how nature's vast variety can emerge from a natural selection process powerfully driven to create a few "most fit" creatures? Why is the ocean filled with remarkably different types of creatures – worms, crabs, stingrays and a thousand others – instead of a few "most fit" species? On the African plains, why are there both lightning fast antelopes and plodding wildebeest?

If natural selection is at work, why don't sharks, with no predators, multiply in the hundreds of millions? The sea certainly holds enough food to support them. Why are there so few lions even though the African plains hold 100 million wildebeest and other animals on which they feed?

Within each animal group, why are there thousands of different types rather than just a few? Why is the Amazon Jungle home to thousands of types of plants and insects, instead of just a few "most fit" species?

Darwinists explained this by saying that each creature occupies a unique niche in the environment. But if that is true, why then, in a 100-square-foot area of a rainforest, do we find thousands of wildly different molds, bacteria, plants, insects and animals instead of one "most fit" type of each?

The "diversity paradox" extended beyond biology to behavior. If natural selection determines what survives, then one "most fit" type of behavior should evolve in each locale. But it does not. On the African savannah, the behavior of mole rats, rhinos, gazelles, elephants and cheetahs is nothing alike. Nor do fish, hippos, crocodiles and water snakes have similar behavior, despite having shared the same watery world for 50 million years.

In 2009, *New Scientist* magazine asked biologists to identify evolution's most important unexplained mystery. One-third of them pointed to the "diversity paradox." "We need to know," wrote Richard Fortey of London's Natural History Museum, "how does evolution produce enormously species-rich genera... many of which can *coexist happily* in the same habitat."[17]

COMPETITION OR COOPERATION?

In natural selection's ruthless, survival-of-the-fittest world, "happy coexistence" should never be seen. But harmony is the rule in nature, and the saga of the Australian rabbit explains why. Australia introduced rabbits in the 1800s as part of the English hunting tradition. However marksmanship didn't match reproduction, says Harvard geneticist Richard Lewontin, and the rabbit quickly became a pest. To halt their spread, a lethal virus was introduced. Although initially devastated, the rabbit population quickly rebounded.

What happened? The virus evolved to be *less* deadly. The reason, says Lewontin, is that a virus which kills its host guarantees its own demise.[16] On the other hand, a virus which learns to live in harmony with the host, survives. Thus, a *less* virulent virus evolved. In nature, it's not survival of the fittest, says Lewontin, but "survival of the nicest."[18]

CHANGING A TRAIT: THE LINKAGE PROBLEM

Victorians also asked, if natural selection is at work, why are the diseases that plagued man and animal in ancient times still the major diseases of today? Modern biologists put the question this way: "Why are infectious diseases still the major causes of animal mortality, despite...pressure to eliminate traits that make individuals susceptible to infection. Why hasn't natural selection weeded out weak immune systems?"[19]

Breeders pointed to yet another problem with natural selection. You can select animals with good traits for breeding, but bad traits always tag along. Breeding pigeons for beautiful white feathers makes them nearsighted. Breeding animals for larger size leads to heart, lung and other problems, and some are fatal.[19] In short, natural selection cannot eliminate *only* the bad traits. Good traits are always mixed with bad, which neutralizes the effects of natural selection.

In some cases, breeding for good traits can have the *opposite* effect on survival that Darwin claimed. Sheep with strong immune systems live longer but have few births, while those with weak immune systems live shorter lives but give birth more often.[20]

Darwin was well aware of this problem, and called it "the mysterious law of correlation." One of the many examples he noted himself was that bald dogs (with no fur) often have bad teeth. Rather than resolve this mystery, Darwin dismissed these "correlations" as irrelevant to evolution.

Today we know these "correlations" hold a central insight into evolution. In the developing embryo, the *same* genes are involved in the early formation of different traits. In dogs, for example, a single gene change early in development leads to both baldness and bad teeth.[21]

DARWIN TRIED TO REVIVE AN OLD RELIGIOUS DEBATE

The second reason Victorians ignored Darwin was because of his obsession over a religious idea that was of little interest to most of them.

Those who have never read *The Origin of Species* are surprised to learn how much it involves religious issues, especially something called "special-creation."

Based on a literal reading of Genesis, "special-creation" is the idea that God created the universe in seven days, and formed each species from clay,, placing them on Earth in a single 24-hour day.

Darwin had been obsessed with special-creation for decades. As far back as the 1830s, he framed his notebook entries "in the form of an argument against creation," said historian Dov Ospovat.[22]

Thirty years later, Darwin was still obsessed with the idea. In *The Origin of Species*, says historian Ron Amundson, the opponent Darwin chose to argue against was "a special creationist, an advocate of the separate divine creation of each distinct species."[23]

This was more than just a theme in *The Origin*. Darwin organized the *entire book* as a refutation of special-creation.[24] Today we are told that Darwin organized *The Origin of Species* as a refutation of special-creation because people believed deeply in the idea, and he feared his discovery of evolution would create a social firestorm. But that is untrue.

On the contrary, for most Victorians Darwin's obsession with special-creation was strange.

Christianity had long taught that Genesis is a metaphor for events which occurred through natural processes. A literal interpretation of Genesis – and the belief in "special-creation" that went with it – were of little interest to most people. A snapshot of how irrelevant the issue was emerges from one of the great achievements of the Victorian era, Charles Lyell's history of the Earth.

The first systematic study of the Earth began in the 1700s. Geologists explained that the rock layers, like those seen in cliffs throughout England, reveal the Earth's history. Starting with James Hutton in 1785, geologists began to extend the age of the Earth deeper and deeper into the past.

In 1830, Charles Lyell, the foremost geologist of his day, pulled these findings together in *Principles of Geology*, providing the first complete picture of the Earth's history. Lyell demonstrated it stretched hundreds of thousands of years into the past – far longer than any biblical calculation. Eventually he estimated the Earth was hundreds of millions of years old.

This direct challenge to a literal reading of Genesis did not hurt Lyell. Quite the opposite. It propelled him to international fame, making him one of the most celebrated men of the age, later honored by Queen Victoria with a Baronet.

Special creation was never a significant issue in the 19th-century. Victorians believed God created the universe. But most of them, including most clergy, believed it occurred through natural processes. Even Darwin acknowledged that most of his contemporaries were not special creationists but instead believed "in some unknown but natural cause of species origins."[23] He wrote, "I found that many naturalists fully accepted the doctrine of the evolution."[25]

Evolution Strengthened Belief in God

When the best-known cleric of the era, the Rev. Charles Kingsley, wrote to thank Darwin for his complimentary copy of *The Origin of Species*, Kingsley noted that the issue of special-creation, on which Darwin had lavished so much effort, was irrelevant, saying:

> "It is just as noble a conception of Deity to believe that He created primal forms capable of *self-development* [i.e. evolution] as to believe that He required a fresh act of intervention to supply [each]."[26]

Kingsley was famous for his *Water Babies* children's stories that ran in *MacMillan's* Magazine. It is a measure of how thoroughly evolution was accepted, that in one story a child says to Mother Carey (mother nature), "I heard, ma'am, that you were always making new beasts out of old." Mother Nature replies, "'So people fancy. But I am not going to trouble myself to make things, my little dear. I sit here and make them *make themselves*."[26]

"The idea that Darwin set science and religion at each other's throats is absurd," says historian Jim Endersby.[27]

"Despite our images of warring factions and angry sermons from parish pulpits, there was no great battle between science and the Church," says Jim Secord.[28] On the contrary, people of Faith welcomed the idea of evolution. In an unexpected way, it strengthened their religious beliefs. It allowed them to join discoveries, like those of Charles Lyell about the age of earth, with a concept many found awe-inspiring – God as a Divine engineer "who had devised natural laws so delicately balanced they would run forever."[27]

"*Vestiges* laid the foundations of how the problem of evolution was tackled in the English-speaking world of the nineteenth century," writes Secord. It "allowed people to link religion and science in a new way, to combine belief in God with the new discoveries about nature."[28] There are many examples of such experiences, he says. One telling instance was Mary Smith, a governess to a group of children at Carlisle:

> "Mary Smith recalled obtaining the book surreptitiously from her employer, after she found it left on a table, and ran up to her garret room to read it. Only when the dawn came and she heard the birds singing, did she reach the last page... She was devout, wrote religious poetry and aimed to teach the truths of Christianity to young children. Deeply concerned about materialism and the role of God in the world, she was 'intensely anxious' to read the book, which had 'most excited

> the wonder and curiosity of the reading world'...
>
> "What did she make of the book?... For Mary Smith, *Vestiges* offered the possibility of linking religion and science in a new way. A belief the origin of humanity through natural causes could be combined with her own evangelical faith. She brought these together in thinking and writing about the book, both in her poems and in her autobiography."[28]

The belief that evolution is not only compatible with religion, but that it confirms God's role in creating the natural world, came to dominate Victorian thinking.

Chapter Five

Darwin's Deadly Social Agenda

THE WAR AGAINST RELIGION

There is a mystery in all of this.

If evolution was well-accepted, and Darwin had little new to say about it, why did he write his book? And why base his theory on the outdated idea of natural selection?

An inherited fortune freed Darwin to devote his days to research. He read the latest journals and books, and corresponded with scientists across Europe. Because of this, he knew natural selection was considered an inadequate explanation for evolution, yet he determinedly persisted with it.

Even more perplexing was Darwin's obsession with special-creation. He knew the idea of evolution was well-accepted, and didn't conflict with people's religious beliefs. Why, then, was his book obsessively organized around special-creation? Why refute an idea that few people believed?

And why was his book aimed at the general public rather than scientists?

The answer to all these questions is ideology. Darwin's focus wasn't science, but society. He had a social agenda. His book was designed not to advance belief in evolution, but to create a view of evolution that would advance a particular social program.

Neither natural selection and the idea evolution is a competitive struggle for existence, nor the idea that evolution is like a branching tree were original to Darwin. More importantly Darwin organized *The Origin* to argue against belief in "special-creation" rather than other scientific explanations for evolution. His goal was to refute the belief that evolution was based on a "Divine Design." Darwin's target was not science, but society.

What was later called "Social Darwinism" was baked into his theory from the beginning. Historians Adrian Desmond and James Moore put it this way:

> "Social Darwinism is often taken to be something extraneous, an ugly concretion added to the pure Darwinian corpus after the event, tarnishing Darwin's image. But his notebooks make plain that competition, free trade, imperialism, racial extermination, and sexual inequality were written into the equation from the start—Darwinism was invented to explain human society."[1]

Darwin's War Against Religion and the Working Class

Darwin was among a group in England's educated upper class with two great concerns: the exploding ranks of the working class (whom they viewed with fear and disdain), and religion (which they saw as a cesspool of superstitions, the great impediment to social progress). For Darwin and his group, the two problems were inextricably linked.

Many reformers at the time believed education and opportunity were the solution to social problems. Darwin and his group saw something else: bad biology. They believed that biology determines ability. The working class were poor because of their inferior biology. Darwin and those around him believed the solution to social problems was improving people the same way farmers improved livestock, through selective breeding, a process later called "eugenics." For them, religion was the great obstacle to social progress through eugenics.

Darwin's goal was to arrange the evidence for evolution in a way that "scientifically" justified eugenics as a solution to the social problems of the working class. He would use the same evidence to make the case that belief in evolution and God – science and religion – are irreconcilable. You can believe in one or the other, but not both. Society would have to choose: belief in evolution and science, or belief in God and religion.

Not everyone who was anti-religion was in favor of eugenics. But among those who held both beliefs, most saw Darwin's theory as their best weapon in the battle to slay religion and pave the way for eugenics.

Science as a Weapon Against Religion

Anti-religious sentiments emerged during the Enlightenment among intellectuals who felt religion infused society with superstitious ideas and irrational beliefs. It was *the* great obstacle to knowledge and social progress. These feelings became increasingly common among intellectuals and by the 1850s their conflict with religion "reached the stage of war to the death." says Yale historian Peter Gay.[2]

Science was the weapon of choice in this war. The intellectuals "had an enormous investment in science as an ally in their war against religion," says Gay.[3] Science was powerful. People believed it determines what is true. Science could be used to change people's beliefs about God.

Science could also be used to validate ideology. Evidence could be selectively organized to support political claims about society and social change. The radical Robert Grant, who taught at University College London (and was an early mentor to Darwin), showed how.[4] Combing through the University's vast specimen collection, Grant pulled out a few items and selectively arranged them to support a view of life that "came out of revolutionary France."[5]

While there was wide support for social reform, with dozens of new programs, it was tempered by a fear of the radical ideas which had animated the French Revolution. Fresh in people's minds was the Revolution's grisly nightmare of violence, followed by decades of blood-drenching warfare across all of Europe. (In 1815, when Napoleon's 20-year reign of war ended, Darwin was already six years old.)

DEATH OR LIBERTY

These concerns were captured in George Cruikshank's famous print "Death or Liberty!" In it, Virtue is buttressed by the Rock of Religion, and Ethics is defended by the Lion of British Tradition. Hiding beneath the

"Cloak of Reform" are the tragic consequences of radicalism – eugenics, starvation, murdering opponents, and the like – which people knew destroyed the fabric of society. (Their fears were realized a few decades later when the same ideas led to the bloody Paris Commune.)

Darwin's Ideas About Religion and God

Darwin was born into what he called a family of "unbelievers." At the age of 16 he left to study medicine at Edinburgh University. "Biographers regularly go back to Darwin's Edinburgh years, convinced the seeds of all his later thinking lie there," says historian Janet Browne, "and to a large degree they are right."[6]

"Edinburgh was a hotbed of radical thinking in the 1820s and the medical school was right in the thick of things," writes Niles Eldredge.[7] "Most of the radical ideas bandied about the medical school were French inspirations." Darwin's closest relationships were with people who shared these beliefs, like the militant atheist Robert Grant, then teaching at Edinburgh.[8]

Within a few years, wrote Darwin, "my disbelief in Christianity was... complete."[6] In his eyes, religion was a fraud – delusional beliefs with no basis in reality – and he surrounded himself with those who thought the same:

> "I can...hardly see how anyone ought to wish Christianity to be true, for if so, the men who do not believe, and this would include my Father, Brother and almost all my best friends, will be everlastingly punished."[9]

Over time, Darwin's feelings about religion deepened from disbelief to contempt. It was a cancer on society that should be torn out root and branch.

Then, in 1849, the unimaginable struck. His beloved eight-year-old daughter Annie –"the apple of her proud father's eye, his favourite child, he confessed"– fell desperately ill.[10] For two excruciating years Annie suffered, and a tormented Darwin suffered every day with her. When Annie drew her final breath, Darwin was blinded with pain. "We have lost the joy of the household, and the solace of our old age," he wrote. "Oh that she could now know how deeply, how tenderly we do still and shall ever love her dear joyous face."[11]

ANNIE DARWIN

As the grief-stricken Darwin struggled to cope, he raged against religion. His great-great-grandson, Randal Keynes, in *Darwin, His Daughter and Human Evolution,* used the family papers to show how, after Annie's death, Darwin's thoughts became dark and deeply brooding.[12] He raged against the "clumsy, wasteful, blundering low and horridly cruel works of nature!" No benevolent God would ever create such a cruel, pain-filled natural world he saw.

The Origin of Species Written to Refute Belief in God

This begins to make sense of why Darwin wrote his book. He poured these feelings directly into *The Origin of Species.* Historian Dov Ospovat said Darwin's idea about evolution "were related to his religious views, his ideas on human evolution, the history of civilization and his hopes for the future."[13]

The Origin of Species was not a science book, written to explain evolution. It was an effort to disabuse people of their beliefs about religion, and to argue for radical social change. Darwin's "science was political," says historian James Moore:

> "Atheists all loved the *Origin of Species.* This was a book that could be used as a weapon to liberate British society from...religion, the state church, the monarchy... The book was extremely useful to that irreligious sector."[14]

From his close friend and mentor, the radical Robert Grant, Darwin learned that scientific facts can be organized to validate ideological claims, and he set out to do just that. *The Origin* contains no new discoveries, or even anything particularly noteworthy. Instead, Darwin took old ideas about evolution and contrasted them with "special creation" to make the case that evolution and religion are irreconcilable. A person could believe in one or the other, but not both. To be "scientific," one would have to abandon religious beliefs, and the ideas about social change they spawned.

This goal – to frame the facts about evolution to make religion and science appear incompatible – has been at the center of Darwinism ever since. This effort animates modern Darwinists, right up to the 21st Century.

Darwin's Dilemma: Direction, Evolution and God

As he began his book, Darwin faced a dilemma. If people already believed that evolution and religion are compatible, how do you make the case they are not?

Victorians accounted for the natural world the same way as Newton and Einstein: God. Newton thought of natural laws as God's mechanics, Einstein called them "God's thoughts." Both believed God crafted a set of laws, then released them to create the universe.

Victorians believed much the same thing. They pointed to natural laws as evidence of God's role in evolution, and as confirmation the world was not formed accidentally, but "by design." In 1802, well-known British scholar William Paley explained this belief.

Imagine you found a watch on the ground, said Paley. Its complex design and perfectly meshing gears would tell you it must have been deliberately crafted. The same argument applies to nature. "Every manifestation of design which existed in the watch...exists in the works of nature," he said.[15] Nature's the perfectly meshing patterns imply a "designer."

Paley did not dispute that creation had occurred naturally, through evolution. He said that natural laws, like those in physics, had guided the formation of everything in nature.[16] Even Thomas Huxley acknowledged that Paley believed in natural evolution, writing:

> "Paley, saw no difficulty in admitting that the 'production of things' may be the result of trains of mechanical dispositions... that is to say, *he accepted the modern doctrine of Evolution.*"[17]

Paley's argument was not to dispute natural evolution, but to make a philosophical point. He said it's an indisputable scientific fact that everything in nature is the *inevitable* result of natural laws. Paley said that God created those laws, which meant He intended the universe to be as it is. Known as the "by design" argument, it said:

1. Natural laws imply a "law maker" (understood by people as God).
2. Laws lead to inevitable changes. (For example, Newton's laws made the future position of planets predictable and inevitable.)
3. These inevitable evolutionary changes mean that God *intended* the natural world to be as it is, thus nature was "Divinely designed."

Scientists agreed that natural laws make evolution predictable. That, by definition, is what scientific laws do. The problem emerged from attributing this predictability to God. Some, like Darwin and Huxley, said that natural laws do not imply a "law giver," and the inevitable outcome of these laws does not mean that a God "intended" the Universe to be as it is.

Here is the problem: Whether evolutionary change is predictable is a claim that can be tested, and thus a question science can answer (and the answer is "yes," evolutionary change is predictable). But whether natural laws and predictable change mean God intended nature to be as it is cannot be tested, and thus is not something science can answer. It is a question for philosophy. Paley believed his analysis was backed by logic (as did Aristotle and later Einstein). Darwin, Huxley and others did not. Either way, it was a philosophical question.

KILLING GOD BY ELIMINATING DIRECTION FROM EVOLUTION

Things would have been fine if left there, a disagreement over philosophical questions. Darwin, however, wanted to use science to refute people's beliefs about God. And this is where the confusion begins – a confusion that haunts the evolution discussion to this day: Darwin tried to use science to answer a philosophical question.

This one idea – predictable evolution implies the existence of God – became central to how Darwin framed his theory. Linking God and predictability forced Darwin to make a series of extraordinary claims that ran counter to science:

- To declare that evolution is directionless and unpredictable.
- To invent fictional processes that ran counter to basic science, and defend them against powerful evidence they were wrong.
- To deny the existing evidence for evolution because it was based on laws, and thus made evolution predictable.

If Victorians held up predictable change as evidence of God, Darwin would claim the opposite – evolution is directionless, in his words with "no more design...than in the course which the wind blows."[18] In 2010 Darwinist Jerry

Coyne put it this way: "The distinguishing feature of [Darwin's] evolution is the complete absence of 'laws' or 'forces' that push it in a single direction."[19]

Darwin achieved his "directionless" evolution with one of the most remarkable claims in the history of science. He announced the existence of a law (natural selection) with no predictability. Evolutionary change is basically random.

This was breathtaking. No one in science had ever made such a claim. Predictable outcomes are the very essence of science, the very definition of scientific laws. Without predictable change, you simply do not have science.

And, by definition, predictability gives direction to evolution. This is how Newton knew the future position of planets, how geologists know the future location of continents, and chemists know the future result of chemical reactions.

In science, repeating an experiment leads to the same result each time. In Darwin's version of nature, repeating an experiment would lead to a different result each time. Science says that if evolution were rerun, nature would produce the same types of atoms, stars, plants and animals as we have today. Darwin said that if evolution were rerun different plants and animals would emerge. In short, "directionless" change ran counter to the whole scientific enterprise.

Despite this, Darwin pressed the idea because it was central to his social agenda. He offered directionless evolution as "scientific" proof that humans are an accident of nature. He, and those around him, believed this idea would destroy belief in religion. In Peter Parnell's play, *Trumpery*, Darwin says: "If I finish the book I murder God."

Directionless evolution also advanced Darwin's second social goal. If true, it would mean humans are just one more biological accident of evolution, and the best way to improve this "accident" was the same way that breeders improve animals, through controlled breeding, or eugenics.

The entire embattled history of Darwinism boils down to the claim that evolution is unpredictable. It was at the core of the bitter disputes that erupted between Darwin and the leading scientists of his day. It is at the center of today's battles over evolution, both within science and between science and society.

As we go forward, remember these two ideas. Whether natural laws make evolutionary change predictable is a question science can answer. It is a claim which can be tested and proven true or false. But whether natural laws and predictable change imply that God "intended" the natural world to be a certain way is not a question science can answer. The distinction between these two ideas is central to untangling the problems that surround modern evolution.

HUXLEY: SCIENCE AS A WEAPON IN THE WAR AGAINST RELIGION

The man who became Darwin's partner in this project was Thomas Huxley, an Englishman destined to become the world's first science czar.

Huxley's family, always financially marginal, slipped into poverty when he was 10 years old, forcing him to leave school and go to work.

In that era, children learned a trade by apprenticing with an adult. Despite his lack of education, Huxley was ambitious, aggressive and a gifted manager. Had he apprenticed at a factory, he almost certainly would have ended up owning the place, and gone on to become a titan of industry. Instead, in 1838, as Darwin moved into a four-story home in a fashionable area of London, 13-year-old Thomas Huxley was apprenticed to a marginal doctor who tended the sick and dying in one of London's poorest districts.

Almost entirely self-educated, Huxley favored languages over science, becoming fluent in German and proficient in Latin and Greek. Capitalizing on his medical apprenticeship, at the age of 17 he began studies at Charring Cross Hospital in hopes of becoming a doctor. But after three years, he declined to take the certifying exam. He simply did not have the education.

With no prospects, Huxley, like Darwin, looked to the Royal Navy. While Darwin sailed as a "gentleman naturalist," in a private cabin with his own bed, expensive books, tailored clothes and nightly dinner with the captain, Huxley shipped out as a "surgeon's mate," sleeping in a crowded quarters, his possessions few enough to roll up in a knapsack.

Huxley was assigned to the HMS *Rattlesnake* for its three-year research voyage to Australia and New Guinea. When the ship's doctor was dismissed for drunkenness, Huxley was tapped to fill the job. Energetic and resourceful, he used his new position to transform himself into the ship's naturalist, amassing a large collection of exotic specimens. Victorians were fascinated by such discoveries, and on his return Huxley began to parlay his work into a career in science.

Success, however, proved a "dreadful struggle."[20] For years, Huxley supported himself by teaching, initially at the South London Working Men's College where he charged a penny a lecture, six pence for a series.[21]

THOMAS HUXLEY

His language skills allowed him to write for popular periodicals – *Pall Mall Gazette, Macmillan's Magazine,* the *Reader, Saturday Review* and others.[22] But he remained so financially strapped that his marriage was postponed for five years running. "I can get honour in science," he wrote a friend, "but it doesn't pay."

Huxley Initially Opposed Evolution

When Huxley began his career, his knowledge was limited. His three years of medical training had focused largely on anatomy and treating diseased and dying people. While he was smart, at this stage of his life, Huxley's days were filled with the demands of earning a living. Because of this, he never learned biology, or the other sciences, in any real depth.

His students complained that while Huxley could dissect dead animals, he did not understand the pulsing chemistry of living things. One of his famous students, Sir Arthur Shipley, wrote, "our obsession was with the dead, with bodies preserved, and cut into the most refined slices."[23] Another student, EW MacBride, said that even late in his life Huxley persisted "in looking at animals as material structures and not as living, active beings."[24] "In a word," said MacBride "he was a necrologist" (a person who makes lists of the dead).

This lack of knowledge meant that Huxley never understood many aspects of science central to unraveling the mysteries of evolution. He did not know the complex chemistry of embryo development, and never fully understood why animals develop as they do and not some other way, nor why some changes are possible in evolution and others not.

There was also no time for the years of study required to learn nature's deeper processes. He did not understand the physics which underlay discoveries like those of Richard Owen. When Owen integrated all vertebrates into a single system, Huxley did not grasp how the laws of physics determine the form a skeleton can take (and thus determine body shape), nor understand why some forms can evolve and others cannot. He did not understand that the way humans walk is determined not by biology, but by the laws of physics. He failed to grasp that it was these laws which allowed Owen to state (correctly) that knuckle-walking apes could not have stood erect and evolved into humans.

Huxley knew the *what* of life, but not the *why.* He knew what exists, but never understood why it exists in that form, and not some other way. This lack of understanding led Huxley, for many years, to oppose evolution. Half a century after Lamarck, and a decade after Owen's convincing demonstration of evolution, Huxley was still arguing against the idea. In 1854, he wrote a venomous review of the new edition of *The Vestiges of Creation.*[25] In a lecture at the Royal Institution the following year, Huxley disputed the entire idea of evolution, or "progressive development" as he called it.[26]

Initially, Huxley also rejected Darwin's theory. But then lightning struck.

HUXLEY'S TWO BIG GOALS

Huxley's two great goals in life were to make the field of science a paying profession for men like himself (who did not come from family money), and to eradicate religion. For him, the two went hand in hand.

Science was then dominated by men of independent means, like Darwin, who could personally finance their efforts, and by a small group of university clerics who controlled Oxford, Cambridge, and many scientific societies.

The first universities were established in the middle ages by the Church to educate the clergy. This was true of Oxford and Cambridge. As the Victorian era opened, Anglican clerics still dominated both universities and were influential in deciding who received scientific appointments.

Huxley was infuriated at the positions given to men whose strengths, as he saw it, were not so much in science but in acquiescing to religion. He was also bitter toward men like Richard Owen who, though not born to money, had mastered the levers of Victorian society and moved into positions of privilege. None of them had to scratch for a living, at one penny a pupil.

Huxley found the entire system suffocating. We can glimpse his anger in an early encounter with Owen. Needing a reference to secure funds for his work, Huxley repeatedly wrote to Owen seeking his endorsement, but never received a reply. When they chanced to meet on the street, Huxley recalled, "I was going to walk past, but he stopped me, and in the blandest and most gracious manner said, 'I have received your note. I shall grant it.'"[27]

Huxley was incensed at what he viewed as Owen's condescension. "If I stopped a moment longer," he said, "I must knock him into the gutter."

Darwin's Theory as a Weapon Against Religion

Huxley's anger toward religion went beyond Oxford clerics. Like some others, he saw religion as a hodge-podge of superstitions that fogged the mind, corrupted education, and impeded social progress. But unlike others, he did not want to just eliminate religion. His second great goal in life was to *replace* religion as the guiding force in society with a new political party based on "scientific rationalism," directed by men like himself.

After initially rejecting Darwin's theory, Huxley realized it could be turned into an effective weapon for his social agenda. It was the battering ram that would break down religion and other obstacles to the "new reformation," and the party of "scientific rationalism" he sought to establish.

Two things are crucial here. First, Huxley thought Darwin's theory was deeply flawed and fundamentally wrong. Second, he didn't care. He would gladly promote bad science if it advanced his social goals. Cambridge historian Jim Secord explains:

> "Significantly, Huxley did not think that natural selection was the right explanation for the origin of species, [nor did he believe Darwin had] correctly explained any of the key topics dealt with in the *Origin*."[28]

Despite this, says Secord, Huxley realized that Darwin's theory could be "the flag of the new party in science" he and his friends sought to establish. "Huxley saw this very strongly," says Secord, "and coined the word 'Darwinism' to encourage the broadest possible implications for what he called a 'new reformation.'"[29]

Huxley Rejects Darwin: Evolution has Direction and 'Design'

Just how wrong did Huxley think Darwin was? Historian Peter Bowler calls Huxley a "pseudo-Darwinist who had little real sympathy for natural selection or the Darwinian approach to life."[30]

Huxley rejected Darwin's three most crucial claims (showing just how unscientific Darwin's ideas appeared even to supporters).

First, he rejected Darwin's claim that evolution is gradual. Darwin excluded any rapid jumps in evolution and Huxley knew this was flatly contradicted by the evidence. Before *The Origin of Species* was published Huxley wrote to Darwin, saying "you have loaded yourself with an unnecessary difficulty in adopting *Natura non facit saltum* [nature does not jump] so unreservedly."[31] Later, in his review of *The Origin* which appeared in *The Times* of London, Huxley publicly rebuked Darwin on this point:

> "Mr. Darwin...embarrassed himself with the aphorism 'Natura non facit saltum.' [nature does not jump]. We believe nature *does* make jumps...and a recognition of the fact is of no small importance."

Huxley also thought natural selection was a dubious idea. Even in his later years he wrote that "until selection and breeding can be seen to give rise to varieties which are infertile with each other, natural selection cannot be proved."[32] (The quote often attributed to Huxley about natural selection -"How extremely stupid of me not to have thought of that"- is a 20th century invention, first appearing only after his death.[33])

Most importantly, Huxley believed that the lynchpin of Darwin's theory – the idea evolution is "directionless" – was wrong. Huxley believed evolution was predictable, and thus had both *direction* and what he called *"design."*

In 1869, a decade on from *The Origin*, Huxley made this belief crystal clear. While he didn't believe in the type of "Divine design" promoted by William Paley, Huxley said "it is necessary to remember" that nature has "a wider Teleology [meaning "design"], based on the *foundational proposition* of Evolution."[35] What was this *foundational proposition* of evolution? Huxley said it is the idea that natural laws make evolution *predictable* and *inevitable*. He put it this way:

> "That [foundational] proposition is that the whole world, living and non-living, is the result of... *definite laws* possessed by the molecules of which... the universe was composed. It is no less certain that the existing world lay potentially in the cosmic vapour, and... from that vapour, [one could] have predicted, say the state of the fauna [animals] of Britain in 1869, with as much certainty as one can say what will happen to the vapour of the breath on a cold winter's day."[34]

Huxley's words are Victorian, but his meaning is modern: *the evolution of everything in nature is predictable and inevitable.* If, at the beginning of the universe, a person knew all the laws of nature they could precisely predict everything that would later evolve, right down to the exact animals that would exist in England in 1869. (This is not a misreading of Huxley's belief. He reaffirms it in print, verbatim, 20 years later.[35])

This belief is the opposite of the claim by Darwin (and modern Darwinists) that evolution has no predictability or direction. Huxley, however, believed evolution has both. He just doesn't want to attribute any of it to God.

This means that everyone, even Darwin's staunchest defenders, understood his theory was both wrong and inherently unscientific. (In his autobiography, an aging Huxley later dubbed himself "Darwin's bulldog," and bemoaned the rejection of Darwin's theory. But, after a lifetime of promoting Darwinism for political purposes, these comments were for public consumption.)

While Huxley privately believed Darwin's theory was wrong, he supported it publicly. Truth was not the issue. Early in his life, Huxley described himself as the "gladiator-general" in the war against religion.[36] Like all good generals, he was organized, ruthless, and saw his first priority not as truth, but winning the war. Darwin's theory was his most powerful weapon:

> "One of its greatest merits in my eyes," he wrote, "is the fact that it occupies a position of complete and irreconcilable antagonism to the vigorous and consistent enemy of the highest intellectual, moral and social life of mankind, the Catholic Church."[37]

(An equal opportunity bigot, Huxley loathed Catholics more than Anglicans, but both were above Eastern Europeans and Jews, whom he despised. In 1858, Huxley sent a note to his friend Joseph Hooker about "a little Jew, of whom I know nothing, and hate as I hate all the chosen people."[38] "Chosen people" was a term commonly used to refer to Jews.)

The *X-Club*: Taking Control of Science

To promote this agenda, Huxley formed a small group called the *X-Club.* "The bond that united us," one member noted, "was devotion to science, pure and free, untrammeled by religious dogmas."[39]

The *X-Club* "soon became an important informal pressure group," writes Harvard historian Janet Browne, "promoting the liberal naturalism associated with Darwin's theory."[40] Darwin's theory was the bludgeon Huxley would use to defeat his enemies – the Oxford-Cambridge elite, Richard Owen, and especially clerics and religious superstitions they promoted.

Although only nine in number, the *X-Club* came to dominate Victorian science for decades, each member organizing their own network to support the others. "One by one," writes Browne, "the members infiltrated every government panel and committee that dealt with scientific affairs."[41] Thomas Hirst became president of the Mathematical Society, George Busk president of the Royal College of Surgeons, Edward Frankland president of the Chemical Society, Joseph Hooker Director of the Royal Botanical Gardens and John Lubbock vice-chancellor of the University of London and president of the Linnaean Society (where Darwin had been ignored years earlier).

In succession, Hooker, Spottiswoode and Huxley served as president of the Royal Society, Britain's most important scientific organization. Huxley, Lubbock, William Spottiswoode and John Tyndall served as presidents of the British Association for the Advancement of Science.

Associates of the *X-Club* broadened the network. On Huxley's recommendations, for example, his protégé William Flower was appointed to a series of increasingly important positions, eventually becoming president of the influential Zoological Society.

This bureaucratic network made Huxley powerful. He discovered early what many others later learned (from Stalin to government clerks): bureaucratic control is power. Huxley seized it with relish and wielded it ruthlessly to achieve his goals. At a time when professional positions were few, Huxley's connections gave him wide influence over who received jobs, funding and professional awards. He became, in effect, the world's first "science czar."

The *X-Club* influenced the institutional levers of science at all levels, advancing favored candidates for high positions in academia and government. The group was systematic and comprehensive. They discussed, for example, whether *X-Club* member John Tyndall should assume the powerful chair in Natural Philosophy at Oxford (once held by Newton), a position which would allow them to control how evolution was taught in universities.[42]

Their network influenced the awarding of prizes, medals and scientific honors, many of which they gave to each other. Together the nine men received 18 honorary degrees and 12 of science's most prestigious awards, including three Copley Medals, five Royal Medals and the Lyell Medal.[43] Two were knighted.

Huxley enforced his agenda with an iron fist. Those outside the *X-Club* network viewed "Huxley and the close-knit band of Darwinians...as a dictatorial, self-regarding clique," writes Janet Browne. "A powerful brotherhood of older men at the summit of their careers who insisted that acolytes ought to adopt their position."[44] "In many ways," she says, this "read the situation accurately:"

> "The inner ring of private clubs and societies which ran scientific London – the X Club, the teaching laboratories and museums in South Kensington, the philosophers... in the Metaphysical Society and Athenaeum Club – were closed to outsiders. The members were influential people who kept a firm grasp on the tiller of scientific progress. Huxley enjoyed his cliques and believed that small groups of 'right-minded men' were by far the most effective way to get things done."[45]

Previously, science had been a disorganized affair, pursued by gentlemen of curious minds and independent means. No one imagined the type of tightly managed organization Huxley created to control what was researched, written, taught and rewarded in science. Once asked what the *X-Club* did, he replied, "Well they govern scientific affairs, and really, on the whole, they don't do it badly."[46]

Huxley accomplished his control with a ruthless politics worthy of the Medici. Unlike the Medici however, who merely sought to control princes, Huxley sought to control Knowledge and Truth.

The X-Club: Promoting Darwin's Failed Theory

Early on, Huxley used this growing power to aggressively promote Darwinism. He wrote three reviews of *The Origin of Species*, two published anonymously, and organized others to do the same.

Unable to generate much interest in Darwin's ideas, Huxley bought two journals, the *Natural History Review* and the weekly *Reader*, replaced their editors and re-launched them as pro-Darwin publications.[47] Despite a booming market for science publications, lack of interest in Darwin's ideas drove existing readers away and both journals soon failed.

Learning from their mistakes, Huxley and his *X-Club* associates founded the journal *Nature* to more effectively promote their social agenda and the Darwinism which justified it. "Far more than any other science journal of the period," says Janet Browne, "*Nature* was conceived, born, and raised to serve polemic purpose."[48] *Nature* became Huxley's megaphone for bashing enemies and promoting his political agenda, especially Darwinism.

"In the first year of its existence," says Browne, "there were six or seven articles using Darwin's scheme, two of which were written by Darwin himself."[49] (This is the same *Nature* that publishes today, and it retains a strong pro-Darwinian bias, inviting devoted Darwinists to review books critical of Darwinism, with predictable results.)

Despite this effort, Huxley still had trouble generating interest in Darwinism. Even those who appeared supportive were sometimes just currying favor with the man who controlled their future. William Flower was typical. A talented anatomist, Flower eventually rose to prominence, but his early positions were entirely dependent on Huxley.

A surgeon in the Royal Navy, Flower met Huxley through an *X-Club* contact. After providing crucial help in one of Huxley's battles with Richard Owen, on Huxley's recommendation Flower was appointed to the Royal College of Surgeons.

Flower continued to provide research for Huxley's ongoing fights with Owen. After a decade, Huxley rewarded Flower by arranging for his appointment to the prestigious Hunterian Professorship in Comparative Anatomy. This was how careers were made in Victorian England.

In later life, some slipped Huxley's noose and made their true beliefs known. Flower, for example, was a committed Christian and in an 1883 speech to the Anglican Conference he made clear that despite Darwin, he believed evolution and religion were not at odds, but fully compatible.

In the three decades after Darwin published, a pattern similar to Flower's was repeated with many others. If you wanted the patronage of the powerful *X-Club*, you would have to support Darwinism, or at a minimum, not oppose it. The price for outright opposition, as you will see, was steep.

Chapter Six

Darwin's Deadly Social Agenda

THE WAR AGAINST THE WORKING CLASS

The second major social goal for those supporting Darwin was to preserve a way of life threatened by the rapidly growing working class.

For a thousand years, England had been home to about four million people, mostly rural peasants led by a small group of aristocrats. In 1750, this abruptly changed. The industrial revolution transformed agriculture, forcing thousands of peasants off the land and into cities where they worked in factories and became a potent new social force – the "working class."

Wealth from the industrial revolution made possible safer water, better sanitation and vaccinations, all of which sharply reduced the death rate of working class children. The four million Britons of 1750 doubled to eight million by 1800, and reached a staggering 17 million by1850. Society wasn't minting any more aristocrats. This tidal wave of humanity was almost entirely working class. The elite were rapidly being swamped.

Mostly hardworking, the working class were nonetheless uneducated and chronically underemployed. They were poor, with all the problems that come with poverty. Many social reformers, especially Christians and socialists, saw knowledge and opportunity as the key to solving these problems. Education and social equality would lift people from poverty.

The Working Class Pollutes Society

The Darwinists saw something different: bad biology. They believed biology entirely determined a person's abilities. Education was useless against its overwhelming power. The "lower orders," as Darwin called them, were prisoners of their inferior biology. No amount of education would ever change that. Yes, a few might become successful, but these were accidents of nature, not the stuff on which broad social progress could be built. (The Darwinists assumed they were part of society's elite because their ancestors possessed superior biology, which they had inherited.)

Darwin viewed entire ethnic groups with contempt, describing the Irish as "careless, squalid, uninspiring... multiplying like rabbits."[1]

Darwin and others like him saw the working class much the same way. They were polluting society and destroying British culture. Darwin was particularly disturbed by the new vaccination programs. It was natural and for the best, he wrote, that "weak in body" die. Vaccination programs, however, meant that thousands of working class children, who previously would have died, now survived, swamping society with their inferior biology. This was "highly injurious to the race of man," said Darwin:

> "With savages, the weak in body or mind are soon eliminated; and those that survive commonly exhibit a vigorous state of health. We civilised men, on the other hand, do our utmost to check the process of elimination; we build asylums for the imbecile, the maimed, and the sick... There is reason to believe that vaccination has preserved thousands who, from a weak constitution, would formerly have succumbed to small-pox. Thus the weak members of... societies propagate their kind. No one who [breeds] animals will doubt that this must be highly injurious to the race of man. "[2]

Darwin continued, "it is surprising how soon... [medical] care wrongly directed, leads to the degeneration of a race." That is how he saw life-saving programs like vaccinations, as misguided humanism that quickly degraded a race.

WORKING CLASS BOYS (MINING COAL) WHO SHOULD NOT BE VACCINATED.

Social Progress By Sterilizing the Working Class

Without intervention, Darwin and others believed the "lower orders" would soon overwhelm the social system, and society would collapse in chaos. The stupid and inferior would inherit the earth. The only rational solution was to control the number of working class children through an aggressive program of what was later called "eugenics."

Eugenics literally means "good breeding." It is the idea that you can improve society the same way breeders improved animals, by weeding out the disabled, sick and slow-witted.

The belief that humans could be improved through selective breeding was as old as Plato. But the exploding growth of the working class brought new attention to the idea. In his influential 1798 book on population, Thomas Malthus wrote, "It does not seem impossible, that by an attention to breeding, a certain degree of improvement, similar to that among animals, might take place among men."[3]

Malthus, like everyone who considered the idea, however, saw an insurmountable obstacle. Eugenics would require sterilizing many people, essentially turning them into servants of those chosen to breed. No one thought this was politically possible. Malthus put it this way:

> "As the human race could not be improved in this way, without condemning all the bad specimens to celibacy, it is not probable [it] should ever become general."[4]

The Darwinists saw things differently. Eugenics was an urgent necessity. Society could be rescued from ruin only by an aggressive effort to eliminate the weak, handicapped, and those working class children who were "highly injurious to the race of man." Eugenics was the only rational solution. "Excepting in the case of man himself," wrote Darwin, "hardly any one is so ignorant as to allow his worst animals to breed."[5]

Science as a Weapon Against the Working Class

The difficulty for Darwinists was how to overcome opposition to an idea most people found monstrous. Their attempt was twofold. The element that had always protected humanity against programs like eugenics was the belief that Man is unique in nature. So, first they would use science to remove this belief. They would use science to redefine humans as nothing more than advanced animals.

Second, they would make the case that elimination of the "unfit" was inevitable and eugenics would make the process more humane. Nature's slow, nasty elimination process would be replaced with rapid, rational programs. Surely this was better than allowing generation after generation to live in squalor and misery.

ARE HUMANS UNIQUE?

Are humans fundamentally different from animals?

For 5,000 years, the answer seemed obvious. Humans build cities, write poetry, play music and sculpt statues from stone. Animals do not. Christianity took this uniqueness a step further, claiming humans have an immortal soul, which was reflected in the unique human mind.

To justify breeding humans like animals, the Darwinists needed to change the belief that humans are unique. They needed to redefine Man as nothing more than an animal, perhaps more advanced than other animals, but not fundamentally different from them.

The question was how to achieve this redefinition.

Claiming that humans evolved from apes wouldn't do the job. One of the great myths about the Victorians is that they were scandalized by this thought, but that is untrue.

The idea of Man as a creature of nature, with animal instincts, was uncontroversial, an established tenant of Western thought, from the ancient Greeks through early Christianity and on up to the Victorians. People believed that strong social training was needed to transform the wild child of birth into something recognizably human and civilized as an adult.

Nor was the idea that humans evolved from apes contentious. The similarity between ape and human anatomy was never in dispute, even small children could see it. The idea humans evolved from apes actually helped society and religion by making sense of human passions and the animal side of Man's nature.

Starting in the 1700s, naturalists like James Burnett wrote openly and without opposition about the possibility humans had evolved from apes, even speculating on how it might have happened.[6] By the early 1800s, the idea was commonplace. French anatomist Jean-Baptiste Lamarck proposed a precise mechanism for such a transition.[7] Prominent scientists like Sir William Lawrence, later head of the Royal College of Surgeons, offered their own explanations for the ape-to-man transition. And the hugely popular *Vestiges of Creation* included the evolution of humans from apes as part of its story.

The ape-to-man idea was not an issue because people assumed that at some point a unique human mind emerged, either through direct Divine intervention or, as many believed, through purely natural processes. People believed that while Man's body may have evolved from an ape, his mind was unique, as amply demonstrated by creations like *Hamlet*, Handel's *Messiah* and those Great Western Railway locomotives that left Paddington Station every hour on the hour.

Science as a Tool to Redefine 'Human'

Darwin understood that to define Man as nothing more than an animal he would have to strip away the unique mind. And he tried. In *The Descent of Man*, Darwin argued that humans are not just physically like apes, but *mentally* like them. "My object," he wrote, is to show "there is no fundamental difference between man and the higher mammals in their *mental* faculties."[8]

Darwin attributed human abilities solely to a *larger* brain. Humans simply have *more of the same* abilities apes possess, not *different* abilities. "The difference between man and the higher animals, great as it is, is certainly one of degree and not of kind," he wrote.[9]

Huxley supported this idea, claiming there was nothing unique about the human brain or mind other than size. The "mental and moral faculties are essentially...the same kind in animals and ourselves."[10]

The claim that humans are not mentally different from apes was unnecessary to make the case for evolution. There was little evidence to support it, and a great deal of evidence to the contrary (such as human civilizations). But this claim was critical to destroying belief in religion and promoting eugenics. Historian James Moore put it this way:

> "If humans were *only* beasts, it meant the overthrow of the established religion of Christianity. All Darwin's friends accepted that political connection."[11]

Darwin and his supporters wove speculation, anecdote and pseudo-science (like Huxley's skull comparisons) together to make a case for which they lacked any real evidence. And they utterly failed to convince society or other scientists that there is no real difference between humans and apes.

(Today, the "unique/not unique" debate is, with a few holdouts, resolved. Modern science shows the human brain is anatomically, genetically and *mentally* unique. "We tend to think of our own species as categorically different", says University of Chicago neuro-geneticist Bruce Lahn, and science backs that idea. "Human [brain] evolution is, in fact, a privileged process," he says, "categorically different from the typical processes of acquiring new biological traits."[12] Michael Gazzaniga, Director of the University of California SAGE Center for the Study of Mind and author of *Human: The Science Behind What Makes Us Unique*, confirms that human thought is unique, different in *type* from that possessed by animals.[13] Other leading cognitive scientists, like Harvard's Susan Carey, say the evidence shows humans have mental abilities *not derived* from any ability in animals. She says "the human conceptual repertoire is a unique phenomenon on this earth."[14)]

DARWINISM: MAKING RACISM 'SCIENTIFIC' AND EUGENICS 'HUMANE'

The Darwinists second line of attack was to make the horrors of eugenics sound both rational and humane.

Evolution works by elimination, said Darwin. In life's struggle, the inferior die. This elimination is inevitable, resistance is futile. Darwin said that soon "an endless number of the lower races will have been eliminated by the higher civilized races throughout the world."[16] In a coming Armageddon, wrote Darwin, "the civilized races of man will certainly exterminate and replace the savage races throughout the world."[15] The "civilized races," he said, would then eliminate the "less fit" *within* their own societies.

Darwin believed these eliminations were natural, unavoidable and necessary for humans to progress.

Darwinists promoted the idea that eugenics is more humane than these natural elimination processes. The first eugenics manual, written by Darwin's cousin Francis Galton, put it this way:

> "Eugenics is the study of [how we] may improve or impair the racial qualities of future generations either physically or mentally. What nature does blindly, slowly and ruthlessly man may do providently, quickly, and kindly. It becomes his *duty* to work in that direction."[17]

For Darwinists, eugenics was not a monstrous travesty, but a social "duty." Euthanasia was not a crime against humanity but, in Galton's word, a "kindly" act bestowed on society's unfortunate masses by the elite.

In the hands of scientific professionals, eugenics programs could be implemented quickly, transforming society within a few generations. Surely this was more humane than perpetuating the misery experienced by the feeble-minded, disabled and poor, generation after generation.

How Would Eugenics Work in Practice?

Darwin argued that some people are demonstrably inferior. He organized individuals and ethnic groups into categories such as "weak," "inferior" and "intellectually superior." In *The Natural Inequality of Man*, Thomas Huxley was blunt: "The doctrine that all men are in any sense...free and equal, is an utterly baseless fiction."[18]

Darwin argued that some entire races are "demonstrably inferior," having evolved at a slower rate than other races. Huxley "proved" this claim by comparing the skulls of different races. He ranked Europeans at the top, then descended down a list that ended with "the Negro or Australian and the gorilla." Biologically and mentally, he said, Africans and Australian aborigines are closer to the gorilla than to the advanced races of man.[18]

Galton and other Darwinists said these rankings "scientifically" ranked people, and determined who had biological value and who did not.

Darwin emphasized that evolutionary progress came not from breeding the best, but from eliminating the "unfit." This meant that the key to a successful eugenics program, and to social progress, was in eliminating other humans. "By proposing that evolution works primarily through the elimination of useless variants," says historian Peter Bowler, "Darwin created an image that could all too easily be exploited by those who wanted the human race to conform to their own pre-existing ideals."[19]

This was the future under Darwinism. Social progress would come not through education, but through a selection process in which an elite decided who in society has biological value and who does not, who can breed and who will be sterilized.

The small amount of social opposition Darwin generated came from those who grasped the full implications of these ideas, and they were horrified.

Opposing Eugenics and Social Darwinism

Opponents of eugenics (and what was later called "Social Darwinism") believed humans are more than their biology. They could point to many examples, such as America, a magnet for exactly the people the eugenicists sought to eliminate: the "poor, downtrodden, huddled masses." By 1859, when the *The Origin* was published, those "huddled masses" had built America into a powerhouse of innovation and industry that would soon roar past England as the world's most prosperous nation, able to provide free universal education for all children, elite and working class alike.

Those who opposed Social Darwinism realized that if eugenics became the solution to social problems, the world would be awash in waves of "scientifically" justified genocide.

One big myth about Victorian evolution involves the concern over eugenics. The story told today is that the Anglican clergy, apoplectic over Darwin's discovery of evolution and the vile claim humans had evolved from apes, organized a debate between the Bishop of Oxford Samuel Wilberforce, and Darwin's defender Thomas Huxley, to vanquish these evil ideas.

As the story goes, in the debate the Bishop railed on and on against Darwin's godless ideas. Having worked himself into a lather, he turned to Huxley and said, "Mr Huxley, please tell us, are you descended from apes on your grandmother's or grandfather's side of the family?"

At this, Huxley whispered to a colleague, "The Lord has delivered him into my hands." Standing, Huxley fixed his gaze on the hushed audience and replied "I would rather be descended from an ape than a dogma spouting Bishop." He then laid out the evidence for evolution, demolishing Wilberforce and the religious critics with a devastating presentation of fact and logic.

Today, this debate is held up as the iconic example of Victorian society's irrational reaction to evolution and the vitriolic abuse heaped on Darwin. It is said to mark the turning point in the war between religion and science, the decisive moment when facts began to prevail over superstition, and reason began to replace dogma.

The story has been repeated countless times, in articles, books and television programs. To this day, it remains a staple of those who say society is inherently irrational and must be brought, kicking and screaming to face reality.

The lack of reality, however, lies not with society, but with those who repeat this story. There *was* a debate at Oxford in 1860, but the real events were quite different.

The Big Debate: What Really Happened

On the Sunday morning prior to the debate, popular and respected cleric Frederick Temple (then Queen Victoria's personal chaplain), preached from the pulpit of Oxford Cathedral that evolution was a welcome idea. Far from conflicting with faith, he said, evolution confirms and strengthens it. The reason, said Temple, is that evolution makes God tangible. God ordained the laws of nature, and evolution allows us to see God's work in action as evolution unfolds.[20] (Later, as Archbishop of Canterbury, Temple confirmed this belief writing "the doctrine of Evolution is in no sense whatever antagonistic to the teachings of Religion."[21])

So if the debate wasn't over religion, what was it about?

The answer: science. Was Darwin's theory scientific? Many thought not. Rather, they saw it as pseudo-science designed to promote an anti-religion, pro-eugenics social agenda. This included Samuel Wilberforce.

As the son of William Wilberforce (whose lifelong campaign led to the abolition of slavery in Britain in 1833), Samuel grew up with a deep sense of protecting society's poor and defenseless. He became a zealous advocate of liberalism and opponent of eugenics.

WILBERFORCE (AGE 40)
THEN DEAN OF EXETER

Educated at Oxford, Wilberforce understood Darwin was attempting to justify his social agenda with science. Wilberforce would defeat Darwin's arguments with logic, facts and the fundamentals of science. "Contrary to the central tenet of the legend," writes Oxford's John Lucas, Wilberforce's argument was entirely scientific, showing Darwin was not supported by the facts, and that the "greatest names in science were opposed to the theory."[22]

Point by point, Wilberforce methodically dismantled Darwinism based on the evidence. For example, Darwin said evolution was gradual. If this is true, said Wilberforce, the history of life should show many tiny changes. But it does not.

Darwin explained this gap by saying the "intermediate" forms existed only briefly and therefore don't leave a record. Thus, says Wilberforce, Darwin uses both the presence and absence of "intermediate" species to prove his theory. He asks, what sort of theory uses both the presence and absence of evidence as proof it is true? This is not only bad science, he said, but it is awful logic.

Wilberforce questioned if Darwin's theory was even real science. Pointing to *The Origin's* many assumptions and unsupported claims (its "utterly rotten fabric of guess and speculation"), he argued that Darwin had violated a cardinal rule of science by substituting speculation for empirical evidence.

Delving deeper, Wilberforce said Darwin had violated science's second cardinal rule. Since natural selection makes no predictions, and thus cannot be tested and proven true or false, can it be considered a "law" in any scientific sense? Wilberforce said not.

His conclusion: by substituting speculation for facts, and by using a "law" that could not be tested, Darwin had violated science's two most sacred tenets, making his theory inherently unscientific.

Wilberforce's speech was later published in the *Edinburgh Review*. After reading it, Darwin wrote to a friend, "it is uncommonly clever; it picks out with skill all the most conjectural parts, and brings forward well all the difficulties."[23]

And what of Wilberforce's question to Huxley about which of his grandparents were descended from apes? It's a myth.

"There is, to my knowledge, no reference to it in the few contemporary reports," writes British historian John Brooke, who has researched the debate in depth.[24]

Huxley's own account of the debate makes no mention of the remark.[25] Nor does Joseph Hooker, who was Darwin's closest friend and was at the debate. Hooker wrote to Darwin immediately afterwards that Huxley was both inaudible and unable to counter Wilberforce's arguments:

> "Huxley...could not throw his voice over so large an assembly, nor command the audience...nor put the matter in a way that carried the audience."[26]

In other words, Hooker thought Huxley had failed in the debate.

More importantly, if Wilberforce had made the comment both Huxley and Hooker would have mentioned it in their letters right after the debate. They did not, because Wilberforce did not say it.[27]

Except for the intelligentsia in London, the debate drew only modest public notice elsewhere in England.[28] Like so many myths about Darwin,

Wilberforce's supposed statement is another 20th century invention. The story first appeared in a letter to Francis Darwin, written many decades later, when all those involved in the debate were dead.[29] It was picked up from there, and in the 20th century the Darwinists turned into "fact."

Darwin's Three Sins Against Science

The reality is that the supposedly superstitious religious cleric, Wilberforce, had succinctly captured the two great sins Darwin had committed against science.

First, Darwin put his theory above the facts. In science, facts determine the validity of a theory, never the other way around. Bacon called this the "bedrock principle of science." Newton chiseled it into the archway to science with the rule "Evidence may *never* be evaded by hypothesis." Putting assumptions ahead of evidence makes a theory ideology, not science.

From the start, however, Darwin demanded to be an exception to this cardinal rule, breaking it over and over. Harvard paleontologist Stephen Jay Gould wrote that Darwinists "sacrificed" the integrity of the evidence to achieve "harmony with their theory," and in the process, he said, science "had paid an awful price."[30]

Darwin's second sin was more egregious, creating an unbridgeable chasm between his theory and science. People were incredulous at his claim that evolution had no predictability. Science's claim to Truth, in the big sense, rests *entirely* on the ability of laws to make predictions. (Many others at the time made these same two points.)

Nobel Prize winning Stanford physicist Robert Laughlin calls Darwinism an "anti-theory." Anti-theories, he says, "have exactly the opposite effect of real theories, they stop thinking rather than stimulate it:"

> "Evolution by natural selection, for instance, which Charles Darwin originally conceived as a great theory, has... come to function more as an anti-theory, called upon to cover up embarrassing experimental shortcomings and legitimize findings that are at best questionable."[31]

"A key symptom of ideological thinking," says Laughlin, "is the explanation that cannot be tested."[33] That certainly is the essence of Darwin's directionless evolution.

CHAPTER SEVEN

DARWIN'S DEADLY SOCIAL AGENDA

THE WAR AGAINST THE EVIDENCE

Darwin ended his life more or less in a sense of confusion.[1]
Simon Conway Morris, Cambridge University, 2003

Society rejected Darwin's ideas. But what about scientists?

Key facts about evolution were well known, even in 1859. Huxley and the *X-Club* were powerful, and worked hard to promote Darwinism. What happened?

As you read on, ask yourself the same two questions Victorian scientists asked. First, did Darwin's theory fit the facts? Did it match what scientists saw in nature? Second, did his theory resolve mysteries, unite the events of evolution into a single coherent picture, and point the way to new discoveries? In short, did it explain, integrate and predict?

DARWIN'S CONFUSION

"The most recent volumes of Darwin's correspondence shed new light on the complex question of the *Origin's* reception," says historian Jim Endersby. They "destroy the myth of Darwin's instant impact."[2] In the years after 1859, Darwin's theory was widely seen as confused, factually flawed, unscientific and of little value in understanding evolution.

"Darwin had not anticipated the vehemence with which even the most respected scientists and philosophers in his day would denounce his efforts as not being properly scientific," says historian David Hull.[3]

Darwin was particularly interested in the reaction of John Herschel, "the most eminent philosopher-scientist of the time." On hearing that Herschel had called his concept of natural selection "higgeldy-piggeldy," Darwin wrote "what this exactly means I do not know, but it is evidently very contemptuous, [and] is a great blow and discouragement."[4]

Similar scathing comments rolled in from across the scientific spectrum. Initially Darwin tolerated the criticisms, writing "my views have often been grossly misrepresented, bitterly opposed and ridiculed, but this has generally been done, I believe, in good faith."[5]

But as the withering criticism piled up, Darwin's feelings changed. "I have got fairly sick of hostile reviews," he wrote Huxley. "Yet having seen what all the Reviews have said against me, I have far more confidence in the truth of [natural selection] than I formerly had."[6]

The problem Darwin faced went beyond his many errors of fact, to the confusion of his basic ideas. "I am conscious that I am in an utterly hopeless muddle," he wrote his friend Asa Gray. "So I must begin at once to work on a corrected edition."[7]

"*The Origin* was still a work in progress long after its first appearance," says Jim Endersby. Darwin was "tinkering constantly with the text, adding and amending, revising and rethinking and, above all, responding to criticism."[2]

Darwin's confusions eventually led him to produced six heavily revised editions of *The Origin*. In each new edition, Darwin furiously reworked his claims in an effort to defend his theory against new criticisms. Over time, 75% of *The Origin* was revised. In some cases, Darwin *entirely reversed* his earlier claims.

Darwin's War Against the Evidence: *The Origin of Species*

Why was Darwin so confused? And why did he *remain confused* for so many years?

The answer is that his theory was at war with the evidence. Darwin's attempt to make evolution directionless forced him to deny established facts in every area of science. From anatomy and botany to physiology and zoology, nature simply did not work as Darwin claimed.

For example, one is tempted to think that a book titled *The Origin of Species* would demonstrate, well, the origin of a new species. But it did not. "In the *Origin of Species*," wrote Darwin, "the derivation of any particular species is never discussed."[8] Never discussed? Darwin, it turns out, had no example of the origin of new species. Despite its title, his book is all about what happens to a species *after* they appear.

The best explanation for the origin of species at the time came from Richard Owen, whom Darwin met as a young man.

During his five-year voyage on the *Beagle*, Darwin amassed a considerable collection of fossils. While good at describing surface features, Darwin often misjudged the underlying biology these features represented. A year after his return, Darwin was introduced to Owen, who agreed to help by reviewing some of Darwin's analysis.

Owen quickly saw that Darwin had misidentified many specimens. This included the bones of a giant rat and other extinct creatures. Although they were unearthed in South America, Darwin was convinced they were related to animals in Africa. Owen showed that despite some superficial similarities, the creatures were ancestors to animals in South America, not Africa. For Owen, the relationships were obvious. For Darwin, they were mystifying, and the experience embarrassing.

By all accounts Darwin was a poor student, with little interest in his studies. He spent a great deal of time riding, shooting birds (a particular favorite) hiking and hunting beetles. His limited understanding of biology and anatomy forced him to abandon his medical studies at Edinburgh University.

Darwin's father, fearing he would end up an "idle gentleman," sent him to Cambridge to study Greek, Latin and Philosophy (at which he did well). But science and math were not Darwin's forte. He took the required elementary science and math courses, but struggled so much he had to hire a private tutor. Even with this help, he passed both only by the slimmest of margins.

Years later Darwin wrote of himself, "I have no great quickness of apprehension... I am therefore a poor critic: a paper or book, when first read, generally excites my admiration, and it is only after considerable reflection that I perceive the weak points." Darwin went on to say,

> "My power to follow a long, abstract train of thought is very limited.... Some of my critics have said, 'he is a good observer, but has no power of reasoning!' [although] I do not think that can be true... On the favourable side, I think that I am superior in noticing things which easily escape attention and in observing them carefully."[9]

Darwin's self-assessment summed up how most scientists saw him: a good observer of details, who failed to grasp the underlying biology of those details. For example, Samuel Haughton, whose work helped revolutionized Victorian biology, dismissed Darwin's ideas for years as the muddled thoughts of someone who didn't understand biology in any real depth. Haughton's work, on the other hand, "made Darwin feel out of his depth and unable to provide an adequate reply."[10]

Darwin's lack of deep biological knowledge led him into repeated errors. For example, he misidentified some of the finches on the Galapagos Islands as other birds. In a repeat of his experience with Richard Owen, the great Victorian ornithologist John Gould later explained to Darwin the differences between the birds.

Darwin's limitations came into sharp relief in 1860, when Richard Owen's review of *The Origin of Species* listed one factual error after another.[11] After all these years, Darwin, it seems, was still misjudging basic biology.

In his review, Owen also dismissed Darwin's obsession with "special-creation," saying there was no sympathy among scientists for such objections to evolution. Owen drove this point home with a list of Biblical events that were generally agreed to be untrue in any literal sense.

Owen then went to the heart of things. He said there is wide agreement that evolution ("the succession of species by natural law") is a fact. He then posed three questions about natural selection, Darwin's explanation for how evolution works.

First he asked, what is the evidence for natural selection? If it is the driving force in all evolution, thousands of examples should be seen in nature. But they are not. While there are many examples that animals adapt to their environment, Darwin offered only speculation that this occurred through natural selection. On the contrary, said Owen, natural selection does not match what scientist see in nature ("the facts of actual organic nature").

Next, said Owen, even if natural selection exists, can it be the *only* law of evolution, as Darwin claims? This is impossible, he said, pointing to other laws of evolution already known to exist.[12] The idea that something as complex as evolution could be explained with a single law seemed absurd. Even the great Newton needed more than that.

Owen then asked, is Darwin's theory real science? Natural selection "either is, or is not, a law of nature," wrote Owen. But Darwin says natural selection has no direction. It is in essence, a random process. This means it cannot be tested and thus cannot be a law in any scientific sense. Directionless natural selection made Darwin's theory inherently unscientific.

The review was devastating and Darwin churned with anger. He later wrote "I used to be ashamed of hating him [Owen] so much, but now I will carefully cherish my hatred and contempt to the last days of my life."

Huxley also saw Owen as an enemy, having spent a decade in vigorous and ceaseless attacks on him in an effort to destroy his career.

For his part, Owen understood this was not a dispute about evolution. He viewed Darwin's theory as pseudo-science cooked up to demagogue religion, redefine what it means to be human and justify the racist ideas that later underpinned eugenics. Earlier, when Darwin and Huxley argued there are no differences between ape and human brains other than size, Owen called it nonsense, saying the human brain has a unique organization not possessed by apes.[13]

Huxley compared the brains of humans and apes to make the case that some races are "demonstrably inferior" to others. He claimed the difference between the brain of a European and an African is as great as between the African and an orangutan.[14] Owen showed this was untrue, demonstrating there was little difference between the brains of the different races.[15] More importantly, argued Owen, there is no difference in the mental abilities of the races, saying "there had been no province of intellectual activity in which individuals of the pure Negro race had not distinguished themselves."[16]

Modern science again proves Owen was right. The human brain does differ significantly from apes, but does not differ by race. More importantly, today we know that human intelligence is not related to brain size but to organization. Small brain or large, it makes no difference. It is how our brain is *wired* that makes us smart.

Darwin's War Against the Evidence: 'Convergence'

Even those who thought Darwin had the germ of a good idea became critical of his cavalier approach to facts. St George Mivart, an up-and-coming young scientist and Fellow of the Royal Society who was initially supportive, came to believe Darwin used unsupported claims to avoid some well established facts about evolution.

It was known that similar traits evolved in unrelated species. Over and over, nature converges on the same processes regardless of the species or environment. Eyes, for example, work pretty much the same throughout the animal world. It was also well known that animals retain their traits despite drastic differences in climate. For example, tropical penguins (yes, there is such a thing) are nearly identical to their cousins in the arctic.

Facts like these contradicted Darwin's claim that evolution was entirely determined by differences in environment and natural selection. Mivart asked Darwin how he reconciled his theory with these facts. Unable to do that, Darwin began a covert war against Mivart, secretly arranging for the publication of essays harshly critical of Mivart's work.[17]

Secret attacks were a long-standing practice of Darwin's. "This indirect defensive technique had served Darwin well in the past," writes historian Janet Browne. "[It] conveniently allowed him to attack with the words of others while maintaining a reputation for non-confrontation."[18]

Darwin and Huxley also launched a campaign to personally smear Mivart, calling his work "the product of bigotry [and] too much Catholicism."[19] "Catholicism was a convenient enemy here," says Browne, "representing to Darwin every outmoded tradition, superstition, and ritual that he felt should be forcibly expelled from modern life."[20]

When a friend said he found the attacks on Mivart extreme, Darwin replied that he was "not as good a Christian" as his friend, "for I did enjoy my revenge."[21]

Others who opposed Darwin came in for similar treatment. Darwin wrote out an enemies list, noting those on "our side" and those who were "outsiders." Outsiders would be punished.

Meanwhile, he called allies like Huxley, my "good and kind agent for the propagation of the Gospel – the devil's gospel."[22]

This comment is a window into Darwin's mind. When combined with his utter unwillingness to deal with important facts about evolution, and his personal attacks on those who asked legitimate questions, it reveals that Darwin saw these conflicts, not as a scientific debate, but as an ideological war to propagate his social agenda.

DARWIN'S CONCEPT OF SCIENCE: IMAGINATION CAN SUBSTITUTE FOR FACTS

In the mid-1800s, science was still young. Many fields were just creating themselves and much was unknown. Facing huge gaps in their knowledge, many scientists "felt perfectly justified in remaining silent on those issues for which they lacked sufficient data," says historian David Hull.[23] Like Richard Owen, they were aware of how much they *didn't* know about nature.

Lack of knowledge, however, is not a hindrance if you believe your imagination is the equal of evidence, which is how Darwin saw it. "Your imagination must fill up *very wide* blanks," he wrote.[24] Darwin repeatedly sought to explain complex processes with simplistic ideas based entirely on his imagination, unconnected to anything in nature.

One of Darwin's inventions –"blending"– provides a window into Darwin's thinking. The picture it presents is startling, revealing how he thought about evidence, ideas, and the whole scientific endeavor. "Blending" is a microcosm of the problems faced by all of Darwin's key ideas.

DARWIN'S WAR AGAINST THE EVIDENCE: DEVELOPMENT & BIOLOGY

Evolution is more than just changing individual traits. Every change, no matter how small, must be integrated into development, the complex process that turns seeds into sunflowers and embryos into elephants. In the 1850s, biologists had begun to unravel some of the intricacies of this unfolding process. As they did, one thing was clear: it didn't match Darwin's claims about evolution.

Darwin faced a dilemma. He could either change his theory to fit the growing knowledge about development, or he could invent his own development process. He chose invention.

Darwin said evolutionary changes are incorporated into development through an imaginary process he called "blending." For Darwin, creatures were like modeling clay. This meant that each of the body's billions of cells were free to move a little bit to form a new trait.[25]

This was a simplistic concept of development – something a child might imagine. There was no evidence of blending, nor could Darwin explain how it worked biologically. But he *needed* blending to be true. It freed his evolution theory from the laws of development.

For biologists, blending was nonsense, and blistering criticism began immediately. Within weeks, Darwin was on the defensive, his ideas about development completely muddled.

There were many problems with "blending," but a few give a snapshot of the immense gulf between Darwin's imagination and nature's reality.

From the University of Edinburgh, well-known biologist Fleming Jenkins wrote Darwin, pointing out a "fatal error." Blending does not lead to slight changes in a trait, but to an entirely *different* trait, just as mixing red and blue creates purple. Jenkins said that what little biologists knew about inheritance indicated it involved "discrete" changes, *the opposite* of Darwin's blending process.

Remarkably, Darwin also did not realize that blending would render natural selection impotent. With blending, the animals in each group would quickly become the same. Natural selection can't act if there are no physical differences between them, rendering it useless.

Blending also contradicted basic biology.

The most celebrated example of natural selection were the small finches Darwin found on the Galapagos Islands, near South America. When the birds first arrived in the area they were all alike. But once isolated on different islands, their beaks changed size and shape to match the different nuts on each island.

Darwin said that adaptive changes like those seen in the finches occur through blending.

For biologists, this was a startling misunderstanding of basic biology. Everything in life is made from cells. When a beak gets bigger, the cells don't get bigger, expanding like tiny balloons. Instead, *more* cells appear and are *organized differently*. Some cells increase the beak's width, some its thickness, while others change how the beak connects to the surrounding muscles and bone.

Plus, you can't just make the beak bigger and stick it in the same spot. Any change requires thousands of other adjustments. If the beak gets bigger, the position of everything on the head changes – eyes, throat, neck, spine – each triggering a cascade of other changes in the body. Some changes are large, some small, but all are necessary for the head and neck to work correctly with the new beak.

Any theory of evolution must explain how all these changes are incorporated into development. How do all the new cells appear at precisely the right time and right place to create these changes? And how does each of the millions of cells know what to do when it arrives? How does it know if its job is to make the beak thicker, or to connect with a muscle?

Blending explained none of this. (As we will see later, genes don't solve the problem either. As geneticist Richard Lewontin told visitors to his Harvard lab, the body doesn't have specific genes for beaks or noses, quipping, "just how nose cells know they are part of a nose, nobody knows."[26])

There was another important problem with blending. In evolution, some parts of the body change while others stay the same. Take those finches. A Waxbill is smaller than your finger and weighs ¼ of an ounce. A Warbler is many times heavier, and fills your palm. Yet both have the same form. If change occurs through blending, where everything is mushed around, how could size change so much while the basic body plan remained nearly identical?

And then there was that perennial problem with Darwin's ideas – no evidence. Each year brought new discoveries about evolution, but nowhere was there any evidence of blending. On the other hand, Lamarck's explanation that new behavior leads to adaptive evolution was visible everywhere, as even Darwin acknowledged.

> "The domesticated duck flies less and walks more than the wild duck," its wings shrinking in size; "the domesticated rabbit becomes tame from close confinement [and] the retriever is taught to fetch and carry."[27]

However, if Lamarck's process played a major role in evolution it would undermine both of Darwin's social goals, showing that:

- Natural selection is not the only force in evolution, and
- Eugenics is not the only method for improving society.

Because of this, Darwin claimed for years that "acquired traits" had only a marginal effect on evolution. But overwhelming evidence showed this was untrue. After a decade of withering criticism, Darwin finally threw in the towel writing, "It appears I formerly underrated the frequency and value of these latter forms of variation as leading to permanent modification of structure independent of natural selection."[28]

"In rescuing natural selection from its critic," says historian James Moore, Darwin "drew repeatedly on acquired traits."[29] In 1868, Darwin published an entire book showing "that a multitude of new acquired characters are often faithfully transmitted."[30] Later editions of the *Origin* incorporate many examples of directly inherited traits, and the sixth edition had an entire chapter devoted to acquired traits.

Importantly, Darwin said Lamarck's acquired traits were the best explanation for adaptive evolution writing, "we may on the whole conclude that inheritance of acquired characters is the rule, and non-inheritance the anomaly."[31] Lamarck had been right all along.

DARWIN CONCOCTS A *NEW* INVENTION: PANGENESIS

While inventing his own development process, Darwin conjured up tiny particles he called "gemmules," which he said carried adaptive changes through the body.

Gemmules had no basis in biology, nor could Darwin explain how they work (nor were they a predecessor to the idea of genes). Nonetheless, he pressed ahead with the idea.

His cousin and close confident, biologist Francis Galton, agreed to help by locating some gemmules. After three years, Galton published his results. He had looked high, low, and in every drawer of biology's cabinet and found nothing. Nature held not the slightest hint of gemmules or Pangenesis ("the doctrine of Pangenesis, pure and simple... is incorrect").[32]

True to form, rather than accept the evidence (from a close supporter, no less), Darwin challenged Galton's research, writing a long rebuttal in the journal *Nature*.[33]

Darwin continued to defend gemmules through years of harsh criticism. When forced to finally concede they don't exist, he was despondent. "I did not sufficiently reflect on the subject and saw not the difficulty of believing in the presence of gemmules."[34]

You may detect a pattern here. "I underestimated the difficulty," "I did not sufficiently reflect" and similar phrases were heard repeatedly from Darwin. Science is straightforward. It starts with facts and looks for an explanation. Darwin did it the other way around. He started with his theory and looked for facts to support it. When the facts didn't fit, he invented his own facts, his own reality, with the same failed results each time.

Darwin's War Against Flowers, Fossils and Fast Evolution

Darwin's war against the evidence extended far beyond biology. Take the crucial question of evolution's speed.

If new species appear abruptly, they could not have evolved though a directionless process. There simply wouldn't be sufficient time for complex traits to emerge from random changes. This saddled Darwin with the need to claim evolution occurs only in tiny increments, which gradually build into complex traits. Darwin said evolution is so gradual that change would be visible only when viewed over many eons.

Like Darwin's other claims, there was precious little evidence for gradual evolution, and a mountain of evidence which showed that change is abrupt and rapid. For example, one of evolution's great mysteries was the origin of flowers. They appear abruptly in life, seemingly from nowhere, showering the world in dazzling new colors. Flowers were a remarkable evolutionary event, and cried out for an explanation. But because they appeared rapidly,

Darwin dismissed the origin of flowers as "an abominable mystery," impossible to explain. (We will see later how this mystery was eventually solved.)

Fossils were a particular embarrassment for Darwin.

The first systematic study of the earth began in the early 1700s. By 1785, James Hutton and other geologists began to dramatically extend the age of the earth. Work by Adam Sedgwick, Roderick Murchison, and others led to important advances every year.[35] By 1830, Sir Charles Lyell was able to assemble a timeline which stretched many tens of thousands of years into the past, and revealed the Earth had gone through several clearly identifiable evolutionary stages.[36]

Inspired by Linnaeus and spurred by the global collection of fossils, geologists began to chart the history of life. By the 1840s, this record revealed that life was almost as ancient as the Earth itself, and had progressed through a series of clearly defined stages.[37]

The fossils revealed something else: life changed in sharp bursts. All across the globe a clear and consistent picture emerged. Starting deep in the ancient past, new species appeared abruptly, fully complex from the start, and then changed only slightly over millions of years. These ancient transitions were so fast that geologists called them "explosions," giving them names like Cambrian Explosion and Devonian Explosion.

Lamarck's process predicted evolution would look like this, rapid and abrupt. Darwin's theory predicted the opposite. Richard Lewontin explains: If Lamarck was right, "then fossils should show distinct jumps separated by long periods of constancy."[38] If Darwin was right, "there should be a slow continuous change in form."

Naturalists examined thousands of species from around the world and found no instances of Darwin's gradual evolution. Like blending and gemmules, it simply did not exist. On the other hand, they saw Lamarck's evolution *everywhere*. All parts of the globe and every period of history clearly showed that evolution occurs in short, rapid bursts. Lamarck was right (again), and Darwin was wrong (again).

Darwin knew for decades that there was no evidence of gradual evolution, acknowledging it many times in private and in print. In *The Origin of Species*, he asked why no "infinitely numerous transitional links" are found in life's history:[39]

> "The number of intermediate varieties which formerly existed [must have been] truly enormous...Why then is not

> every geological formation and every stratum full of such intermediate links? Geology assuredly does not reveal any such finely graduated organic chain."

Darwin understood the profound implications of the evidence, writing, "this perhaps is the gravest objection which can be urged against my theory."[40]

Darwin Declares His Theory is Superior to Evidence

Despite the evidence, Darwin was unmoved. He said it is the facts that are wrong, not his theory. "The geological record is extremely imperfect," he wrote. "This fact will to a large extent explain why we do not find interminable varieties, connecting together all the extinct and existing forms of life by the finest graduated steps:"[40]

> "Species...sometimes falsely appear to have come in abruptly; and I have attempted to give an explanation of this fact, which if true would have been fatal to my view. But such cases are certainly exceptional."[41]

The evidence was clear. Darwin was wrong. His staunchest supporters warned that this threatened his entire theory. His cousin Francis Galton conceded the criticism, agreeing evolution occurs in what he called "a series of jerks."[43] Even Thomas Huxley, in his 1860 review of *The Origin of Species*, wrote "Mr. Darwin's position might...have been stronger if he had not embarrassed himself with the aphorism 'Natura non facit saltum.' [nature makes no leaps]. We believe nature *does* make jumps."[42]

"Huxley...warned Darwin that his strict and unwarranted adherence to gradualism might undermine his entire system," wrote Stephen Jay Gould. He impressed on Darwin that "the fossil record, with its abrupt transitions, offers no support for gradual change, and the principle of natural selection does not require it – selection can operate rapidly."[43]

For supporters, gradual evolution was an embarrassment. To others it was just another instance of Darwin's cavalier disregard for facts, one more example of his contempt for science's cardinal rule: facts determine the validity of a theory, not the other way around. It reinforced the belief that both Darwin and his theory were deeply unscientific.

Darwin had always understood, that on this point, the evidence was entirely against him. Privately he worried, writing in the margin of his 1844 *Essay*, "If species really [are] created in showers...my theory [is] false."[44]

(A vast expansion of fossil evidence in the 20th century, coupled with new dating technology, decisively resolved the issue. Modern paleontologists explain that while new species do evolve from old, the process is rapid. "The fossil record flatly fails to substantiate...finely graded change," write paleontologists Niles Eldredge and Ian Tattersall. New species appear "with a bang... often with no evidence of prior evolution."[45]

(This means new species *are* created in abrupt "showers." And as Darwin correctly understood, it also means his theory is false.)

Darwin Ended His Life in Confusion

Blending, gemmules, gradualism – they were all part of a pattern Darwin repeated over and over. Rather than explain, integrate and predict, he took another approach – attack, invent, and exclude. When facts conflicted with his theory, he attacked them as flawed or inaccurate. When this failed, as it always did (facts are stubborn things), he invented imaginary processes to make the facts fit his theory. When his inventions failed, Darwin complained that he had been misunderstood ("my conclusions have lately been much misrepresented") or acknowledged his thinking was flawed ("I underrated...," "I did not sufficiently reflect...," "I saw not the difficulty...."[46]

But even after these admissions, rather than accept the evidence as proof his theory was wrong, Darwin simply declared the facts irrelevant to evolution. In Darwin's world, his theory was superior to any evidence.

The denial of the facts, the attempt to substitute speculation for evidence and the resulting confusion continued throughout Darwin's life. In 1860, Samuel Wilberforce pointed to the "utterly rotten fabric of guess and speculation" that filled *The Origin of Species*. In 1871, *The Times* of London made the same point in its review of Darwin's *The Descent of Man*, writing that Darwin's ideas will, "in the minds of the learned [create an] enormous and painful improbability by the amount of speculations on which Mr. Darwin relies." *The Times* assessment was scathing:

> "We wish we could think that these speculations were as innocuous as they are unpractical and unscientific, but a man incurs a grave responsibility who advances the disintegrating speculations of this book. He ought to be capable of supporting them by the most conclusive evidence of facts. To put them forward on such incomplete evidence, such cursory investigation, [and] such hypothetical arguments as we have exposed, is more than unscientific – it is reckless."[47]

While Darwin was acknowledged for his work on niche topics like barnacles and pigeons, his evolution theory continued to be seen as "speculative," "unscientific," "confused" and, for those concerned with his unwarranted claims on serious social matters, simply "reckless."

Darwin Discredited

Because he was constantly at war with the evidence, Darwin spent much of his later years fighting with critics and offering increasingly muddled explanations for things. "Darwin ended his life more or less in a sense of confusion," says Cambridge paleontologist Simon Conway Morris.[48]

It is a measure of how widely Darwin's theory was rejected, that despite Huxley's great power and Herculean efforts, it remained marginal throughout the Victorian era. A snapshot of how the scientific community viewed Darwinism is seen at the 1894 meeting of the British Association for the Advancement of Science.

An aging Huxley arrived at the meeting to find himself still defending Darwin from attacks on all sides. Scientists of every stripe repudiated Darwinism, including leading physicists, like Lord Kelvin, and many biologists. Even Darwin's supporters had "to admit that some of the objections appeared to be valid and that some of the alternative mechanisms would have to be taken seriously," says historian Peter Bowler.[49]

In a major speech, Association president Robert Cecil told the group that while evolution was well-accepted, Darwin's natural selection was another matter. Despite decades of effort, said Cecil, no one had shown that natural selection produced new species, or even major changes in species.

Echoing a problem that had haunted Darwinism from the beginning, Cecil went on to say that Darwin's theory seemed impossible to test experimentally, and thus was of little value to science in understanding evolution.

Peter Bowler says that as the century came to a close, "those who opposed the selection mechanism had no doubt about the overall trend. Darwinism was on the decline and would soon be eliminated altogether as a major evolutionary theory." A confidence, says Bowler, "best illustrated by [a book entitled] *At the Deathbed of Darwinism*."[49]

After Huxley's death, a group of ardent eugenicists continued to promote Darwinism as the "scientific" solution to social problems. But within science it remained, as it had always been, a relatively marginal idea that, without Huxley's great effort, would have long ago disappeared.

The evidence for evolution, however, continued to mount. New examples of Lamarck's acquired traits, Owen's laws of organization, and the laws of development appeared yearly.

Lamarck was particularly influential. "The inheritance of acquired characters had great breadth of appeal in the last decades of the nineteenth century, cutting across both national and disciplinary boundaries," says prize-winning historian Richard Burkhardt.[50] It explained a wide range of evolutionary facts – the fossil history, geographic variations in plants and animals, nature's rapid adaptive changes, the degeneration of useless organs, and the origin of many instincts and behaviors.

Lamarck's theory integrated evolution with all areas of biology, and made predictions that could be scientifically tested. "The Lamarckian position was supported in England, France, Germany, Austria, Switzerland, Italy, Russia, and the United States", says Burkhardt," by embryologists, paleontologists, physiologists, bacteriologists, and plant geographers."[50]

PART 3

HOW DARWIN CAME TO DOMINATE

Redefining Evolution

Today, it is hard for us to grasp just how comprehensively scientists rejected Darwinism well into the 20th century. But they did.

So the question is, with little supporting evidence, its unscientific denial of predictability and its wide rejection in science, how did Darwinism (on its "deathbed" in 1900) come to dominate current ideas about evolution?

Beyond that, where did the false history of evolution come from – all those myths about the Victorians, evolution, and Darwin?

And if evolution was widely accepted by Victorian clergy and people of faith, what led to the 20th century religious battles over evolution?

In the 20th century, while Darwinism remained a marginal idea in science, it became popular among those who sought to impose eugenics and a biological view of social progress on humanity. This kept Darwinism alive for decades.

Then, in 1946 an unexpected turn of events catapulted Darwin's theory over other ideas about evolution, and eventually made the name "Darwin" synonymous with "evolution."

Immediately after this shift, evolution, which had been on the verge of a revolution in thinking, suddenly lurched back to 19th-century ideas.

In the years that followed, as Darwinists came to control how evolution was defined, they fabricated our current beliefs about Victorian evolution and Darwin's place in it.

In those years, however, new scientific discoveries continued to contradict Darwin's explanation for evolution. Modern Darwinists dealt with these discoveries as Darwin had done. In each case they first attacked the evidence, then invented imaginary explanations for it, and finally dismissed it as irrelevant to evolution.

This pattern, begun by Darwin, was repeated dozens of times in what today are called the "Darwin Wars." Beneath these bitter wars *within science*, was a continued attempt by Darwinists to use science to redefine what it means to be human, and to advance an anti-religion, pro-eugenics social agenda. For Darwinists, the human future was defined by biology.

Chapter Eight

The Makeover

REDEFINING EVOLUTION TO MATCH DARWIN

Those who can make you believe absurdities can make you commit atrocities.
Voltaire

In September 1906, New York City's Bronx Zoo paired an orangutan with a young Congolese pygmy, Ota Benga, and displayed them together in a cage in the zoo's Monkey House. The exhibit was a sensation.

On Sunday, September 16th, when *The New York Times* ran the story "Bushman Shares a Cage With Bronx Park Apes," 40,000 people showed up that day alone. The Times reported that everywhere Benga went he was pursued by crowds, "howling, jeering and yelling... Some poked him in the ribs, others tripped him up, all laughed at him."[1]

When a group of local African-American clergy objected, they were told the exhibit was based on Darwin's theory, which was an accepted scientific fact that showed "the 'lower races' were... closer to pigs and dogs than to human beings, and that a different value should be put on their lives."[2]

This was the terrifying center of Darwin's evolution. Everyone's life had a different value – and the Darwinists would do the ranking. Astonishingly, *The New York Times*, *New York Globe* and other media agreed. In defending the idea that Ota Benga was not fully human, *The Times* editorialized:

> "We do not quite understand all the emotion which others are expressing in the matter...It is absurd to moan over the imagined humiliation and degradation Benga is suffering.

> Pygmies...are very low in the human scale, and the suggestion Benga should be in a school instead of a cage ignores the high probability school would be a place...from which he could draw no advantage whatever. The idea that men are all much alike except as they have had or lacked opportunities for getting an education out of books is now far out of date."[3]

Ota Benga

The New York Times continued that "as for Benga himself, he is probably enjoying himself as well as he could anywhere in his country." However, what *The Times* saw as "imaginary degradation" was felt differently by Ota Benga, who later committed suicide.

Darwinism 'Scientifically' Justifies Racism and Eugenics

Although a scientific failure, Darwinism became increasingly popular as the pseudo-scientific justification for racist beliefs and eugenics programs. Held up as the "scientific" solution to social problems, eugenics programs spread rapidly in the early 20th century, to France, Germany, the Soviet Union, and dozens of other countries. In the United States, tens of thousands had their children taken away and were sterilized, lobotomized (used on Rose Marie Kennedy and film star Francis Farmer), and worse.

The connection between Darwinism and eugenics cannot be dismissed as the abuse of Darwin's ideas by others. Those who created modern Darwinism were the same people who created the modern eugenics movement.

The eugenicists were the Darwinists.

Darwin laid the foundation for eugenics with his ideas of social progress through eliminating the "unfit." His cousin and confidant, Francis Galton, took up the movement's leadership. In *The Comparative Worth of Different Races*, Galton wrote that each race had a different biology which could be scientifically ranked to determine their value.[4] The differences between savages and civilized people, he said, are explained by the biological differences between their races. For example, Galton wanted to ship thousands of Chinese to Africa, in the belief, he wrote:

> "That the Chinese would...supplant the inferior Negro race. I should expect the large part of the African seaboard, now sparsely occupied by lazy, palavering savages... might in a few years be tenanted by industrious, order loving Chinese."[5]

Galton hoped the Africans would be managed ("tenanted") by a superior race much the way farmers managed livestock. Galton assumed the Chinese, in turn, would be managed by an even more superior race, North Europeans, like himself. That is how the Darwinian hierarchy of races worked.

After Galton, leadership of the British Eugenics Society was picked up by Charles Darwin's son, Leonard (1850-1943), and his grandson, Charles Galton Darwin (1887-1962).

Leonard was called "the embodiment and authoritative mouthpiece of eugenics," authoring such articles as "The Cost of Degeneracy" and "Quality Not Quantity."[6] He organized the First International Congress of Eugenics in 1912, at which Darwinism and eugenics were woven into a single explanation for social progress. In a rousing speech, Leonard told the assembly "the unfit amongst men are now no longer killed off by hunger and disease, but are cherished with care, thus being enabled to reproduce their kind, however bad that kind might be."[7] He said it was a "paramount duty" that eugenics be imposed on society, saying "countless future millions" would benefit.

Charles Galton Darwin bolstered this work by popularizing key Darwinist ideas, like the "comparative worth of different races," in journals like *Mankind Quarterly*.

THE 20TH CENTURY EUGENICS MOVEMENT

The belief that such ideas are scientific was cemented by Karl Pearson, a brilliant mathematician who, in the 1890s, revolutionized science with the use of statistics. "His laboratory was a world centre in which men from all countries studied," says mathematician Helen Walker. "Few men in all the history of science have stimulated as many other people."[8]

For many intellectuals, the spirit of the new century was pro-socialist and anti-religion. This included Pearson, who became a socialist over what he saw as Christianity's hypocritical opposition to social progress. "The old faith no longer provided a moral compass for society in this age of capitalism and consumption," says historian Ted Porter. "Nor could it give direction to his own life."[9]

Like Huxley and those in the *X-Club*, Pearson saw himself as a political radical and viewed science as the tool of liberation. The scientific "frame of mind seems to me an essential of good citizenship," he wrote.[10] In science, Pearson discovered a sense of historical and personal purpose, says Porter. "Science defined the next, decisive phase in what he conceived as a grand saga of human progress."

We think of socialists as being opposed to eugenics, but this is untrue. For many on the left it was the easy, rapid path to social progress. This was true for Pearson, whose belief in Darwin's natural selection led him to eugenics, which became his new religion. He became a zealot, turning eugenics into his life's work, eventually serving as Professor of Eugenics at the University of London.

Pearson believed that his statistics revealed nature's deepest truths. In the 1890s, he became a protégé of Francis Galton and sought to justify eugenics with statistics. Pearson's group correlated hundreds of physical, mental and social traits, (called "biometrics") which they said were the result of natural selection.[11] Galton called this the "great extension of my work by Professor Karl Pearson."[12]

Pearson himself was delighted with the results, believing they confirmed the idea that social progress is best accomplished through eugenics.

EUGENICS: AN ELITIST, COERCIVE CONCEPT OF SOCIETY

Pearson believed natural selection confirmed racist beliefs, and made eugenics both natural and necessary. In *National Life from the Standpoint of Science*, he passionately argued:

> "History shows me one way, and one way only, in which a high state of civilization has been produced, namely, the struggle of race with race, and the survival of the physically and mentally fitter race. If you want to know whether the lower races of man can evolve to a higher

> type, I fear the only course is to leave them to fight it out among themselves."[13]

Pearson said the "lower races" have inferior biology because they did not evolve in the harsh North European climate "on which so much of the Aryan [race's] success depended."[14] Hitler would later use this and similar Darwinism inspired eugenics ideas to justify his programs.

Eugenics, by its very nature, is elitist and coercive. Those of "high rank" (in Francis Galton's words) should breed more and the lower classes less, voluntarily if possible, by force if necessary. Galton reflected the widespread belief among eugenicists that forced sterilization was not just a civic duty, but a "kindness" bestowed on the inferior masses by their social betters.[15]

Pearson sought to scientifically justify this coercion. "My view – and I think it may be called the scientific view of the nation," he wrote, is that social progress occurs by expanding the "better stocks." He saw education as useless. "No degenerate and feeble stock will ever be converted into healthy and sound stock by the accumulated effects of education, good laws, and sanitary surroundings."[16] For Pearson, the biologically inferior *must* be managed by those of "high rank."

'Scientific' Racism

In the early 20th century, Pearson's work was pointed to as scientific proof that some races are inferior to others, an idea the Darwinists called "scientific racism" (popularized by Leonard Darwin and Charles Galton Darwin in his *Mankind Quarterly* articles).

Supporters of "scientific racism" included Madison Grant, a trustee of the American Museum of Natural History, director at the American Eugenics Society and co-founder of the Bronx Zoo (where he arranged pairing Ota Benga and the orangutan).

In his influential 1916 book, *The Passing of the Great Race*, Grant glorified the Nordic race and argued that culture is not a product of ideas and beliefs, but of a person's biology:

> "The view [was] that the Negro slave was an unfortunate cousin of the white man, deeply tanned by the tropic sun and denied the blessings of Christianity and civilization... It has taken us fifty years to learn that speaking English, wearing good clothes, and going to school and to church do not transform a Negro... Nor was a Syrian or Egyptian transformed into a Roman by wearing a toga and applauding

> his favorite gladiator... Americans will have a similar experience with the Polish Jew, whose peculiar mentality...is being engrafted upon the stock of the nation."[17]

Grant's impressive credentials made these ideas seem scientific. No one reading Grant's book thought a man of his stature would make such claims if they weren't true. "Scientific racism" must be true. Darwin discovered it, Pearson proved it, and *The New York Times* confirmed it.

Promoted by Darwinists like Grant, the belief that cultural differences are the result of biological differences became widely accepted as scientifically true. It was this belief that led Ota Benga into the Monkey House, and became the justification for much 20th century racism and genocide.

The Discovery of Genes

Despite Darwinism's appeal as a justification for "scientific racism" and eugenics, most biologists continued to view it as scientifically wrong. Any lingering doubt ended in the early 1900s with the discovery of genes. Genes categorically proved Darwin's ideas wrong.

In 1900, work by the Austrian monk, Gregor Mendel, was re-discovered. In the 1850s, using hundreds of careful experiments, Mendel showed that many traits don't change gradually, a bit at a time, but abruptly in whole units. They appear or disappear in a single generation. He also discovered that changes in traits are governed by laws, which meant that changes in inheritance – and evolution – are predictable.

Dutch botanist Hugo de Vries confirmed these discoveries, and said traits are inherited in units he called "pangenes," later shortened to "genes."

Mendel's laws were a sensation. They explained inheritance and meshed perfectly with developmental biology and Lamarck's idea of evolution.

Well-known British biologist William Bateson wrote, "The monk had succinctly described the transmission of elements governing heritable traits." Mendel had uncovered the secret to inheritance.

Genes and their "laws of interaction" led to an explosive burst of discovery. In the following years, hundreds of studies showed Mendel's laws at work in every type of plant and animal, even in humans.[18] They confirmed that traits arise not gradually, but abruptly. Nor are these changes influenced by natural selection.[19] To the contrary, gene changes appeared immune to natural selection.

In England, William Bateson used Mendel's laws to show that genes often work in groups to change a trait (a process called "linkage").

Bateson, who coined the term "genetics," became the world's first professor of genetics in 1908 at Cambridge University, where he established genetics as an independent discipline in biology.

The two towering figures of early genetics – de Vries and Bateson – emphatically rejected Darwinism. Evidence from laboratories around the world showed precisely the reverse of Darwin's slow, blended change. Traits came in discrete units, which don't gradually blend, but abruptly appear and disappear. These hard facts became part of the continued rejection of Darwinism well into the 20th century.

THE SECOND MAJOR DARWIN WARS: OPPOSING GENETICS

Mendel's laws also threatened Pearson's eugenics efforts.

If evolution comes from rearranging genes, it was impossible to improve society simply by eliminating people with "bad" genes. The new genetics showed there are no good or bad genes, just good or bad *combinations* of genes. And these combinations could happen to anyone.

This meant brilliant minds and great athletes would appear among the working class, while fools and weaklings would emerge among the elite. For those seeking the rapid transformation of society through eugenics, this was a devastating blow.

Pearson and the Darwinists unleashed a ferocious assault on the new genetics. If the battles Darwin waged against his many opponents were the first major Darwin Wars, the attack on Mendel and the geneticists was the second. Darwinists ridiculed the new genetic discoveries, denied they had broad application in biology, and tried to discredit the idea that Mendel's laws had any influence in evolution.[20]

Like Darwin himself, Pearson and the modern Darwinists chose to simply dismiss all the conflicting evidence. This was easier than you might imagine. In 1900, when Mendel's laws were discovered, technology was primitive. Gas lamps lit most homes, and horses were still the staple of transport. No biologist had ever seen a gene, or could explain how they created changes. If statistical proofs from the great Karl Pearson showed Mendel's genetic laws were irrelevant to evolution, who would say he was wrong?

The Darwinists succeeded for two decades in muddling the role of genes in evolution. By the 1920s, however, technological advances and new research confirmed Mendel's laws and their role in heredity. Thomas Hunt Morgan discovered that genes link in organized ways. His later discovery of the chromosome confirmed the new gene theory of heredity.

Morgan's discoveries transformed biology. He became one of the world's leading scientists, won the Nobel Prize in 1933 and established the California Institute of Technology's biology department, which produced seven more Nobel Prize winners.

Morgan and those like him, known as the "Genetics School," rejected the Darwinian model. "Like most embryologists and many biologists" at the time, "Morgan found the Darwinian theory of evolution lacking in plausibility," says Washington University historian Garland Allen.[21] Genes work in coordinated groups, and change according to precise laws. Morgan found it difficult to imagine how such complex interactions could arise, as Darwin claimed, "simply by an accumulation of slight chance variations," says Allen.[21]

The fact that genetic change occurs in abrupt leaps based on internal processes, rather than gradually based on natural selection, confirmed for Morgan and others that Darwin had gotten the evolution process backwards.[22] "The neo-Darwinians," wrote Morgan, "seem less concerned with the advancement of the study of evolution than with expounding Darwinism as a dogma."[23]

Beyond this there was the nagging problem of whether Darwinism was real science. "Although Morgan believed that evolution itself was a fact," says Allen, "the mechanism of natural selection proposed by Darwin... could not be put to an experimental test."[21]

Redefining Evolution to Match Darwin

By 1930, the evidence that genes are not effected by natural selection was overwhelming. For their theory to survive, the Darwinists would somehow have to deal with this. The solution began with Ronald Fisher, who rescued Darwin's theory from Mendel's laws.

From an early age, Fisher had been deeply troubled by the direction of British society. He firmly believed that civilizations collapse when the upper class has few children and the masses have many. Britain's situation had become even more dire than in Darwin's day. Eugenics offered the only hope. Fisher made it a crusade, devoting his life to eugenics.

In 1911, while still in his teens, Fisher and Charles Darwin's son, Horace, organized a eugenics society at Cambridge University. Fisher later worked with the Eugenics Education Society in London (founded by Francis Galton). In 1933 he was appointed Professor of Eugenics at University College London. For decades, Fisher used this position as a platform to campaign for government-run programs to sterilize the "unfit."

Like Karl Pearson, Fisher saw natural selection as the "scientific" justification for eugenics and believed he could harness statistics to prove it. A formidable mathematician, Fisher picked up Pearson's work, creating a direct link between Darwin's original eugenics ideas and himself.

Fisher's daughter, Joan Box, writes that when her father began his work "Darwin's evolutionary theory had largely given way to theories involving genetics, and most biologists assumed the effects of [natural] selection to be negligible."[24] Fisher's fertile mind, however, saw a path to survival. In a remarkable move he would separate the inheritance of traits from evolution.

Fisher said that how traits are *inherited* (via genes) is unrelated to how they *survive* (via natural selection) and evolution. This meant that how a gene changes (via Mendel's laws) will have no effect on evolution. Only natural selection determines which genes survive, thus only natural selection determines the course of evolution.

In effect, Fisher said, the laws of genetic change are irrelevant to evolution. If he could prove this idea, it would return natural selection as a potential force in evolution. Fisher went to work.

(In Darwinian terms, Fisher said "variation"- how a gene changes - is irrelevant in evolution. Only "selection" - which genes survive - matters.)

Making Mendel's Laws Irrelevant to Evolution

Fisher's first problem was that Darwin's theory had never matched any process seen in the real world. The constant refrain from Darwin's opponents was "nature doesn't work that way." Fisher solved this by creating an alternate version of nature - an alternate reality.

In real plants and animals, changes appear and disappear abruptly. But when Fisher statistically averaged the changes in a large group, these abrupt shifts vanished. When viewed over long periods, this statistical averaging made evolution look gradual, as Darwin claimed.

Fisher also dismissed the cause and effect connections central to biology as irrelevant in evolution. Statistics is about finding correlations. In Fishers world, the flesh and blood cause of events is unimportant. All that matters are the patterns these events create.

Fisher took his statistical plants and animals (none of which existed in nature), then placed them in a statistical world (which also did not exist), and then watched how they statistically change over time. In this nonexistent "statistical world," Fisher said Darwinism is true.

By 1933, Fisher had developed enough "statistical evidence" to attribute the gradual changes he saw *solely* to the effects of natural selection. Fisher called this proof of natural selection "population genetics."[25]

Fisher said his work also provided a new scientific foundation for eugenics. He claimed statistics allowed him to track "good" and "bad" genes as they spread through a group. He said the ability to track genetic changes in society made eugenics scientific, and devoted fully *one-third* of his magnum opus (*The Genetical Theory of Natural Selection*) to explaining the urgent need for society to impose eugenics on the masses.[26]

The 'Modern Synthesis': A Scientific Sleight of Hand

Other Darwinists added to Fishers work. Geneticists JBS Haldane, for example, said random differences in births and deaths led to the random changes in which new genes survive over time. This further marginalized the role of Mendel's laws in evolution. Haldane's work and similar ideas were grouped under the "population genetics" umbrella.[27]

Julian Huxley (grandson of Thomas Huxley) took this a step further and said "population genetics," integrated Mendel's genetic laws with Darwin's natural selection. The two were no longer at odds, said Huxley, an idea he called the "Modern Synthesis."

Fisher, however, had made natural selection "compatible" with genetics only by excluding Mendel's laws from any role in the evolution process. A "synthesis" that *excludes* is more like a sleight of hand, or a magic trick, than science, and no one was fooled. In the 1930s "population genetics" was widely rejected. Developmental biologists in particular opposed it, said Ernst Mayr. "In Germany, in France, in England...and in the United States they were opposed to the 'synthesis' on a Darwinian basis...They were all Lamarckians."[28]

The Modern Synthesis Makes Eugenics 'Scientific'

While many scientists rejected "population genetics," it was embraced in the 1930s by a new generation of eugenicists who saw it as a way to make eugenics modern and more "scientific."

Julian Huxley was a driving force in this effort. Civilized societies, he wrote, "are doomed to decay unless they enforce adequate measures for regulating human reproduction, for controlling the *quantity* of population, and at least preventing the *deterioration of quality* of racial stock."[29]

Darwin worried that vaccinations debased society by saving working class children who would otherwise die. Julian Huxley shared this concern, arguing in 1941 that "the lowest strata are reproducing too fast. Therefore... they must not have too easy access to...hospital treatment lest [this] should make it too easy for [their] children...to survive."[30] This was Huxley's solution, deny health care to working class children so more will die.

This did not mean the entire working class should be done away with – someone still had to mine coal, build roads and clean the streets. Rather, said Huxley, society should aim for "the virtual elimination of the few lowest and most degenerate types," then strictly limit births among the rest.[31]

The Nazis and Natural Selection

These ideas, backed by Fisher's "statistical proofs" of natural selection, became central to the Nazi justification of their programs. "National Socialism is nothing but applied biology," wrote Hitler's deputy, Rudolf Hess.[32]

The Nazi belief in their intellectual superiority, and the inferiority of other races, was grounded in the idea that a struggle to survive in an oppressive environment determined a person's abilities and intelligence:

> "The [ancient] tribes that emigrated early to the north [of Europe] and there gradually became white, had to develop all their intellectual powers and...perfect all the arts in their struggle with...misery... brought about by the climate... And out of it all came their high civilization."[33]

Contrary to popular belief, the Nazis did not focus on blonde hair or fair skin. Hitler, most Nazis and many Germans had dark hair. Rather, they focused on pseudo-scientific ideas about the different blood types of ethnic groups. "*Blut und Boden*"– "blood and soil"– was the popular expression.

In the 1930s, members of eugenics societies in America and England became advisors to Ernst Rudin, Hitler's eugenics architect and author of the Nazi "racial purity" laws.[34] They explained to Rudin and other Nazis the "science" underlying eugenics. They showed how to apply it to blood types, how to rank people based on their "biological value," and how to create effective sterilization programs for those at the bottom of the ranking.

The Nazis were appreciative of the help they received. One day, in the mid 1930s, Leon Whitney, president of the American Eugenics Society, burst into the office of Madison Grant waving a personal letter from Hitler that praised his work. Grant smiled, reached into his desk drawer and pulled out his own personal letter from Hitler. In it, Hitler called Grant's racist book, *The Passing of the Great Race*, his "bible."[35] Nazis were particularly taken with Grant's argument that:

> "A mistaken regard for what are believed to be divine laws and a *sentimental belief in the sanctity of human life* tend to prevent both the elimination of defective infants and the sterilisation of... adults...of no value to the community. The laws of nature *require the obliteration* of the unfit."[36]

"Human life," continued Grant, "is valuable *only* when of use to the community or race." Hitler agreed completely.

Huxley, Fisher and other Darwinists made nearly identical arguments about the value of human life and race in the 1930s. They agreed that the laws of nature require eliminating (or in Grant's words "obliterating") those judged to be "defective." As the Darwinists repeatedly pointed out, natural selection and population genetics "scientifically" proved this is true.

These ideas were directly incorporated into Nazi eugenics laws, including the 1935 "Law for the Protection of German Blood and Honor," which focused on "racially distinctive" minorities with "alien blood" (Jews, Gypsies, Africans), forbidding their marriage to Aryans, and eliminating their civil rights.

Paradoxically, many eugenicists were also avid conservationists. Social Darwinism led directly to the belief that some animals are more valuable than some humans. Julian Huxley, for example, co-founded the World Wildlife Fund, advocating for animals the care he felt should be denied "lower order" children. In Huxley's view there was no real difference between wildlife and the working class, except the latter were not in danger of extinction.

Nazi leader Hermann Göring thought similarly. While writing laws that sent "unfit" humans to extermination, he also banned animal traps, boiling lobsters and the like to, as he said in a 1933 radio broadcast, "show sympathy for their pain."[37]

(This Darwinist mindset continues today. Princeton University philosopher Peter Singer, infamous for his belief that disabled human infants should be killed, in 1993 launched the Great Ape Project, designed to provide "basic rights" to apes. For Social Darwinists like Singer, biology determines your rights – disabled human infants have none, healthy apes apparently do.)

Darwinist Praise for the Nazi Programs

In 1935, Clarence Campbell, president of the American Eugenics Research Association, appeared before the Nazi "World Population Congress" in Berlin, in a role *The New York Times* described as "a champion of Nazi racial principles." *The Times* reported that Campbell praised "the leader of the German nation, Adolf Hitler, [who] has been able to construct a comprehensive racial policy of population development and improvement that promises to be epochal in racial history."[38]

Covering the same event, *Time* magazine headlined its story "Praise for Nazis" and said "Germans heard gratefully last week a warm, approving speech from Clarence Campbell." *Time* wrote that,

> "Long before Adolf Hitler was anybody, Dr. Campbell [was] teaching that if 'Americans' do not speed up their birth rate, the result after a few more generations will be to leave control of their country in the hands of some more pushing race."[39]

Time reported that Campbell's "boldest dicta" to those at the Nazi Population Congress was that "the difference between the Jew and the Aryan is as unsurmountable as that between black and white... Germany has set a pattern which other nations must follow."

A year later, as Nazi "racial purity programs" were put into practice, American and British eugenicists accepted honors from Nazi eugenic institutes and universities. In 1936, Harry Laughlin, author of *Eugenical Sterilization in the United States,* was given an honorary degree by the University of Heidelberg in recognition of his service to eugenics. Laughlin wrote that he was happy to accept,

> "Not only as a personal honor, but as evidence of the common understanding of German and American scientists of the nature of eugenics... and the practical application of those biological principles which determine [the] racial endowments... of future generations."[40]

In the face of the escalating brutality against Jews and other minorities, American eugenicists remained enthusiastic about the Nazis programs. In 1937, Frederick Osborn, secretary of the American Eugenics Society, remarked that "the German sterilization program is apparently an excellent one. Taken altogether, recent developments in Germany constitute perhaps the most important social experiment which has ever been tried."[41]

So pleased were Darwinists with the Nazi programs that in 1939, as all across Europe Jews were corralled into ghettos and branded with numbers like cattle, Julian Huxley and a group of Darwinists penned a "Eugenics Manifesto" to promote similar programs worldwide.[42]

They wrapped the "Eugenics Manifesto" in humanitarian language ("economic equality," "human betterment"), but the thinking and the programs they advocated were identical to those established by the Nazis. Human improvement, the Manifest author's wrote, can occur "only as a result of some kind of selection...either through conscious choice, or as an automatic result of the way in which [people] lived."[42]

The Eugenics Manifesto argued every person should be tested to provide a "valid basis for estimating and comparing the *intrinsic worth* of different individuals." Once people are ranked based on the "intrinsic worth" of their biology, the "genetic improvement of mankind" can then be achieved through sterilization, artificial insemination and similar programs.

Huxley and his group urged governments to form a "federation of the whole world" to implement these programs worldwide as quickly as possible. Although the authors said the program should be voluntary, the evidence in every country was that they were imposed by force. As Darwin himself noted, no one ever considers themselves unfit.

HUXLEY, HITLER AND WORKING-CLASS CHILDREN

The claim promoted by Huxley, and the other Darwinists, that people should be "ranked" based on their "biological worth" led inexorably to some people being ranked as sub-human. Walter Reich, at the Woodrow Wilson Center, explains the consequences:

> "The Nazi animating vision for the racial reordering of Europe [was] its conviction that...Slavs were subhuman, and Jews weren't human but beasts in human form, and had to be exterminated... The SS carried out that vision...Having been indoctrinated by Nazi propaganda, they murdered and brutalized Slavs and... methodically exterminated Jews. Gypsies, too, became targets for mass murder. In the name of genetic purity they even killed their own handicapped citizens."[43]

So convinced were people that natural selection scientifically justified such acts, that for a month after Germany's surrender, the staff at the Kaufbeuren-Irsee clinic in Bavaria continued to kill handicapped people, stopping only when ordered to by American army officers.[44]

Children were a special target of these programs. The concern Darwin and Julian Huxley had, that too many of the "wrong" children were surviving, was shared by Hitler. After a 1982 visit to Auschwitz, historian Robert Lifton wrote: "I was shown the many exhibits maintained there, exhibits that leave nothing to be added concerning the evil human beings can do to other human beings. But the one that left the most profound impression on me was the simplest of all: a room full of shoes, mostly baby shoes."[45]

CHILDREN JUDGED TOO UNFIT TO REMAIN ALIVE.

NATURAL SELECTION AND SOCIAL PROGRESS

It would be unfair to focus just on Germany. During those years, Darwinism played a role in atrocities around the world. A wide range of eugenics programs, many quite brutal, were implemented in dozens of countries.

Dr. Osamu Sakura, at Yokohama University, says Darwinism's primary impact in Japan was not in biology, but "as a kind of social scientific tool [for] eugenics."[46] During World War II, the Japanese implemented an array of eugenics programs in the vast areas of Asia they conquered.

The United States implemented similar programs. In one particularly notorious example, the US Public Health Service for decades experimented on African Americans with syphilis, in the infamous Tuskegee project.

Many socialist countries also implemented eugenics programs. "A popular misconception," says University of Maryland historian John Glad, is that eugenics "was exclusively a product of political conservatism. In point of fact the movement had its roots in the left as much as in the right."[47]

In 1927, for example, Joseph Stalin established The Institute of Experimental Pathology and Therapy to support biologist Ilya Ivanov's efforts to breed humans and apes.

"Ivanov started the experiments in Africa and continued them here," says the Institute's current director, Vladimir Barkaya.[48] "He took sperm from human males and injected it into female chimpanzees." Ivanov planned to inseminate women with ape sperm, and hoped eventually to breed a human with super strength.

A central rationale for the program was to provide conclusive proof that science had superseded religion. In supporting Ivanov's request for funding, the Commissariat of Agriculture wrote to Stalin, "The topic proposed by Professor Ivanov...should become a decisive blow to religious teachings, and may be aptly used in our propaganda and in our struggle for the liberation of working people from the power of the Church."[49]

Eugenics in the Early 20th Century

Today, Darwinists attribute the eugenics atrocities to non-Darwinists who abused his ideas. But as we have seen, that is not true. Not all Darwinists were eugenicists, nor all eugenicists Darwinists. But the leaders of the eugenics movement and of modern Darwinism were the same people. In this sense the eugenicists were the Darwinists.

The Social Darwinism which justified eugenics had been central to Darwin's thinking from the beginning. As historians Adrian Desmond and James Moore wrote, "Social Darwinism is often taken to be something extraneous, an ugly concretion added to the pure Darwinian corpus after the event, tarnishing Darwin's image. But his notebooks make plain that... [issues like] racial extermination, and sexual inequality were written into the equation from the start—Darwinism was invented to explain human society."[50]

"Darwinism was not 'responsible' for social Darwinism or eugenics in any simple way," writes historian Peter Bowler. "[But] it has to be admitted that, by making death itself a creative force in nature, [and] by proposing that evolution worked primarily through the elimination of useless variants, Darwin created an image that could all too easily be exploited by those who wanted the human race to conform to their own pre-existing ideals."[51] Bowler says:

> "Darwin exploited the idea of the struggle for existence in a way that was unique....The potential implications were drawn out even more clearly when Galton argued that it would be necessary to apply artificial selection to the human race in order to prevent "unfit" individuals from reproducing and undermining the biological health of the population. This was the eugenics program, and in its most extreme manifestation at the hands of the Nazis, it led not just to the sterilization but also to the actual elimination of those unfortunates deemed unfit by the state."

There is no doubt that Darwin's theory was widely rejected in science from 1859 well into the 1940s (except by eugenicists).

The question is, after nearly 100 years of broad scientific rejection and a history of justifying eugenic horrors, like those of World War II, how did Darwinism become the dominant theory of evolution? How did it come to prevail in the scientific community?

Chapter Nine

The Takeover

HOW THE DARWINISTS TRIUMPHED

Darwinists hijacked the field of evolutionary biology.[1]

Lynn Margulis, National Medal of Science winner, 2009

As the genetics revolution unfolded in the early 20th-century, three discoveries placed evolution on the cusp of a conceptual revolution.

Glimpsing the "Hidden Plan of Life:" Three Big Discoveries

One of biology's big mysteries is how the body develops. How does a seed unfold into a flower, or a tiny embryo become an elephant? The process is complex. When cells start life, they're all identical. Yet some become teeth, others become toenails. How does a cell know which it should be? Genes are involved, but how do they manage this?

In the early 1930s, British biologist Conrad Waddington began to unravel the mystery when he discovered "organizer" genes. These specialized genes guide cells through the development process, telling them what to be when they grow up.

In a surprise, Waddington discovered that the *same* "organizer" genes direct this process everywhere in the body. This meant that a gene is not a blueprint for making body parts. Rather, it's a chemical signal. Genes often work in groups which create a unique chemical signal, similar to how letters combine to form different words, each with a unique meaning. These changing signals are what guide a cell to become a tooth or toenail.

A second key discovery emerged in the 1940s when geneticist Richard Goldschmidt found that the whole chromosome can abruptly reorganize, with genes moving to new places. This alters when and where genes send their signals during development.

Goldschmidt realized this is how new species arise – not by gradually accumulating many small changes, as Darwin claimed, but by abruptly rearranging their genes. Goldschmidt put it this way:

> "The change from species to species is not a change involving more and more additional atomistic changes, but a complete change of the primary pattern...into a new [form]."[2]

This discovery explained why new species appear abruptly, with radical new traits, and integrated genetics, development, and evolution into a single system. It was a formidable achievement, widely seen as a major advance in understanding evolution. Even Darwinists like Theodosius Dobzhansky acknowledge it was the first truly new idea about evolution in 100 years.[3]

The third major discovery came from Barbara McClintock, a young woman of piercing intellect and immense energy.

In the early 1930s, barely out of graduate school, McClintock was already making breakthroughs. She amazed colleagues by discovering that genes change in groups, and then connected this to specific chromosome changes and alterations in traits.[4]

YOUNG BARBARA MCCLINTOCK

Other discoveries followed. In a science overwhelmingly dominated by men, the brilliant McClintock rapidly rose to the top, becoming the first female vice president of the Genetics Society of America in 1939, and assumed the presidency a few years later.

In the 1940s, McClintock made a discovery that revolutionized the entire field of genetics. While studying Indian corn (the type with multi-colored kernels), she asked why some kernels have unexpected color combinations. Although this seemed of little significance, these combinations couldn't be explained with existing ideas about genes. Just as Einstein revolutionized our thinking about the universe by solving a few "loose ends" in physics, McClintock's solution to this little mystery transformed our thinking about genes, genetics and evolution.

McClintock found that these unexpected color changes occur because genes are "mobile." They move around – a lot. She initially called them "jumping genes" because they moved so much.

McClintock also discovered that, in a "cut and paste" process, these mobile genes can cut themselves into pieces, which then "paste" themselves inside another gene, creating a new gene. Some combinations include "small genes entirely contained within larger genes."[5] These gene fragments later cut themselves out of their temporary home and recycle themselves.

McClintock's Brilliant New Theory of Evolution

McClintock's discoveries had profound implications. They meant genes aren't static, but dynamic, with the ability to rapidly reorganize themselves. All prior ideas about what a gene "is" instantly became obsolete. They also made genetics vastly more complex than previously imagined.

Her findings also meant that all the genes needed for development are not contained in the genome. Some genes are made "on the fly," when and where they're needed. McClintock grasped that this meant the body must have some "higher" organization process, one that sits above genes, controlling them and orchestrating what they do.

These discoveries led McClintock to rethink all of genetics and evolution. Working at a feverish pace, she integrated her own findings with Waddington's discovery of "organizer" genes (which regulate biological processes) and Goldschmidt's discovery that reorganizing genes can lead to new species. When done, she had created a comprehensive new vision of development and evolution.

McClintock published her theory in a landmark 1951 paper with three groundbreaking ideas.[6]

First, the mobile genes she had discovered manage the development process. As development unfolds, they move from place to place, telling other genes what to do and when to do it. (These were Waddington's "organizer" genes, what she now called "controller" genes.)[7]

Second, during development, the job of a "controller" gene is not fixed, but can change. At each new location it can combine with other genes to create unique signals that create different body parts. Working in ever-changing groups, the "controller" genes create unique signals that transform general cells into the specialized cells (tooth or toenail) that form body parts.[8] (Today we know these "controller" genes make up 95% of the genome.)

Third, because a gene's job is not fixed, it does not carry the instructions for making a specific part. Rather, the instructions for making each part emerges from a "higher" organization process in the body, the same process that guides "controller" genes, telling them where to go and what to do. She said evolution occurs when this higher organization process changes.

McClintock's theory brilliantly integrated the discoveries of the prior half-century, and added crucial insights into the evolution process. It was a tour-de-force worthy of Einstein. Her ideas later transformed all of genetics, and set the stage for the modern Genome Revolution that began in 2000 and dominates genetics today.

McClintock's discoveries led to many honors, including a 1983 Nobel Prize. The US Postal Service, in its *American Scientists* series, placed McClintock alongside Josiah Gibbs, America's greatest 19th-century scientist, John von Neumann, the genius whose work transformed the social sciences, and physicist Richard Feynman (often compared to Einstein).

These recognitions, however, lay far in the future. In the 1950s, the conceptual revolution about to emerge from her work and from other discoveries was abruptly halted. After 100 years, the Darwinists were about to have their day. The hand of old Thomas Huxley reached out from the grave and wrapped itself around the throat of modern evolution.

How Darwinism Came to Dominate Evolution

Thomas Huxley's position had made his family prominent. The grandchildren were expected to follow suit, and they did. Grandson Andrew won the Nobel Prize in Medicine, David became the youngest Queen's Counsel in the British Empire and Aldous became world famous as the author of *Brave New World* and other books. Gervas Huxley won Britain's Military Cross for heroic actions in World War I and went on to a distinguished military career, and Lawrence Huxley became a prominent diplomat and ambassador.

Julian Huxley was another matter. Growing up on his grandfather's knee, Julian listened to tales of past battles, splendid victories and the family's rise to power and distinction. The old "bulldog" (a nickname Huxley bestowed on himself in old age), was the glory of Julian's youth, and he set out to imitate his grandfather. He would become a man of science.

THOMAS AND JULIAN HUXLEY

Julian's career began well enough. But plagued with mental problems that began in his youth, Julian suffered a series of mental breakdowns that recurred throughout his life. In 1913, then a 27 year-old assistant professor of biology, he had a breakdown severe enough to require months of nursing home recovery. The following year, another mental collapse forced him to return for a long convalescence.[9]

JULIAN HUXLEY

Over the years, these bouts became increasingly severe. In 1927, barely past 40, Julian resigned from academia, and stunned his family by taking a job as an assistant to the science fiction writer HG Wells. All thoughts of a career in science were over. Julian had failed his illustrious family.

Julian later traveled and became an avid conservationist. He was also fascinated by social engineering, spending time in the Soviet Union, where he admired the centralized social planning. On his return in 1932, he joined a group called "Political and Economic Planning," which developed centrally run social programs, then campaigned for the government to adopt them. Here his life took a dramatic turn.

Julian's work on centralized social planning took him back to his grandfather's vision of government-run eugenics programs as the key to social progress. He studied how the old "bulldog" had masterfully come to dominate Victorian science. An idea began to form – a path to redemption. Julian would take up his grandfather's battles, and he would win.

The ghost of the old Thomas Huxley now infused Julian's mind. He would succeed by using every lesson in his grandfather's playbook: gather a small group of promoters, create journals, orchestrate publicity, grab control of the committees that fund research, and all the rest.

Huxley did not do this alone, but he began the process and guided it with a small group of loyal followers. At every step of the way, the same tiny group provided the ideas and leadership to accomplish the transition: Julian, his protégé Hermann Muller, Theodosius Dobzhansky, Ernst Mayr, and a half-dozen others. They went on to accomplish in ten years what Thomas Huxley could not in a lifetime. They made Darwin dominant.

1. A Tiny Group Gained Control over Defining Evolution

Margaret Mead once said, "A small group of thoughtful people could change the world. Indeed, it's the only thing that ever has." That is what happened in evolution.

In the 21st century, with millions of scientists and the vast information that flows from the Internet, it is hard for us to imagine how a few people could gain control of an entire area of science. But in the 1940s, things were different.

At the time, few scientists supported Darwin. Harvard paleontologist Percy Raymond said of his colleagues: "Probably most are Lamarckians of some shade... To the uncharitable critic it might seem that many out-Lamarck Lamarck."[10]

This was true of science in general. But like Percy Raymond, the focus of most scientists was not evolution, but some other field – genetics, biology, paleontology. Evolution was an adjunct to what they did, not their central concern. At the time, evolution, as a *standalone* discipline, had few scientists dedicated solely to it.

Understanding how his grandfather had used a small group to take control of Victorian science, Julian Huxley began laying the groundwork for a similar effort with modern evolution.

Huxley's first step was to organize a small group that would gain control over how evolution was defined. In 1939, Huxley asked Ernst Mayr and Theodosius Dobzhansky to form an informal network to promote Darwinism.[11]

In 1942, Huxley provided an intellectual foundation for his budding group with *Evolution: the Modern Synthesis*, which defined evolution as Darwin's natural selection.[12] Huxley brought together scholars "who had little or no previous contact," says Princeton professor Alan Mann, and organized evidence in ways that supported Darwinism.[13]

In 1945, Huxley's network became a formal group patterned after the *X-Club*. Called the Society for the Study of Evolution, and led by Mayr and Dobzhansky, it pulled together the small group of Darwinists in an ambitious effort to promote their ideas.[14]

Just as the *X-Club* created the journal *Nature* to promote Darwinism, the new group founded the journal *Evolution* with the same goal. "The aims of the Society, through its journal and otherwise," wrote Ernst Mayr, "reflect the conviction that the [Darwinists] evolutionary approach...will provide common goals and mutual comprehension among all the life sciences."[15]

With the founding of the Society for the Study of Evolution and the inception of the journal *Evolution*, Darwinists came to dominate the evolution discussion. As the only journal of its type, *Evolution* allowed the Darwinists to define the evolution debate, guiding it toward their ideas.

The group filled their new journal with articles supporting Darwinism, Fisher's "population genetics" and Huxley's "Modern Synthesis" as the explanation for evolution. Again imitating the *X-Club*, they favorably reviewed each other's books, and bashed opponents as "anti-evolution."

Along with the new society and the journal, the Darwinists planned a conference. This became the pivotal event – the moment everything changed. At the January 1946 Princeton Evolution Conference, a small group of 57 attendees declared that all biologists had reached "a common point of view" on evolution.[16] This small group declared that Darwinism was now the "consensus" theory of evolution in biology and science.

They went further. It was not enough for Darwinism to be the "consensus" theory. It must be the only theory of evolution. Historian Michael Dietrich says that although the Darwinists could not agree among themselves on a single model of Darwinism, they were unified in the claim that every other theory of evolution should be eliminated from discussion and research.[17]

The conference proved pivotal. It gave an academic imprimatur to the idea that Darwinism was now the accepted theory (the *only* theory) of evolution among biologists. This was untrue, but truth was irrelevant. Darwinists now controlled how evolution was defined.

2. Gaining Control over Evolution Research

Taking another lesson from Thomas Huxley, the group next set out to gain control over the committees that approved funding for research on evolution. It proved astonishingly easy.

In the aftermath of World War II, France, Germany, Italy and Japan, all previously strong in science, struggled simply to feed their people. All faced a decade of rebuilding. Even England teetered at the edge of bankruptcy, surviving only with a loan from the United States.

On the heels of this destruction, a "Cold War" erupted between the United States and the Soviet Union. The developments of World War II – radar, missiles, the atomic bomb, the first jet planes – made it clear that the military weapons of the future would emerge from science. As the rest of the world lay in ruins, the still economically powerful United States began an unprecedented expansion of scientific funding. Federal spending on basic research skyrocketed a staggering tenfold in just a few years.

This massive effort was centralized in a few federal agencies, and administered by bureaucrats required to justify their decisions. Research, however, is often hard to explain. As Einstein said, "if we knew exactly what we're doing, then it wouldn't be called research, would it."

Evolution posed a particular problem. Evolution rarely, if ever, had its own dedicated department, as physics, chemistry and the other sciences did. Bureaucrats asked who in academia was "in charge" of evolution? With their Society, journal and conferences, the Darwinists appeared to be.

Fisher's "statistical evolution" sounded scientific and came with mathematical proofs, which made it easy for bureaucrats to justify. Plus, they could point to academic journals and conferences which confirmed Darwinism was the "consensus" theory of evolution among experts.

This led bureaucrats to put a handful of Darwinists in charge of determining which evolution projects would be funded. Just as the *X-Club* gained control of the groups that determined success in Victorian science, modern Darwinists now controlled the committees that determined success in evolution research.

Today science is a vast affair, but in 1946 it was comparatively small. Within that small world, evolution was a minuscule discipline, easily dominated by a small group sitting on a few committees deciding what research would receive funding.

Lynn Margulis explained the effect. Recalling a lecture on genetics she attended, she says the speaker presented his work in Darwinist terms, but at the end said none of the Darwinist ideas had been proven. Margulis said "I later asked him, if none of it could be shown experimentally, [and] he himself was pointing to serious flaws...why did he teach this nonsense?" He said if he didn't describe his work in "Darwinist terms...he wouldn't be able to obtain grant money."[18]

This was the reality faced by everyone who wanted to study evolution. In science, without funds for research and publication, a scientist becomes invisible. Control over grant approval gave Darwinists the power to marginalize anyone who challenged their theory, stamping them with career ending labels such as "anti-evolution" or "creationist."

It's worth remembering the effect the nine member *X-Club* had on science just a few decades earlier. Those on the outside viewed "Huxley and the close-knit band of Darwinians…as a dictatorial, self-regarding clique... who insisted that acolytes ought to adopt their position," writes Harvard historian Janet Browne. "In many ways, this read the situation accurately," he says. "Huxley enjoyed his cliques and believed that small groups of 'right-minded men' were by far the most effective way to get things done."

In the 1950s and later, those who failed to tow the Darwinist line saw their careers abruptly stall. The more people acquiesced to Darwinism, the harder it became for others to criticize it without being marginalized. (As we will see, this problem continues right up to today.)

Gaining Control over Evolution Education

As the authorities on evolution, Darwinists were also put on the education committees that decided how evolution would be defined in textbooks and taught in schools. In the early 1950s, for example, Ernst Mayr joined the National Research Council's Biology group "to chart the course of biology." This became a powerful platform which Mayr used to separate what he called "functional" or regular biology from "evolutionary" biology, and to "secure the place of evolutionary biology" in textbooks.[19]

Mayr defined evolution as Darwinism, dissolving any distinction between the two.

With funding from the National Science Foundation, evolution became a regular part of science instruction and biology textbooks.[20] Darwinists dominated this process, determining the definition and description of evolution used in textbooks at all levels. For example, the evolution taught in universities was based largely on Fisher's Darwinist population genetics. Lynn Margulis says that's what "I was taught, and so were my contemporaries and so were the younger scientists. Assumptions were taught as fact."

Those who objected to all this couldn't counter the institutional machinery now under Darwinists' control. UCLA's Everett Olson spoke for many when he wrote in 1960:

> "There exists a silent group...engaged in biological pursuits who...disagree with much of the current thought, but say little because they...are so strongly in disagreement that it seems futile to undertake the monumental task of controverting the immense body of information and theory that exists in the formulation of modern [Darwinian] thinking."[21]

Within a decade, other theories of evolution disappeared for the new generation of scientists, now educated largely in Darwinism. By the 1959 University of Chicago conference on evolution, Darwin had become synonymous with evolution. What they could not establish factually, the Darwinists had achieved bureaucratically.

3. Inventing the Darwinian Myths

The third factor in the takeover was publicity and propaganda.

The close collaboration between Darwinists and Nazis in the 1930s had shown Huxley the power that propaganda has to change people's perceptions. In 1939, Julian's cousin, Gervas Huxley, helped establish the British war-time Ministry of Information to counter Nazi propaganda. This gave Julian a further window into the Nazi propaganda success, especially how they rewrote history, then backed it with a media deluge – films, newspaper stories, magazine articles, "scientific" reports – to create a new reality in people's minds.

With these lessons in hand, Darwinists in the 1940s used their new control over evolution to rewrite history. During those years an aggressive "Darwin industry" emerged, fabricating the stories about Victorian evolution and Darwin's role in it that are common today.

This is when the image of Darwin as a brilliant but persecuted visionary, valiantly fighting religious intolerance and social ignorance in an effort to reveal a great scientific truth was invented. History was rewritten to portray an epic struggle between dogmatic religion and rational science, a frightened, superstitious society versus the enlightened genius of Charles Darwin.

In re-writing the past, the Darwinists waged a scorched-earth war against their old opponents, depicting them as creationists in disguise, resolutely opposed to evolution. Ernst Mayer, for example, dismissed Richard Owen's brilliant discoveries as "metaphysical" nonsense. "With the Darwinian 'victors' writing history," says historian Adrian Desmond, "Owen, arguably England's greatest comparative anatomist... was still a hate figure."[22]

It was all untrue. Even Darwin acknowledged that most of his contemporaries were not creationists but instead believed "in some unknown but natural cause of species origins."[23] But the stories nonetheless stuck.

Modern opponents also had their discoveries denigrated, and were personally ridiculed.

Barbara McClintock was a particular problem. Having shown that genes operate in highly synchronized groups, she demonstrated that genetic change cannot occur through random mutations or "chance events," as the Darwinists claimed. Random changes in such a system would lead quickly to biological chaos and death.

McClintock was so effectively marginalized that she stopped publishing, ceased giving lectures, and retreated into isolation at the Cold Spring Harbor Laboratory in New York. McClintock later wrote,

> "[It] became painfully evident to me in my attempts during the 1950s to convince geneticists that the action of genes had to be and was controlled... It is painful to recognize the *fixity of assumptions* that many persons hold... One must await the right time for conceptual change."[24]

(Purging McClintock's ideas was particularly tragic, denying evolution what could have been an Einstein-level revolution in thinking. The effort to marginalize her was so successful, that her discovery of mobile genes disappeared entirely from scientific awareness, and was not rediscovered for a generation. Her brilliant insight that genes are part of a higher organization process in biology was delayed half a century, reemerging only after 2000.)

In true ideological form it was not enough that Darwin's opponents be wrong, they must be demonized.

For example, unable to disprove Richard Goldschmidt's evidence of rapid evolution, the Darwinists launched a vicious personal attack, turning Goldschmidt into an object of ridicule. Stephen Jay Gould recalls, "when I studied evolutionary biology...during the mid-1960s, official rebuke and derision [were heaped] upon Richard Goldschmidt, a famous geneticist who, we were told, had gone astray."[25]

Goldschmidt became an object of scorn, said Gould, his name evoking "snickers" and "laughter" when mentioned in classrooms. Gould compared the attacks on Goldschmidt to what happened in the totalitarian novel *1984*, where every day Big Brother directed "two minutes of hate" against Emmanuel Goldstein, "enemy of the people."[25]

After an aggressive campaign, these fictions became the conventional history of evolution passed around today. It was an ideological rewriting of history worthy of *Darkness at Noon* or *1984*. And the transformation was complete in less than a decade.

Browbeating Opponents into Submission

In the years that followed, with control over funding Darwinists browbeat opponents into submission, and threatened the careers of anyone who contradicted their theory.

The effort to bully opponents into submission continues today. Biologist Jerry Coyne, for example, through his web site "Why Evolution Is True," regularly bashes scientists who explain evolution outside of Darwinian orthodoxy. Richard Dawkins and other Darwinists run similar operations.

This is unique to evolution. There are no websites called "Why Geology is True," or "Why Cosmology is True," which try to enforce ideological adherence to old ideas. Calling Coyne a "grand inquisitor, one historian asks "Why is it that of all the sciences, evolutionary biology alone requires these counterparts to the Spanish Inquisition to keep the faithful in line?"

Even leading scientists come under attack if they question Darwinism. In 2008, Harvard's Edward Wilson and David Sloan Wilson (no relation) at Binghamton University, showed that no evidence existed for a key Darwinian process (something called "kin selection"). Despite 40 years of research, they said, no evidence for "kin selection" had ever materialized. The attack from Darwinists like Richard Dawkins and Jerry Coyne was swift. David Sloan Wilson said the assault included:

- Misrepresenting the lack of evidence supporting kin selection and the substantial evidence against it.

- Misrepresenting the number of scientists who oppose kin selection, calling Wilson's article "a futile one man crusade" when in fact kin selection is dismissed by many scientists.
- Questioning Wilson's personal competence, saying he was "losing it again," was" totally over the waterfall," and his ideas were "sheer madness."[26] (Unable to win a discussion with facts, Darwinist frequently retreat to personal attacks, a tactic, we saw, that began with Darwin himself.)

Wilson says these attacks have a powerful effect on those who study evolution. "For every colleague who dares to use [alternate ideas], there are others who strip the terms from their manuscripts to avoid harassment by people like Dawkins and Coyne."

Science, says Wilson, is supposed to result in "progress rather than endless gridlock," but the controversy over kin selection "seems to persist forever." He asks,

> "Is that because the topic is so very complicated?... Or might it be the kind of futile gridlock that occurs when some of the participants stop functioning as scientists?"[26]

Wilson had put his finger on the problem. Darwinists aren't "functioning as scientists." That's because the issue isn't science. Underlying these never ending disputes is the effort to twist the evidence for evolution to support their war against religion.

Wilson (an atheist) understands that the obsession with religion corrupts evolution as a science. An example, he said, "concerns the recent atheists attacks on religion by Dennett, Dawkins, Hitchens and others."

> "In this case the problem is...departures from responsible scholarship to portray religion as bad, bad, bad in every respect. This includes... extremely poor scholarship with respect to [the] scientific understanding of religion, from an evolutionary or any other perspective. It doesn't matter that these authors have distinguished reputations... They need to be roundly criticized... This is not repression. It is fighting to prevent responsible scholarship and scientific inquiry from becoming a carnival."[27]

How this obsession with religion corrupts our understanding of evolution is best seen in the abusive attacks on Stephen Jay Gould.

One of America's leading evolutionary scientists, Gould was also an atheist. However, he was a scientists first, and presented the evidence on evolution without regard to how it affected the Darwinist war on religion. For this he was relentlessly assaulted, culminating in a 1999 attack in *The New Yorker,* "Why Stephen Jay Gould is Bad for Evolution," by Robert Wright. Wright "indicts" Gould, not for bad science, but for being insufficiently ideological and anti-religion.

> "This indictment of Gould will no doubt surprise his large reading public," said Wright. "After all, in addition to being America's unofficial evolutionist laureate, Gould is a scientist of sterling credentials – a Harvard paleontologist and president of the American Association for the Advancement of Science. In what more capable hands could the defense of science rest?
> In truth, Gould is not helping the evolutionists against the creationists and the sooner the evolutionists realize that the better."[28]

Here we have America's "evolutionist laureate," a scientist of "sterling credentials," being attacked simply because he's more concerned with getting the evidence right, than with the Darwinist effort to portray all their opponents as "creationists." But Gould was not "bad for evolution," as Wright claimed, he was simply bad for Darwinists and their attempt to deny or twist facts to support their ideological agenda.

Many Academics "Want" Darwin to be True

The attack on Gould points to another important factor in the triumph of Darwinism. Many intellectuals support Darwinism because they deeply *want* it to be true. It is the lynchpin of their view of life. New York University scholar Thomas Nagel explained the issue with exceptional honesty:

> "In speaking of the fear of religion, I don't mean to refer to... the association of many religious beliefs with superstition. I am talking about something much deeper – namely, the fear of religion itself... I want atheism to be true and am made uneasy by the fact that some of the most intelligent and well-informed people I know are religious believers. It isn't just that I don't believe in God and, naturally, hope that I'm right in my belief. It's that I hope there is no God! I don't *want* there to be a God; I don't *want* the universe to be like that."[29]

Darwinism is crucial to many who don't want the world "to be like that." For 150 years, they have viewed Darwinism as the ideological acid eating away at belief in religion. Such a passionately held idea spills into their research, writing, teaching and, most unfortunately, their willingness to defend Darwinism even when it contradicts established scientific facts.

4. Darwinism was Promoted by the Popular Press

Darwinism's domination also came in part from the popular press. Cornell University communications professor Bruce Lewenstein said that in the 1930s and 1940s, many newspaper and magazine journalists believed it was their job to persuade the public that science could greatly improve society. He describes this as a continuation of the Progressive Era belief among intellectuals "that society was perfectible and that the wonders of science and technology would lead civilization towards this ideal."[30]

The 1950s was the beginning of the era of big government, big science and mass media. All these groups, said Lewenstein, had a vested interest in promoting science to the public:

> "After World War II, the term 'public understanding of science' became equated with public appreciation of the benefits that science provides to society. This equation was the result of the independent, but parallel...needs of four different groups with an interest in popularizing science: commercial publishers, scientific societies, science journalists, and government agencies."[31]

In the 1950s, Huxley leveraged the desire of journalists to convey the benefits of science to promote Darwinism. In the popular press, the concept of natural selection was easy to explain. The idea that genes are little blueprints for traits, while wrong, was also easy to convey, and it captured the popular imagination.

Over time, everything, including people's behaviors and beliefs, were reduced to the effect of genes. The press trumpeted a gene for "addiction," "intelligence," being "gay" and hundreds of other traits. There was even a "God" gene that explained why people believe in religion. Evolution was a breeze to explain when all you needed to do was point to a gene.

It was during these years that the "Darwin industry" used the press to promote the image of Charles Darwin as a "persecuted visionary." Julian Huxley popularized *The Origin of Species* as a "book that shook the world."[32] This was untrue, but journalists loved the story Huxley told: Darwin, the brilliant man of science who made a great discovery, paid a terrible price, battling religious intolerance and social ignorance his entire life.

The story of this epic struggle between a superstitious society and religious dogma versus rational science was repeated by journalists so often that it became embedded in the public mind as reality. Lost was the true story of a Victorian society that embraced evolution as a physical manifestation of God at work and confirmation of their religious beliefs.

Huxley Takes Control of UNESCO

Huxley took this publicity effort to a level never before seen in science – the entire world. In 1946, aided by his cousin Gervas, who sat on the board of the British Council (a group promoting international education), Julian was named the first Director of UNESCO, the United Nations Educational, Scientific and Cultural Organization.

Huxley filled UNESCO with eugenicists, and used it as a platform to promote Darwinism and eugenics worldwide. Huxley embedded eugenics in UNESCO's founding mission statement (!) writing:

> "Even though... any radical eugenic policy will be for many years politically and psychologically impossible, it will be important for UNESCO to see that the eugenic problem is examined with the greatest care, and that the public mind is informed of the issues at stake so that much that now is *unthinkable* may at least become *thinkable*."[33]

For Huxley, UNESCO was just one more propaganda outlet for his eugenics agenda. While most people considered eugenics a source of unimaginable atrocities, Huxley believed after they were "properly educated" on it's benefits by groups like UNESCO, it would again become acceptable.

After leaving UNESCO, Huxley went on a worldwide media blitz – books, articles, radio, television and an Oscar-winning wildlife film – to popularize his ideas. In 1953, UNESCO awarded him the Kalinga Prize in recognition of this "public education" work.

By 1959, with the new education curriculum, extensive mass media program, and validation by the Federal government and United Nations, all the public knew about evolution were Darwin's ideas.

In people's minds, Darwin and evolution were synonymous. Julian had fulfilled his grandfather's dream, and had done it in little more than a decade. He even managed to get the United Nations to promote eugenics. The old "bulldog" would have been proud.

Chapter Ten

The Tragedy

DELAYED DISCOVERIES, RACISM AND EUGENICS

So what were the effects of the Darwinist takeover of evolution? What happens in science when ideology replaces fact?

- It delayed scientific progress. While the other sciences rapidly advanced, evolution spiraled into what some called a "second-class" status. For decades important research was misdirected, depriving humanity of lifesaving discoveries that affected millions.
- Darwinism "scientifically" justified racism, claiming that racial differences are a product of genetic differences ("Today we use genetic instead of racial" wrote Dobzhansky.[1]) And it "scientifically" justified eugenics. Programs begun in the 1930s in Europe, America and other countries were reconfirmed and ran well into the 1970s.
- It led to moral crisis. In the 1950s, Darwinism's justification of racism and eugenics led many people to reject it, and in the process to reject the idea of evolution. Modern Darwinists had created the false choice between evolution and religion which Darwin never could.

False Facts Delayed Important Discoveries

With their new power, Darwinists obstructed any discovery that contradicted their theory, delaying important medical advances for decades.

The discoveries of immunologist Ted Steele are a particularly tragic example. Geneticists knew that viruses evolve quickly. Steele believed the immune system must also evolve fast, otherwise viruses would quickly kill us.

He set out in the 1970s to understand how this rapid evolution works, and harness that knowledge to save lives.

Steele discovered that after our body is attacked and fights off a virus, the immune system stores part of the virus in our own DNA. If the virus ever attacks again this fragment allows the immune system to rapidly recognize and respond to it.[2] Steele also discovered that the rapid response can be directly inherited by the next generation, giving them protection from the virus, even if they've never been attacked by it.[3]

This was an extraordinary breakthrough, poised to transform immunology. The BBC ran several programs on Steele's work and *The Times* of London ran an article calling it an "astonishing experiment which seems to challenge Darwinism and resurrect Lamarck." In *The Time's* article, Nobel Prize winning immunologist Peter Medawar said that while the discovery needed to be confirmed, "I hope Steele is right." Medawar understood that if Steele was correct, it could save millions of lives.

Ordinarily a discovery of this magnitude would be showered with research funds. But as he later explained in his groundbreaking book, *Lamarck's Signature*, Steele said immune system changes are inherited through the process Lamarck had described.[3] This contradicted Darwinism, and a blistering assault was unleashed. The Darwinist establishment battered Steele for decades, attempting to discredit his discoveries and obstructing his access to research funds.

Steele spent the better part of his life fighting these attacks. Years later in a radio interview he said that "to be branded a heretic and a pariah meant that my career [and] opportunities to keep doing this research in this area were extremely limited so I had to survive the best way I could."[4]

Steele's discovery that immune protections can be directly inherited is now widely confirmed.[5] Peter Medawar later called Steele's work "one of the landmarks in the history of biology."[6] Unfortunately, the immunology advances from Steele's discovery were delayed by decades, leading to countless deaths that would otherwise have been prevented.

At the other end of the medical spectrum, one of the most important breakthroughs of the last decade is reprogramming cells to grow new body parts (like those used on Claudia Castillo mentioned in the Introduction). The ability to reprogram cells, central to modern regenerative medicine, was first discovered by John Gurdon in the late 1950s, but his work was sidelined for nearly 50 years.[7] Gurdon, now 80, won the 2012 Nobel Prize in Medicine for his discovery. He said of those early years, "I was a graduate student flying in the face of [established] knowledge."[8]

A "petrified form of evolution emerged from the [Darwinian] modern synthesis," said the great micro-biologist Carl Woese.[9] "Dogmatic thinking prevailed all too often, with disastrous consequences for progress [in] microbiology, molecular biology, and the study of evolution."

During those years Darwinism became, in effect, a scientific serial killer. Dozens of lifesaving discoveries – from Lamarck to McClintock, Steele and a dozen others – disappeared for half a century, purely for ideological reasons. Each was eventually rediscovered, proven correct, and now plays an important role in understanding evolution.

Even Richard Goldschmidt, the person most demonized by Darwinists, has in recent years seen a dramatic reversal in fortune. Decades ago, Stephen Jay Gould predicted that "Goldschmidt will be largely vindicated in the world of evolutionary biology."[10] He was right. Goldschmidt's work is now being revived and the journal *Nature* says "many researchers have welcomed the return."[11] Canadian biologist Brian Hall says the ridicule heaped on Goldschmidt is today "giving way to an appreciation of the important role he played as the first to integrate genetics, development and evolution."[12]

Eugenics Through Other Means

Over time, however, these delays became a form of eugenics, leading to the death of millions around the world – primarily the malnourished, infirm and working poor. What Darwinists were not able to accomplish through direct eugenics programs, they achieved by obstructing life-saving research.

The effect of this obstruction continues today.

For example, increased resistance to antibiotics is one of the best-documented examples of rapid (non-Darwinian) evolution, with direct life-and-death consequences. However a group at the University of Virginia shows that the word "evolution" is rarely used in medical research papers.[13] Instead, antibiotic resistance is said to "emerge" or "arise," rather than "evolve." The University of Virginia scientists write that:

> "The failure to use the word "evolution" may reflect the mistaken sense that evolution implies processes that are slow, and imperceptible... It has [also] been repeatedly rumored (and reiterated by one of the reviewers of our article) that

> both the National Institutes of Health and the National Science Foundation have in the past actively discouraged the use of the word "evolution" in titles or abstracts of proposals to avoid controversy."
>
> "Nowadays, medical researchers are increasingly realizing that evolutionary processes are involved in immediate threats associated with not only antibiotic resistance but also emerging diseases. The [rapid] evolution of antimicrobial resistance has resulted in a 2 to 3 fold increase in the mortality of hospitalized patients, has increased the length of hospital stays, and has dramatically increased the costs of treatment."

In short, the idea of rapid evolution is avoided in medical research to "avoid controversy," despite the fact that the rapid evolution of microbes has led to a dramatic increase in deaths.

The Battle for Eugenics: The Scopes "Monkey" Trial

Another tragedy of the takeover was that Darwinism gave new "scientific" credence to eugenics and racism. Let's step back to the Scopes Monkey Trial to see how this happened.

In 1925, high school biology teacher John Scopes was arrested in Dayton Tennessee for violating the state's Butler Act, which disallowed teaching certain claims about human evolution.

At the trial, virtuoso defense attorney Clarence Darrow argued for Scopes. Inspired orator William Jennings Bryan argued to uphold the law. Today the trial is seen as a classic example of the triumph of rational thought and enlightened science over religious fundamentalism and backwater biblical superstitions.

Historians tell a different story.

For years Tennessee high schoolers were taught biology from George Hunter's *A Civic Biology*, which described evolution as "the belief that simple forms of life on the earth slowly and gradually gave rise to those more complex and that thus ultimately the most complex forms came into existence."[15] And for years, no one in Tennessee objected to teaching this idea.

The problem was not evolution but eugenics, and what it means to be human. This is why it was called the Scopes "Monkey" Trial, not the Scopes "Evolution" trial.

University of Missouri historian Douglas Linder says that three-time Democratic candidate for President William Jennings Bryan "cared deeply about equality [and] worried that Darwin's theories were being used by... a growing eugenics movement advocating sterilization of 'inferior stock.'"[14] To protect real people from real harm, Bryan campaigned for state laws that would bar Darwinism and eugenics from the classroom.

Tennessee's Butler Act reflected this specific concern. The law prohibited teaching "any theory that denies the story of the Divine Creation of man as taught in the Bible, and to teach instead that man has descended from a lower order of animals."

Few today would view this as the best way to stop eugenics. But in 1925, it seemed an effective way to make clear that in Tennessee human beings were seen as having a God-given uniqueness that made them more than mere animals and subject to eugenics programs.

That goal led to a conspiracy.

At Fred Robinson's drugstore in Dayton, Tennessee, a small group of local men often met for a soda and socializing. A member of the group, George Rappalyea arrived one afternoon in 1925 carrying an ACLU announcement that offered support to anyone who challenged the Butler Act. Rappalyea explained that another Tennessee law required teachers to use a textbook which explicitly taught the idea that humans had evolved from animals. Thus, one way or the other, teachers were required to break Tennessee law.

Linder writes that "listening to Rappalyea, the others... became convinced that the publicity generated by a controversial trial might help their town, whose population had fallen from 3,000 to 1,800." This small group then proceeded to gin up a case over the new law.

The trial emerged, not over religious issues, but from an anti-eugenics movement attempting to protect society's most vulnerable, and a foolish effort at local promotion.

TWO SALIENT POINTS: THE TRIAL WAS *NOT* ABOUT RELIGION

A lot happened surrounding the trial, but two facts are salient. First, in a trial that lasted eight full days, religion entered the picture only briefly, when introduced by Clarence Darrow, itching to make it a conflict between science and religion.

For days the case was argued on purely factual grounds ("is eugenics scientific?"), with no mention of religion. On day four, Darrow tried to introduce religion by asking that the Butler Act be declared unconstitutional on religious grounds. Judge John Ralston denied the motion, saying the Act "gives no preference to any particular religion or mode of worship." Ralston added, "our public schools are not maintained as places of worship, but on the contrary, were designed, instituted, and are maintained for the purpose of mental and moral development."[16]

On day eight, with the trial ending in mere hours, Darrow had still found no way to inject religion into the case. At the last moment, he called Bryan to the stand and began relentlessly questioning his religious views.

Bryan described his anti-eugenics campaign as a defense of humanity, based on the Judeo-Christian values of the Bible, from which he quoted.

Darrow, attempting to make the issue religious fundamentalism, asked about the six days of creation in the Genesis story. Bryan replied, "I do not think it necessarily means a twenty-four hour day." Darrow pressed him, asking if they were literal days. "My impression is that they were periods," said Bryan.

An unrelenting Darrow rephrased and repeated the same question over and over: how long did creation take? An exasperated Bryan finally said it made no difference if it took "six days, or six years, or six million years or six hundred million years... I do not think it's important whether we believe one or the other."[16] Hardly the words of a man promoting biblical literalism.

Judge Ralston eventually brought the questioning to an end, ruling that Bryan's religious views "can shed no light upon any issues that will be pending before the higher courts."[17]

Two Salient Points: The Trial *Was* About Eugenics

The second point is that the eugenics problem which Bryan opposed was real. The first witness for the defense was Maynard Metcalf, a member of the American Eugenics Society. From the start, the defense intended to define evolution in terms that would justify eugenics.

Indeed, that was the problem with the textbook used by Scopes, which taught biology in a way that promoted eugenics.[18] Here are a few items Tennessee teenagers learned from this textbook (emphasis added). As you read them, ask yourself how would you react if your children were taught this material today? Would you object to this being presented as "science"?

- "There exist... **five races** of man, **each very different from the other**... These are the Ethiopian or negro type, originating in Africa; the Malay or brown race, from the islands of the Pacific; the American Indian; the Mongolian or yellow race, including the natives of China, Japan and the Eskimos; and finally, **the highest type of all, the Caucasians**, represented by the civilized white inhabitants of Europe and America."

- "If the stock of **domesticated animals can be improved**... the health and vigor of the future generations of **men and women... might be improved** by applying to them the [same] **laws of selection**. This improvement [is a] science called eugenics."
- "Studies have been made on... families in which mental and moral defects were present in one or both...parents. The "Jukes" family is a notorious example. The first mother is known as "Margaret, the mother of criminals." In seventy-five years the progeny of the original generation has cost the state of New York over a million and a quarter dollars, besides giving over to the care of prisons and asylums considerably over a hundred feeble-minded, alcoholic, immoral, or criminal persons."
- "Hundreds of families such as those described above exist today, spreading disease, immorality, and crime to all parts of this country. The cost to society of such families is very severe. Just as certain **animals or plants become parasitic** on other plants or animals, **these families have become parasitic** on society. They not only do harm to others... they are actually cared for by the state out of public money...**They are true parasites.**"
- "If such people were lower animals we would probably **kill them off to prevent them from spreading**. Humanity will not allow this, but we do have the remedy of separating the sexes in asylums...and various [ways of] preventing intermarriage and... perpetuating such **a low and degenerate race**. Remedies of this sort have been tried successfully in Europe and are now meeting with some success in this country."

It was against these ideas that Bryan and the people of Tennessee fought. They lost.

The Darwinists won. Eugenics won.

The price was soul shattering.

Two years later, in 1927, the United States Supreme Court approved the forced sterilization of Carrie Buck, a 17-year-old single mother judged to be feeble-minded and morally delinquent. Summing up the court's thinking, Justice Oliver Wendell Holmes Jr. wrote "It is better for all the world if instead of waiting to execute degenerate offspring for crime, or to let them starve for their imbecility, society can prevent those who are manifestly unfit from continuing their kind."[19]

It was now a crime simply to be born "their kind."

Some states had earlier passed eugenics sterilization laws, but the Supreme Court now approved it for the nation. In the years that followed, hundreds of eugenics programs were unleashed on thousands of people all across the country. By 1941, California alone had sterilized 14,955 people.[20]

Other programs, like the infamous Tuskegee experiment, were even more insidious. Begun in 1932 and run by the US Public Health Service, the Tuskegee experiment used poor black men as guinea pigs to study the effects of syphilis. Lasting forty years, the men involved, and the families who loved them, lived their lives suffering in pain.

These programs ran well into the 1960s, and some, like the Tuskegee experiment, continued into the 1970s. The damage – physical, mental, emotional – was profound. A 2011 report on NPR told the story of the deeply scarred lives of those still living who had survived these programs.

Inherit the Wind

Why do people today see the Scopes Trial as a battle between religion and science rather between society and the eugenicists?

The confusion stems partly from the press coverage at the time, especially HL Menken, whose nationally syndicated column daily called the people of Dayton "yokels" and "morons," and described Bryan as a "buffoon" spewing "theologic bilge."[21] Like Darrow, Menken wanted to make the trial about religion, which he called "a curse to mankind."

Despite this media portrayal, the trial would likely have faded from memory except for one thing, *Inherit the Wind*, which retold the story.

Historian Edward Larson, who wrote a Pulitzer Prize-winning history of the Scopes trial (*Summer for the Gods*), says *Inherit the Wind* fundamentally distorts what happened.[22] Focusing largely on Bryan's time on the stand, it portrays his biblical quotes as opposing science. *Inherit the Wind* also misrepresents the larger social context of the trial, framing it as a struggle between those who are educated and enlightened, versus ignorant hillbillies mouthing Biblical superstitions.

Inherit the Wind's version of events was repeated over and over until it became cemented in the public mind – first as a 1955 Broadway play, then a 1960 movie, a 1965 television film, two more movies in 1988 and 1999 and multiple book printings. Its portrayal led millions of Americans to ridicule opponents of Darwinism as religious fundamentalists. Lost was the true issue – eugenics – and the very real tragedies that followed.

The Post World War II Push for Eugenics

Another tragedy spawned by the takeover was the emergence of a new type of "scientific racism."

The events of World War II discredited eugenics. But having won the battle for eugenics in the 1920s, and seen these programs spread worldwide in the 1930s, Darwinists set out in the 1940s to rehabilitate eugenics. They did this by attributing racial traits to genes. If racial differences are the result of having different genes, this made racial distinctions - and racism -"scientific." Others did not misuse Darwinism to make such claims. From the start, it was the Darwinists who invented and promoted them.

In 1946, as a stream of horrific Nazi eugenics programs were revealed, the Darwinists aided former Nazi eugenicists. This included Otmar von Verschure, who trained the infamous Josef Mengele (director of the appalling eugenics experiments at Auschwitz) and had used human body parts sent by Mengele from Auschwitz in his own research. Verschuer wrote to a former student, geneticist Hermann Muller, expressing his commitment to restoring the reputation of "our science" - eugenics.[23]

Hermann Muller, a Darwinist and protégé of Julian Huxley, had taken up residence at the Kinsey Institute in the US where, although criticizing those who had "brought odium on the concept of eugenics," he worked to restore the reputation of eugenics.[24]

Where the Nazis had talked about blood types, the Darwinists began talking about genes. Eugenic zealot Theodosius Dobzhansky spearheaded the Darwinists effort to equate race with having different genes. "Today," he wrote, "we would use the word genetic in the place of...racial." In 1946, in *Heredity, Race, and Society,* Dobzhansky explained that the races have different genes and this accounts for the differences in their biology, behavior and abilities.[25] "Race can be defined as populations which differ in... genes," he wrote.

Dobzhansky later penned the foreword to a book by America's most prominent eugenicist, Frederick Osborn, which made the same point: differences in genes explain differences in race.[26]

"Redefined in... genetic terms, race was not relegated to the margins but remained at the center of the study of human evolution in the 1950s," says historian Lisa Gannett. Whenever "Dobzhansky details his vision of future research," she says, "the evolutionary origins of racial differences are always front row and center."[27]

"The most violent eugenicists, and in some ways racists," said Richard Lewontin, "were geneticists who said genes dominate everything, and if there are differences between groups and how rich they are, how much power they have, it must be in the genes."[28]

Supporters argued that by equating race with genetic differences, Dobzhansky hoped to make race a matter of science, and remove it as a social issue. Dobzhansky did wrap his "racial genetics" in humanitarian language, for example saying the "artificial selection [of humans] constitutes the sound core of eugenics, the applied science of human betterment."[29]

Detractors, however, said this was the old "scientific racism" of 1900 dressed up in new genetic terms. Dobzhansky clearly understood that basing race on genetic differences would "scientifically" justify racial prejudice. He knew it would give "scientific" credence to Madison Grant's old claim that "speaking English, wearing good clothes, and going to school and to church do not transform a Negro."

And in fact, racists, arguing for segregated societies, rushed to use Dobzhansky's racial genetics as "scientific" confirmation of such claims.

Darwinists like William Hamilton went further, saying that racial conflict and race wars were inevitable. Why? Because each race represents an extended family who all share the same genes. Hamilton said it was natural, even desirable, to discriminate against people from inferior races because they all have the same inferior genes.

"These geneticists and their scientific allies, ultimately helped to preserve the concept of race in science [in a way that] would quickly be exploited and manipulated by racists – scientists and non-scientist alike," says Michael Yudell at Drexell University. [30]

Using Genetics to Prove Racial Inferiority

As we will see, the claims about genetics and race by Dobzhansky and the Darwinists are false. The ability to decode the human genome in the 1990s showed that humans have few genetic differences, and those that exist involve minor issues like lactose tolerance, resistance to local diseases or skin color. (People have about the same number of genetic differences with someone from another race as they do with a third cousin.)

In the 1950s, however, Dobzhansky's claims about natural selection, race and genes became widely accepted. Darwinists used them for decades to promote both eugenics and racism.

In 1951, Ronald Fisher (one of the eugenicists brought to UNESCO by Julian Huxley), was asked to help prepare the United Nation's "Statement on the Equality of the Races." Fisher protested vigorously, saying science had established that the races are *not equal* – some races are clearly inferior.

Fisher prepared an opposing statement, writing that the races differ greatly "in their innate capacity for intellectual and emotional development...This problem is being obscured by entirely well-intentioned efforts to minimize the real differences that exist."[31]

Many assumed Fisher's assessment must be correct. If natural selection is true, then important genetic differences must exist between the races and ethnic groups, "scientifically" justifying eugenics.

It's a measure of how widely accepted these ideas were at the time, that the United Nations appended Fisher's racist document to its "Statement on Equality," where it remains to this day (available online).

Natural Selection and Social Progress

The idea that natural selection is the path to social improvement, and justifies killing the "unfit," extended deep into the 20th century.

In *To the End of Hell: One Woman's Struggle to Survive Cambodia's Khmer Rouge*, Denise Affonço describes how this belief fueled genocide in her native Cambodia.[32]

In 1975, the Khmer Rouge, a revolutionary communist army, took control of Cambodia. Their first action was to round up the country's entire urban population and march them deep into the jungles. A Khmer officer explained why to Affonço's group. "In principle we should shoot you, but there are so many of you, the ammunition is too expensive.... So [we are] going to make a selection to eliminate the bad elements by work and hardship."[32]

The Khmer subjected everyone to a brutal regime: long marches, little food, no rest and no shelter from the sun, rain and elements. The Khmer were planning a new society. The idea was that those who were unfit for the new society would die. And they did, in the millions – the old, the disabled, the sick and the very young. Especially the young.

While they might not have approved of the method, Darwin, Huxley, Fisher and other Darwinists would certainly have approved the outcome, believing society was much better off for it. (The 1984 Academy Award-winning film *The Killing Fields* tells the story of this genocide.)

Ideas like those of Dobzhansky and Fisher appear whenever human beings are viewed though a Darwinist lens. An updated version came in 2005 when UCLA's Jared Diamond used natural selection to justify his claim that New Guineans are genetically superior in intelligence to Westerners. "New Guineans [have] impressed me as being on the average more intelligent, more alert, [and] more expressive...than the average European or American."[33]

The reason for this higher intelligence, says Diamond, is that incessant tribal warfare allowed only the smartest, most cunning New Guinea men to survive each generation (natural selection at work). On the other hand, "most Western infants survive fatal infections and reproduce themselves, regardless of intelligence and the genes they bear."[33] (Darwinists just can't get over the fact that children whom they judge unfit or unintelligent are able to survive because of modern medicine.)

Provoking a War with Religion

The third tragedy of the takeover was that Darwinists finally created the false choice between science and religion they had long sought.

Although today this is presented as a battle between science and religious fundamentalism, like the Scopes trial, history tells a different story.

Richard Lewontin points out that while Genesis says God created the universe in three days, physicists, cosmologists and geologists say it took billions of years. "So," he asks,

> "Why aren't Kansas schools under extreme pressure to change the curriculum in physical science courses? Why should physicists be allowed to propagate, unopposed, their godless accounts of the evolution of the physical universe? Something more is at stake than a disagreement over the literal truth of biblical metaphors."[34]

The "something more at stake" was the issue that had always driven Darwinism: defining what it means to be human. Darwin said religion and evolution led to irreconcilable answers, and society would have to choose between them. Victorians in 1859, knowing that Darwin had shown no such thing, rejected this as a false choice.

But a hundred years later the situation was different. In newspapers, magazines and textbooks people were repeatedly told that Darwinism is true, a claim confirmed by major scientific organizations. Many scientists also said Darwinism made eugenics "scientific" and, although previously discredited, eugenics began to make a resurgence.

Modern Darwinists tied their hopes for eugenics closely to a campaign against religion. Francis Crick, for example, was both a staunch atheist – resigning his position at Churchill College at Cambridge to protest their building a new chapel – and a committed eugenicist, telling an interviewer:

> "[Eugenics] is not a subject at the moment which we can tackle easily because people have so many religious beliefs... I think it would be risky to try and do anything in the way of eugenics. [But] I would be astonished if... society did not come round to the [eugenics] view."[35]

The Scopes trial had been an object lesson to Darwinists in how religious beliefs could be portrayed in ways that advanced eugenics. The 1959 Chicago Evolution Conference expanded Darwinism to cover the "evolution of mind and behavior." Historian Janet Browne says Julian Huxley gave a "secular sermon in the style of his grandfather, and provocatively declared that religious belief was merely a biological feature of evolving mankind."[36]

Huxley believed that religious beliefs may have served some evolutionary purpose in the past, but they were now harming society.

People wondered, would religious beliefs soon be labeled a genetic defect to be eliminated as a matter of public health? Darwinists said as much. "I believe," said Huxley, that a "drastic reorganization of our thought is now becoming necessary, from a God-centered to an evolution-centered pattern."[37] Richard Dawkins was more direct: "I think a case can be made that [religious] faith is one of the world's great evils, comparable to the smallpox virus but harder to eradicate."[38]

The Choice People Faced

In the 1950s, many people felt they faced a real choice about the future of society.

On one side was Darwinism, which said biology defines a person's worth, some races are genetically inferior to others, and sterilizing people like Carrie Buck ("their kind") was the path to social progress.

It was Darwinists (Huxley, Fisher, Dobzhansky, Muller, others) who fueled the campaign for eugenics. As in the past, this was not Darwinism being misused by eugenicists. *The Darwinists were the eugenicists.*

On the other side was the belief that human life has an intrinsic value, that "all men are created equal," and the path to social progress is through education and equal opportunity for all classes and races.

The choice people confronted can be seen in the fate of three children.

In the novel *Sophie's Choice*, at the entrance to a concentration camp, a young mother with two small children is told she can take only one child with her. The other child will be sent immediately to a death chamber. She is also told that she must decide which child will die. If she does not choose between them, both will be killed immediately. The air is thick with the acrid odor of burning flesh As she holds each child's small hand tightly in her own, her mind convulses, her stomach heaves, her soul collapses.

The other type of society was described by the writer Calvin Trillin, who began his career in the 1950s. Writing about his wife Alice, Trillin said that "one summer, working at a camp for children with genetic disorders, Alice wondered how little L. could be so cheerful even though she neither grew nor could digest food." Alice later saw a letter from L.'s mother that read "if God had given us all of the children in the world to choose from, L., we would only have chosen you." When he arrived to pick her up, Alice pulled her husband aside and said, "Quick. Read this. It's the secret of life."[39]

The question was, would people live in a society like Sophie's, that forced them to choose between their children, or a society where children like Little L. were loved and nurtured?

This choice was not a matter of education or intelligence, but went to the core values about the belief in human equality central to Western society. "As Darwin clearly saw," says Oxford scholar John Gray, "his idea represents a fundamental moral challenge to our Western world-view."[40]

Nor was this a choice that could be made passively, intellectually rejecting Darwinism while doing nothing more. A stream of stories about the horrors of World War II eugenics programs splashed across the pages of *Life*, *Look* and other magazines.

Especially ghastly were the experiments of Josef Mengele. He would stand on the arrival platform of the Auschwitz concentration camp, deciding who would die immediately, who would be worked to death in the camp, and who would be sent to his clinic to have their arms and legs amputated, eyes removed, and similar experiments, to see how the body responded. Children were experimented on sexually. Pregnant women were brought in and dissected alive, along with their unborn babies, before being sent to a death chamber.

Many American soldiers had seen the concentration camps first hand. Others saw the photographs. They were horrific in ways never imagined possible. (A key revelation of the just released biography of *Catcher in the Rye* author JD Salinger, is how the trauma of seeing the concentration camps haunted him throughout his life.)

The consequences of not actively opposing such ideologies was all too clear. World wide, over *seven million people* were killed in these Darwinian justified eugenics programs. Writing about the years of Nazi rule, German theologian Martin Niemöller expressed the problem in words that echoed in people's minds:

> "First they came for the communists, and I did not speak out because I was not a communist...then they came for the trade unionists, and I did not speak out because I was not a trade unionist. Then they came for the Jews, and I did not speak out because I was not a Jew. Then they came for me, and there was no one left to speak for me."

This moral dilemma led many people to reject Darwinism. Seen through their eyes, who could do otherwise? Darwinism was both scientifically flawed and ethically reprehensible. Opposing Darwinism was about preventing real atrocities, not in far off countries but in eugenics programs in America right then. Little L. was precisely the type of child eugenics programs were designed to eliminate.

In addition to the moral dilemma, there was a vast trove of scientific evidence that proved Darwinism wrong – from paleontology and geology, to biology, genetics and a dozen other sciences. People asked, how can Darwinism be true, if it is contradicted by all this scientific evidence? If Darwinism is scientifically untrue, and Darwinism and evolution are the same thing, then by definition evolution must also be *scientifically* untrue? The logic was irrefutable ("ipso, facto"), even Aristotle would have agreed.

For many people preserving society meant rejecting the scientifically flawed Darwinism. If this also meant rejecting the idea of evolution, then so be it. What rational person would put the conflicting scientific claims of Darwinism ahead of protecting real people, like Carrie Buck and little L?

In making evolution synonymous with a theory contradicted by so much scientific evidence, Darwinists provided the scientific ammunition that perpetuates today's anti-evolution arguments. (This seems to serve their purposes. The more people argue about whether evolution is true, the less they focus on the fact that Darwinism is fundamentally wrong.)

Over time, the rejection of Darwinism became a social war and then morphed into the special-creation versus evolution debate we have today.

Despite all the rhetoric about science and religion, however, the true conflict in this war was (and is) the dispute central to Darwinism from the start, defining what it means to be human, and what is permissible for one human being to do to another.

These Ideas Pervade Current Thinking

If you think this choice wasn't real, or is not present today, think again. These ideas are widespread, appearing in places you might not expect.

Princeton philosopher Peter Singer became known for advocating that disabled human infants should be put to death. In 2012 Alberto Giubilini and Francesca Minerva took this idea to a new level. In "After-birth Abortion: Why Should the Baby Live?," they argued that even *healthy living infants* have no right to life, and should be killed if they merely prove inconvenient to their family:

> "Both a fetus and a newborn... are...potential persons but neither is a 'person' in the sense of... an individual who is capable of attributing to her own existence some basic value such that being deprived of this existence represents a loss to her... It is not possible to damage a newborn by preventing her from developing [into] a person...What we call 'after-birth abortion' (killing a newborn) should be permissible... including cases where the newborn is *not disabled*."[41]

Like others who view humans solely in biological terms, the authors justify this idea by saying that "to bring up such children might be an unbearable burden on the family and on society as a whole."

What is remarkable is not that this claim was made (people say ridiculous things all the time), but that it appeared in Oxford University's *Journal of Medical Ethics*. Julian Savulescu, the journal's editor and director of Oxford's Centre for Practical Ethics defended child killing as a topic worthy of consideration. "The goal is not to present the Truth or promote some single moral view," he said. "It is to present well-reasoned argument based on widely accepted premises."[42]

When did child killing become an idea "based on widely accepted premises"? When did behavior previously linked only to psychopaths become worthy of social consideration?

The Nazi rationale for experimenting on Jews was that they were not really human beings. If infants are not really human, as the Oxford group claims, does this mean they can be experimented on, as Josef Mengele did to Jews?

This was exactly the future under Darwinism that people feared, and why so many rejected it and, unfortunately, evolution along with it.

Chapter Eleven

A New Darwinism

MAKING GENES MAGICAL

The Darwinists succeeded in the 1940s in part because much about biology was still unknown. This left them free to explain important problems with simple models and speculation. Despite a lack of evidence, some claims were repeated so often they entered textbooks as established facts.[1]

A wave of discoveries in the 1960s, however, shed new light on biology and made the old Darwinist claims untenable, forcing them to develop a New Darwinism, one which turned their old claims upside down.

The Second Genetic Revolution

In the 19th century most scientists believed that all processes in biology reduce to chemistry, which in turn reduces to the behavior of atoms. This idea, called "reductionism," said that everything in nature can be explained simply by adding up the activity of atoms. Human awareness, for example, was called an illusion. Because humans are made entirely of atoms, they can't have a property that doesn't exist in atoms. Atoms don't have awareness, thus what feels like awareness to us must be an illusion.

Those who created the 20th century revolution in physics showed that "reductionism" has nature fundamentally wrong. They showed that while physics, chemistry and biology obey the same laws, as nature becomes more complex new properties emerge which *cannot be reduced* to the properties that exist at a lower level. For example, iron and oxygen are made from particles called quarks, yet both possess unique properties that don't exist in quarks.

The man who led the revolution in physics, Neils Bohr, call this discovery "complementarity." He said the higher levels of nature are compatible with the lower levels. But as nature becomes more complex, unique properties emerge at the higher levels which can't be reduced to any property at the lower levels.

In a famous example, Bohr showed that light, which is central to all life, behaves differently at the level of biology than it does at the level of physics.[2]

Neils Bohr and his band of revolutionaries (all Noble prize winners) understood the profound implication of this discovery. It meant biology *could not be reduced* solely to chemistry or physics. Life is something more – something *different* – than the activity of atoms.

Bohr's associate, Nobel physicist Max Delbrueck, took these revolutionary ideas to Linus Pauling, a young biologist at the California Institute of Technology. Pauling used the new ideas to unravel one of biology's biggest mysteries: how atoms form the complex "molecules of life" called proteins.[3] In the 1940s, Pauling's groundbreaking work revolutionized biology and won him a Nobel Prize.

Another of Bohr's revolutionaries, Nobel physicist Erwin Schrödinger, triggered a similar revolution in genetics.

Although biologists knew that genes existed, they had never seen one. In a lecture, "What is Life?," Schrödinger said the evidence indicated genes must be a special type of molecule that allows them to "carry information," and thus hold the "code of inheritance."[4]

The furious hunt began for Schrödinger's "information molecule." In 1944, an American-Canadian group, led by Oswald Avery, discovered that genes are made of DNA, for which they received the Nobel Prize.

In the years that followed much was learned about how DNA works. But its exact structure remained a mystery. In 1953, Rosalind Franklin, who pioneered the use of photography to reveal the structure of molecules, produced the first blurry photos of DNA. They revealed crucial aspects about DNA, including its spiral staircase shape, the famous "double helix."

Using Franklin's photos, biologist Francis Crick made a guess at the composition of the double helix, and rushed his guess into print (without getting Franklin's permission to use her work, or giving her any real credit).[5]

(Many people today believe Crick and his lab partner, James Watson, discovered DNA. They did not. At the time, Crick's insightful guess at the *rough structure* of DNA was seen as important but not transforming. It explained nothing about *why* genes act as they do, or how their complex actions build living things. And those were the burning questions of genetics.)

'Genetic Clocks' Reveal Evolution's History

As Linus Pauling continued to make discoveries, he found that some molecules change in a predictable sequence. Belgian chemist Marcel Florkin used these predictable changes to begin outlining a new, more precise history of life and evolution (what he called "biochemical evolution").[6]

Pauling and geneticist Émile Zuckerkandl took this to a new level when they found that some molecules change at regular intervals, making them, in effect, "biological clocks."[7] This allowed geneticists to begin mapping an even more precise time-line of evolutionary events.

By the early 1960s these developments were creating a revolution. The new "molecular biology" explained the basic processes of life, and the new "molecular genetics" explained inheritance and revealed the history of life.

Since both used the same underlying mechanisms, this "molecular revolution" was poised to unite biology, genetics and evolution into a single, coherent picture of life.[8]

This revolution created two problems for Darwinists. First, the new discoveries changed the history of life. Previously, plants and animals were organized based on their physical traits. If they shared the same trait it was assumed they were related. Molecular biology shoved this idea aside, showing it was often untrue. A more acculturate history of evolution emerged and the story it told was at odds with Darwinism.

Second, powerful atomic forces govern how molecules (like genes) form and thus how they change and evolve. This meant that genetic evolution was not determined by natural selection, but by these atomic forces. Natural selection was again relegated to (at best) a minor role in evolution.

The Darwinist War Against The New Evidence

Darwinism had always been at odds with the facts of nature, and these new discoveries were no exception.

Dartmouth historian Michael Dietrich says that "in the 1950s and 1960s, molecular biology was developing into a field of its own that was often perceived to be in direct competition or conflict with [Darwinian] evolutionary biology."[9]

In what he calls the "Molecule Wars," Dietrich says that "as molecular biologists began to ride roughshod over the [Darwinian] evolutionary synthesis, [Darwinists like] Theodosius Dobzhansky, Ernst Mayr, and G. G. Simpson, began an unprecedented counter attack on the place of molecular biology" in evolution.

In essence, Darwinists declared the new discoveries irrelevant to evolution. Opposing what he called the molecular "bandwagon," Ernst Mayr said the new discoveries had no bearing on "a correct understanding of the evolutionary forces."[10] He contemptuously labeled the new genetics "beanbag science."

Darwinists Invent Their Own Biology

But new discoveries kept coming, and many contradicted Darwinism. Unable to discredit the evidence, the Darwinists made an audacious move. They claimed nature had two types of biology – one for evolution and one for everything else in life. The Darwinists argued there is regular or "functional biology," which explains how the body works, and "evolutionary biology," which explains how traits evolve.[11]

"As the molecular wars heated up," says Dietrich, "this distinction was deployed by Mayr, Dobzhansky, and Simpson as a means of negotiating the division of the biological landscape."[9] Essentially, the Darwinists wanted to separate biology and evolution, and make biology irrelevant to evolution.

A second wave of discoveries created new problems for the Darwinists.

They had achieved their 1940s "Synthesis" of Darwinism and genetics only by excluding important parts of biology from any role in evolution.[12] When it was discovered in the 1960s that developmental biology and evolution are intimately connected, the Darwinists excluded it from any role in evolution.[13] Ernst Mayr wrote that anyone "who uses findings from [development] to prove one or another evolutionary theory proves that he completely misunderstands the working of evolution."[14]

University of California biologist John Gerhart and Harvard's Michael Kirschner says that "the Modern Synthesis made [life] like modeling clay, and remolding of the clay meant that each of the billions of little grains was free to move a little bit in any direction to generate a new form."[15] With this approach, they says, the problem of how development creates living things "could be largely dismissed, further segregating evolutionary biology from its peer disciplines."

One after another, as other discoveries contradicted their theory, the Darwinists excluded those areas of biology from any role in evolution. Leading scientists like Niles Eldredge complained that "a major effect of the [Modern] synthesis has been the removal of most biological disciplines... from the active ranks of investigation into the evolutionary process."[16]

Others like Dudley Shapere became alarmed that Darwinism looked unlike any other theory in the history of science, advocating the quasi-magical idea that biology could be isolated into separate spheres that have no relationship to each other.[17]

At the heart of this spectacular denial was the never-ending attempt by Darwinists to define what it means to be human. Theodosius Dobzhansky said the new discoveries were unimportant to "understanding man and his place in the universe."[18] In his 1959 declaration that nature has two types of biology, Ernst Mayr put it this way:

> "We live in an age that places great value on molecular biology. Let me emphasize the equal importance of evolutionary biology. The very survival of man on this globe may depend on a correct understanding of the evolutionary forces and their application to man. The meaning of race, the impact of mutation... of competition – all these evolutionary phenomena are of the utmost importance for the human species."[19]

THE NEED FOR A NEW DARWINISM

Despite Darwinist claims, nature has only one type of biology. And with each new discovery their alternate "evolutionary biology" began to collapse.

In 1959 one of the world's leading geneticists, Motoo Kimura, made a remarkable discovery. He found that bacteria actively exchange genes with each other! This explained how antibiotic resistance spreads so rapidly.[20]

Kimura also discovered that genetic changes are unaffected by natural selection. Kimura's colleagues say that his "detailed studies of the evolution of individual molecules cast into doubt... Darwinian concepts of what determines evolutionary change."[21]

Kimura resolved the difference between his evidence and Darwinism by simply removing natural selection as a factor in genetic evolution.

Darwinists, unsurprisingly, saw this as a potentially fatal threat and attacked Kimura at every turn. "The theory was rejected out of hand by many evolutionists," says James Crow, and "Kimura spent most of his energies for the rest of his life defending the theory."[22]

Later research, however, proved Kimura right.

In a landmark 1969 study, "Non-Darwinian Evolution," University of California biologists Jack King and Thomas Jukes said that in every plant and animal they examined, genetic evolution was unaffected by natural selection, just as Kimura had said.[23] Thus, they concluded, "we cannot agree" with the Darwinists:

> "Evolutionary change is *not imposed* upon DNA from without [by natural selection]; it *arises from within*."

Whatever force was driving genetic evolution was coming from inside creatures, not outside them. Other discoveries showed similar results. The Darwinists could no longer deny the evidence building against them.

The New 'Gene-centered' Darwinism

Beginning in the late 1960s a tiny group of Darwinists salvaged the situation by inventing a New Darwinism, and a new story about evolution that turned Darwin's original ideas upside down.

While it was known that genes behave according to Mendel's laws, no one knew what caused genes to change in the first place. Some speculated it was the result of random mutations. In the 1920s genes were bombarded with radiation to alter them, but no one knew if the changes were random.

Besides this there was no evidence this is how genes change in nature, and important evidence it was not. In 1932 Barbara McClintock showed that cells have a natural repair process that fixes broken genes, like those created by radiation. The technology to even test the idea of random change did not even exist until the 1970s (and showed random change is untrue; see Chapter 16).

Despite all this, in 1937 Theodosius Dobzhansky declared in *Genetics and the Origin of Species* that *all* genetic change is random. "The *only* known method of *genetic differences* is through mutation," and he then defined a "mutation as a *random* change of the gene structure."[24] In the Darwinists world, all genetic change is random.

Although his book is filled with facts, Dobzhansky offers no evidence for this claim. In true Darwinist form, he simply asserts it is true. Despite a lack of evidence, in the late 1940s, under Darwinists domination, the idea that genetic change is random became textbook orthodoxy.[24]

In the 1960s, group of young Darwinists salvaged their theory by taking the idea of random change and making it the center of a New Darwinism.

First, they said each gene makes a unique protein, which then forms a unique trait. In simple terms, eye genes make eye proteins, which then form eyes.

Next, they said that genes change randomly. This made the origin of new traits a random process. (And, as Darwinists ceaselessly repeated, this made evolution directionless, which meant humans are an accident of evolution, belief in God and religion are superstitions, etc.)

Third, they said natural selection can target and eliminate individual genes and their trait if they prove unhelpful ("unfit").

Called the "gene-centered" view of evolution, New Darwinism was formalized in 1966 by George Williams, and popularized by Richard Dawkins' book *The Selfish Gene*, based on those three ideas: [26]

- Genes make unique proteins, which then form unique traits.
- Random gene changes create new proteins and new traits.
- Natural selection eliminates the "unfit" genes and traits.

When the new "gene-centered" evolution was introduced it had two compelling features. First, it was easy to understand. Random gene changes create new traits. Natural selection determines which genes and traits will survive. Voilà, evolution in two simple steps.

It also had great popular appeal. Over the next decade, scientists claimed to find individual genes for hundreds of traits: being tall, short, fat, thin, smart, gay, there was even a gene for being a compulsive shopper.

The media found these stories easy to report and popular with readers. The new gene-centered Darwinism told people that genes explained everything about themselves and their lives: beauty, talent, personality, shyness, alcoholism, intelligence, athletic ability and so on.

The idea of random gene change quickly went from baseless claim to textbook truth. Pure speculation had once again become scientific fact.

GENE-CENTERED EVOLUTION REVERSED DARWIN'S ORIGINAL IDEAS

But for all its appeal, New Darwinism had the same problem as old Darwinism. First, it failed the evidence test. Going back to Mendel, geneticists knew genes aren't "eliminated," as the Darwinists claimed.

It was also clear that genes don't work in isolation. They link with each other in complex interdependent groups. Looking at individual genes, says UCLA geneticist Michael Oldham, would be like "reading every fifth word in a paragraph."[27] These overlapping relationships mean that a gene couldn't just appear or disappear without creating biological chaos.

The new gene-centered evolution had another problem – it *reversed* Darwin's original theory of how evolution works.

Darwin wrote an entire book showing that the origin of new traits is *not random*. In many cases, he said, new behavior leads to new traits (through

Lamarck's "acquired characters" process). But, he said, the *origin* of new traits was irrelevant. All that counts in evolution is which traits *survive* the natural selection process. (In Ronald Fisher's words, evolution occurs through "random survival" or "random extinction."[28])

The New Darwinists reversed Darwin's original formula for evolution, saying it is the *origin* of new traits (via gene mutation) that is random, *not* the natural selection process that follows.

Darwin also said that creatures live and die as a whole, thus natural selection can't target individual traits (or genes) for elimination through natural selection.

The New Darwinists now claimed the opposite: natural selection works *only on individual genes* and traits. "The fundamental unit of selection," wrote Richard Dawkins, "is not the species, nor the group, nor even, strictly, the individual. It is the gene."[29]

In effect, New Darwinism turned Darwin's original theory of evolution upside down. And the old Darwinists were incensed. It's "a step backwards," thundered Ernst Mayr.

> "People like Dawkins who think the gene is the target of selection are evidently wrong... Dawkins' basic theory is totally non-Darwinian," wrote Mayr. "In the case of [George] Williams, I have come to the unhappy conclusion that not many of the proposals of his are valid."[30]

Similar criticisms flew back and forth between the old and New Darwinists for a decade. "Dawkins explicitly abandons the Darwinian concept of the individual as the units of selection," wrote Stephen Jay Gould.[31]

If this leaves you confused, don't worry. It left most scientists confused, too. Darwinism had now been used to prove *two diametrically opposite* versions of evolution.

- Darwin said the origin of new traits is not random, New Darwinism said the process is entirely random.
- Darwin said arbitrary environmental changes and natural selection made evolution random. New Darwinism said the natural selection process was not random.

Science had never seen anything quite like this: a theory which could prove one thing in the 1950s and its opposite a decade later. (Imagine if Newton has woken up one day and said, "Wait, I got gravity backwards. It doesn't pull things together, it pushes them apart. All my equations should be reversed.")

How could Darwinism reverse itself without notice?

Although this reversal was dramatic, it drew little scrutiny in the popular press. The technical complexity of the issues made it impossible for mere reporters to evaluate the conflicting claims.

Even scientists felt ill-equipped to make judgments about complex disputes outside their own narrow area of expertise. Plus, few wanted to wade into a battle with the Darwinists (new or old) having seen the fierce assaults on powerful scientists, like "America's evolution laureate" Stephen Jay Gould, and even Nobel Prize winners like Linus Pauling. In this battle, what chance did a no-name professor have against the Darwinist attack machine.

Beyond these issues, the scientific truth of Darwinism was irrelevant for many scientists. Darwinism was the "acid" eating through believe in religion. However defined, either you were on the side of Darwinism (and science, reason and enlightened thought), or you were on the side of the religion (and "creationism," superstition, ignorance and irrational beliefs). As always, Darwinism was about ideology, not science, and thus could change as needed to fit the times.

Chapter Twelve

New Darwinism

MAKING EVOLUTION "EVERYBODY'S TOY AND PLAYTHING"

Science is based on the idea that events are predictable.

What happens when predictability is replaced by random change? You get magical explanations. That is what random genetic change did to evolution. Having set out to eliminate "the hand of God" from evolution, Darwinists replaced it with magical stories.

Random Change Makes Genes Magical

Under New Darwinism, the origin of every form, feature or trait, no matter how complex, was easily attributed to a random gene change. They became the magical explanation for everything. How did life first appear? The right chemicals randomly assembled into the first cell. How did fish move from the sea to land? Millions of random gene changes. How did hands, eyes and brains evolve? Millions of more random changes.

Modern Darwinism "reduced evolutionary biology to everybody's toy and plaything," says geneticist Janis Antonovics. Explaining evolution with "simplistic speculation...seemed to qualify anyone as an evolutionary biologist," he says. "Worse it led [people] to imagine that this is what professional evolutionary biologists do... From eight o'clock to ten o'clock real biologists work collecting data and doing science, but then for relaxation they run to the evolutionary playground [for the] gay abandon of speculation."[1]

SUBSTITUTING 'JUST SO STORIES' FOR SCIENCE

With the New Darwinism, the cause and effect explanations of science were no longer required in evolution. All you needed was a random gene change and a good story. "The method," said biologists Steven Rose and Leon Kamin, "consists essentially of contemplating the trait and then making an imaginative reconstruction of a...history that would have made the trait adaptive."[2]

"Imaginative" is the operative word. Darwinism became infamous for such explanations, reaping scorn from other scientists which ridiculed them as "just so stories"– tales which neatly explained things, *after the fact.*

If this sounds more like children's stories, thats because it is. The name "just so stories" comes from Rudyard Kipling's 1912 book, *Just So Stories for Little Children,* which told fantastic tales about the origins of thing – "How the Leopard Got His Spots," "How the Camel Got His Hump," and the like. In the other sciences, Darwinists' explanations were on par with Kipling's stories.

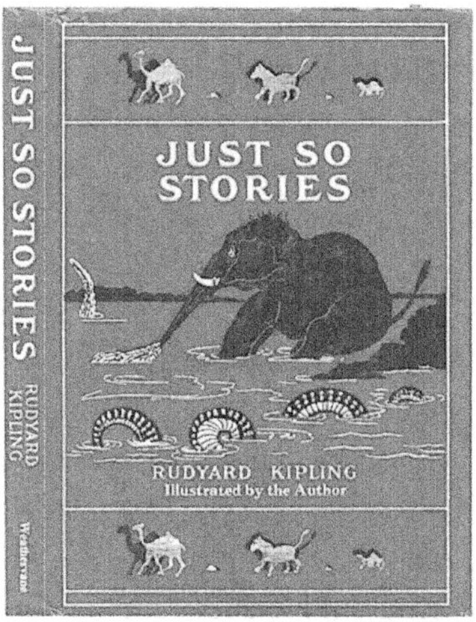

While Darwinists had always used such stories, in the 1960s they began to appear in droves. The evidence, however, often failed to match these stories.

For example, when asked to explain the small stature of African pygmies, Darwinists said random gene changes led some people to have small bodies. This made it easier to move through dense jungle vegetation, stay cool and consume less food, all of which aided survival.

Anthropologists, however, showed that many pygmies live in cool dry areas, some completely outside forests, and have plenty of food. They also have short lifespans (less than 20 years), leading to fewer children than larger people. Thus their size is not an adaptive adjustment that aids survival.[3]

And some of the tallest peoples, like the Maasai of East Africa, have suffered severe food shortages, and live in a hot climate, yet have remained tall. Why didn't natural selection make them smaller to conserve food and stay cool?

Darwinists explained large bodies the same way as small ones, a random gene change made some people bigger and stronger, making it easier to hunt for food and escape predators, aiding survival. Whatever trait people had, it was explained by a random gene change and "selection" for adaptation and survival.

JUST SO STORIES: THE PEPPERED MOTH

Even the "classic" examples of natural selection turned out to be "just-so-stories." The most famous may be the tiny Peppered moth, about the size of your thumbnail, with wings that are either mostly white or black.

Both types were common in England before the Industrial Revolution, but by 1900 the white-winged moths had largely disappeared. Darwinists called this a classic case of natural selection. The story they told is that the coal-burning factories which sprang up in the 1800s left a layer of soot on the landscape. This camouflaged black moths, but made their white-wing cousins easy pickings for predators, so they were soon eliminated.

PEPPERED MOTH

In the 1950s, the passionate Darwinist Bernard Kettlewell put this idea to the test. The results, he said, confirmed natural selection led to the change and called it "the most striking evolutionary change ever actually witnessed in any organism, animal or plant."[4] For the first time in 100 years, Darwinists had some experimental proof of their theory.

The Peppered moth was hailed as "the clearest case in which...evolutionary process has actually been observed."[5] The story was embedded in textbooks as "the canonical example of [Darwinian] evolution in action."[6]

Like other classic examples of natural selection, the Peppered moth story was easy to explain, sounded so plausible and... was untrue.

From the beginning, many biologists felt Kettlewell's study was wrong. In 1998, one of the skeptics, biologist Ted Sergeant at the University of Massachusetts, conducted a meticulous review and concluded Kettlewell's experiment was indeed "deeply flawed." "There is little persuasive evidence in the form of rigorous...observations and experiments to support this explanation," he wrote.[7]

That same year, Darwinist Jerry Coyne was asked by *Nature* to review the facts. Finding major errors and "many other flaws in the work," Coyne concluded "we must discard [the moth] as a well-understood example of natural selection."[8] Coyne was distraught. "My own reaction," he wrote, "resembles the dismay attending my discovery, at the age of six, that it was my father and not Santa who brought the presents on Christmas Eve."

Despite these results, the Peppered moth still appeared in textbooks as a "classic example of natural selection."[9]

Just-so-stories: Early Humans as Killers and Cannibals

Fictional tales about moths are harmless, but a false picture of human evolution corrupts our ideas about human nature and ourselves, with serious consequences.

Archaeologist Raymond Dart believed that human evolution was driven by a passion for killing. He said human evolution took off only after we began to use bones as weapons to kill, especially our fellow humans.

In true Darwinist style, Dart fabricated a story of human nature, painting Man as violent and cannibalistic. He said early humans were:

> "Confirmed killers, carnivorous creatures that seized living quarries by violence, battered them to death, tore apart their broken bodies, dismembered them limb from limb, slaking their ravenous thirst for the hot blood of victims and greedily devouring their writhing flesh."[10]

Dart based this gruesome picture on a handful of human fossils which he said bore the marks of man-made weapons. Asked by anthropologist CK 'Bob' Brain why he used such lurid imagery, Dart said "that will get them talking!... Never let the facts get in the way of a good story!"[11]

The image Dart created was wildly at odds with the evidence, especially the picture assembled by famed fossil hunter Louis Leakey during 30 years of careful research. Leakey's work showed early humans to be "gentle, loving, and kind, living in peaceful cooperation with their fellows."

But Dart ignored all this. He was more interested in using what he called "the blood-spattered archives of humanity" to promote a Darwinian view of human evolution.

Careful research by Bob Brain, then Director of South Africa's Transvaal Museum, showed that the wounds Dart said came from weapons actually matched the puncture holes made by leopards, the most lethal predator of early humans.[11]

The holes found on infant skulls, which Dart said were evidence of child killing, perfectly matched the talons of eagles, a common predator of young primates. Microscopic analysis of the scratches around the holes showed they were indeed produced by bird talons. Anthropologist Robert Sussman, an expert on early humans, says the evidence shows that a staggering one of every twelve perished from these animal attacks.[12]

"Our early ancestors bore little resemblance to killer apes," says Donald Johanson, who discovered the famous "Lucy" fossils.[13] He says they used bones and horns to find food, not to kill. "Microscopic wear on the tips of fossilized antelope horns indicates they were used as digging tools, not lethal weapons," says Johanson. "Without protection, our ancestors were easy prey. They were not the hunters, they were *the hunted.*"

Dart had human evolution backwards. Early humans weren't cannibalistic killers. Murder is not in our genes.

Unfortunately, the research that demonstrated the many flaws in Dart's lurid story came years after it made international headlines and became embedded in people's minds as the scientific truth about human nature.

JUST SO STORIES: HOW FABRICATIONS BECAME 'FACT'

Dart's idea that a thirst for killing drove human evolution, and which continues to lurk just below the surface of modern man, triggered a cascade of popularizations which are still going strong. The "killer instinct" myth, says Sussman, was "repeated in the 1960s in anthropology texts and popular literature [and] in the 1970s it was adopted by socio-biologists to explain human nature."[14]

In the bestselling *Territorial Imperative*, Robert Ardrey expanded Dart's idea, claiming our brains became bigger only as we developed more complex killing tools.

TERRITORIAL IMPERATIVE

The belief that an urge to kill drove human evolution was further popularized by Arthur Clarke and Stanley Kubrick in *2001: A Space Odyssey*. In the opening scene, an ape, with a large bone in hand, suddenly realizes that it can be used as a weapon to kill. The bone spirals high into the air and morphs into a space station. The message: A desire to kill drove human evolution and led to the modern world.

Today we now know that compared to other species, humans are unusually cooperative, and we have been throughout our entire evolutionary history. Nonetheless, the Darwinist view of human evolution remains ingrained in our ideas about ourselves. Another version of this myth appears in the recent book, *Demonic Males*, which argues that rape, violence and murder are deeply embedded in human nature.

This pattern – dramatic Darwinian story, great publicity, story disproved, no publicity – was repeated many times in the 20th century.

JUST SO STORIES: SELFISH GENES

The biggest just-so-story of all is Richard Dawkins "selfish gene." A zoologist by training, Dawkins was interested in animal behavior (ethology) and trained under well-known ethologist Nikolaas Tinbergen.

The study of animal behavior was then dominated by the idea of innate traits popularized by Konrad Lorenz, particularly an "innate aggression." In

1976, in his book *The Selfish Gene*, Dawkins combined a bit of genetics with the idea of innate aggression to claim that Darwin's "survival of the fittest" struggle operated at the level of genes.[15] Unseen by us, genes are slugging it out with each other for survival.

Long on rousing prose, but short on testable science, the book made Dawkins famous and became a key to establishing the new gene-centered view of evolution. Like so much of Darwinism, *The Selfish Gene* is less about genetics than making the case humans are an accident of evolution. In Dawkins' story, the sole reason humans exist is to serve genes:

> "We are...robot vehicles blindly programmed to preserve the selfish molecules known as genes. They created us, body and mind; and their preservation is the ultimate rationale for our existence."[15]

The Selfish Gene said that natural selection and the "survival of the fittest" struggle are fought out between individual genes, trying to "maximize their own survival." (In Dawkins' description, genes are given human motives, abilities and awareness. If you changed the words "selfish gene" to "Mafia gangster" or "Wall Street banker," the book would still read just fine.)

The Selfish Gene has been popular for 37 years. Much has been discovered over those four decades, but Dawkins has never updated the book. The reason, he says, is that his publisher advised against "a conventional, comprehensive, page by page revision."[16] This seems disingenuous. Dawkins is modern Darwinism's most ardent promoter. If new evidence supported his ideas, one assumes he would rush to use it.

I believe the truth is that he can't revise the book. Like other "just so stories," the evidences shows the selfish gene idea is simply untrue. From the start Stephen Jay Gould and many others called the self gene idea "fatally flawed." Creatures live and die as a whole, thus individual genes can't be eliminated, as Dawkins claimed. "Natural selection simply cannot see genes and pick among them directly," wrote Gould, "it must use bodies as an intermediary."[17] Dawkins eventually conceded the point, saying individual genes are "not selected directly."[18]

Plus, if genes don't cooperate then complex life can't evolve. But if genes do cooperate, then "selfishness" is not driving evolution. To resolve this dilemma, Dawkins' pal Mark Ridley invented the "cooperative gene."[19] Ridley said nature had evolved a process that "sabotages" selfish genes, allowing "cooperative genes" to emerge. How did this cooperative process arise in Darwin's ruthlessly competitive evolution? Ridley didn't know. What did it look like in biology? Ridley couldn't say. It all sounded like more Darwinists story telling.

In addition, as we will see, modern genetics shows that genes operate almost entirely in large, overlapping networks where cooperation is not just helpful, but an absolute necessity.

It is hard problems like these, not pleas from his publisher, that explain why Dawkins cannot revise his book. As geneticist Richard Lewontin wrote, "of all the vulgar errors about biology presently circulating, [Dawkins'] notion that we are 'lumbering robots blindly programmed' by our genes which 'control us body and mind'...is surely the most popular."[20]

JUST SO STORIES: COOPERATION AND KIN SELECTION

Richard Dawkins says he wrote *The Selfish Gene* in part to deal with what Darwinists call the "problem" of cooperation. Cooperation is a "problem" because in Darwin's hyper-competitive world cooperation should never be seen. But it occurs widely throughout nature.

The difficulty cooperation posed for Darwin is crystallized in insect colonies. Among social insects, like bees and ants, most in the colony don't breed and thus don't pass on their genes. If creatures are competing for survival, why do most colony members become sterile, toiling selflessly to nurture the queen and her offspring?

To explain cooperation in what should be a brutally competitive world, William Hamilton invented "kin selection." Hamilton said evolution is all about perpetuating genes. Since you share many genes with your relatives, their survival is almost as good as your own. The closer the relative, the more genes you will share, and the more you will help them survive.

Despite a lack of evidence, by the 1970s "kin selection" was embedded in textbooks as proven fact and evolutionary dogma.[21]

Kin selection, however, never made sense in the real world. For example, it predicts that creatures will exchange their lives to save close relatives because they share the same genes. But African army ants share 99% of their genes with the ants they attack and cannibalize. Blue Jays routinely eat the eggs of birds with genes nearly identical to their own. Chimpanzees share 99% of their genes with other chimps, but are clearly unwilling to sacrifice themselves for them. Geneticist Steve Jones once said "humans share 50% of their genes with bananas, but that doesn't make us half banana."

Contrary to "kin selection," cooperation is widespread in nature, *regardless of genetic relationship*. On the other hand, close relatives who don't cooperate are expelled from the group. Bacteria live in colonies with others nearly identical to themselves, yet they expel any member if for any reason they don't collaborate.[22] Fire ants will even kill their queen if she fails to behave cooperatively.

Cooperation is important even *between* species. Yes, there is a food chain and some animals eat others. But like Richard Lewontin's Australian rabbit and its virus, most animals live in a state of balance with each other.

All across nature, what matters is not genetic relatedness, but willingness to cooperate. Martin Nowak, director of Harvard's Evolutionary Dynamics Program, says cooperation is so crucial in life it amounts to one of evolution's "basic forces."[23] "In the Wimbledon championship, you must defeat your opponent to move to the next round," says Nowak, "but everyday life is not like a tennis tournament. Overall, success is proportional to the amount of cooperation."

Kin selection also suffered from the flaw that plagued all Darwinist creations: no evidence. While vast amounts of mathematics were churned out to justify kin selection, few tangible facts ever materialized. In 2010, Martin Nowak and Edward Wilson said that after 40 years as the dominant Darwinist explanation for cooperative behavior, kin selection had produced "only hypothetical explanations."[24] They went on to say,

> "During the same period, in contrast, empirical research... has revealed the rich details of... communication, colony life cycles, and other phenomena. Almost none of this progress has been... advanced by [kin selection], which has evolved into an abstract enterprise largely on its own."[24]

Kin selection is a microcosm of every process Darwinists put forward to account for evolution – a "hypothetical explanation," spun into an "abstract enterprise," for which there was never any evidence.

DELAYING SCIENTIFIC DISCOVERIES

The consequences of abandoning the cause and effect explanations required in the other sciences for random gene changes and just-so-stories, sent researchers down blind alleys and dead-end streets, delaying important discoveries for decades.

For example, in 1974, renowned cancer researcher Lewis Thomas wrote, "the capacity to blunder slightly is the real marvel of DNA. Without this special attribute, we would still be anaerobic bacteria."[25] The idea that random gene mutations were at the heart of evolution led Thomas and other researchers to assume that cancer also originates with a random mutation that turns normal cells into crazed mutants rampaging through the body. Getting cancer was like the *Night of the Living Dead*, a terrifying struggle against zombies that were impossible to kill.

We now know this is untrue.

The overwhelming majority of cancers, say University of Chicago pathologists Hans Schreiber and Donald Rowley, are caused by "genes from cancer-causing viruses."[26]

Viruses float around in our body all the time, but don't harm us. They are inactive or we've developed a defense against them (this is what Ted Steele discovered). Occasionally, however, a virus will "cut and paste" its DNA into the DNA of one of our cells, taking control of them.

Precisely because this process is not random, it can be researched and targeted for removal by specially designed molecules. "[This] is one of the great hopes in the fight against cancer," say Schreiber and Rowley. "Therapeutically targeting such molecules should eradicate cancer without harming normal tissue"[26] We won't need to blast the entire body with caustic doses of chemotherapy to kill a few cancer cells. Instead, we will harness nature's own processes to safely break down the cancer-causing viruses.

This new understanding is allowing researchers to make real progress to fight cancer, even harnessing our own immune system to do the job.[27] Had the work of Barbara McClintock, Ted Steele and others not been sidelined, we would be two generations further along in this process.

Genetic Determinism, Fate, Racism and Eugenics

"Our fate is in our genes."- John Watson

The idea that genes determine our behavior also had a malignant effect on our concept of human nature, one that remains deeply embedded in our thinking today.

Most ancient societies – Greece, Rome, China, India, others – believed life was governed by "fate," which could not be changed. One's fate was often foretold in the movement of the stars. People were counseled to accept their fate in life, not to fight it.

Western Judeo-Christianity overturned this idea. Based on the belief that each person's choices determine their future, it placed their destiny in their own hands. In Shakespeare's immortal words, our fate is "not in our stars, but in ourselves." People were no longer a prisoner of forces beyond their control. They could alter their future – the concept was awe-inspiring.

The idea that behavior is largely determined by genes (called genetic determinism) returned people to a life governed by fate. In the words of geneticist James Watson, "We used to think our fate was in the stars, but now we know, in large part, that our fate is in our genes."[29]

Genetic determinism instantly erased 2000 years of social progress.

The ancient idea of fate and genetic determinism lead to the same belief: forces beyond our control determine our destiny. Choice is an illusion.

In this view of life "politics becomes a branch of molecular biology, and our social and political institutions are as immutable as the chemicals of which we are made," said Richard Lewontin. "Now that's rubbish, but that's what's pushed. We are governed, body and soul, by our genes. The whole notion is that the genes inside us are independent of us."[29]

The idea that genes control us, and determine our behavior, spiraled into a justification for every type of atrocity.

In *A Natural History of Rape,* University of New Mexico biologist Randy Thornhill and anthropologist Craig Palmer argue that rape is "a natural, biological phenomenon," part of "human evolutionary heritage."[30]

In an NPR interview Thornhill said that like all behavior, rape is a product of natural selection and served a useful evolutionary purpose. In this case, he said, rape allowed a man to spread his genes to unwilling partners.

Thornhill said the drive to rape is now deeply embedded in our genes. As he put it, rape is to be expected. All men are walking rape machines, held in check only by a thin veneer of social control. We should realize this and accept it. For Thornhill, the logic is inescapable. If natural selection is true, then rape, by definition, must be a beneficial trait. "It's not a debatable matter," he told the NPR interviewer.

FROM EUGENICS TO VIOLENCE: THE COLUMBINE KILLINGS

Just as it was a short step from Huxley's sterilization to Hitler's extermination, it is a short step from defining rape as a helpful adaptation to justifying other types of violence.

In 2009, Dennis Sewell wrote in *The Times* of London, "You wouldn't know from the celebrations of Charles Darwin's life, that the amiable

Victorian gent portrayed in those TV drama-docs...has been fingered as a racist, an apologist for genocide, and the inspiration of a string of psychopathic killers. There is hardly a mention that his name has been associated with some of the most infamous crimes of modern history... fostering moral nihilism and scientific racism, ultimately expressed in the Holocaust."[31]

Sewell, a 20-year veteran investigative reporter for the BBC, said that "most startling of all, a connection has now been drawn between Darwin's theories and a rash of school shootings." Going into deep investigative detail, Sewell's report showed that teens like Eric Harris and Dylan Klebold, who perpetrated the 1999 killings at Columbine High School, were obsessed with Darwinism and the idea of natural selection.

In the documents Eric Harris left behind he (like Darwin) railed against vaccination programs, which he believed interfered with evolution's natural elimination process. Harris said "let natural selection take its course. All the fat, ugly, retarded, crippled dumbass, stupid f***heads in the world would die...Maybe then the human race can actually be proud of itself."

In their videos, Harris and Klebold said they felt they should help the natural selection process by eliminating those who are weak and inferior. Harris fantasized about putting his classmates into a violent computer game in which only the fittest would survive. On the day of the shooting, Harris wore a "Natural Selection" t-shirt.

Barry Arrington, an attorney for some of the families whose children died, told Sewell that after reviewing the documents "it became evident to me that Harris consciously saw his actions as logically arising from what he had learnt about evolution. Darwinism served as his personal intellectual rationale for what he did. There cannot be the slightest doubt that Harris was a worshipper of Darwin and saw himself as acting on Darwinian principles."[31]

Sewell reports that Harris and Klebold became heroes to a group of teens who frequented an internet social networking site called "Natural Selection's Army." In 2007, Pennsylvania police, acting on a tip about a potential school shooting, discovered that the alleged perpetrator was a frequent participant at the "Natural Selection Army" website. So, apparently, was Finnish student Pekka-Eric Auvinen.

In 2007, Auvinen entered his high school in Tuusula, Finland and methodically shot his teacher and seven students, then killed himself.

Sewell says other students recounted that as Auvinen went through the building, he would point his gun at people's heads, study them carefully, and then either squeeze the trigger or "suddenly turn away and let his terrified target go free." Auvinen was making a decision, one by one, to weed out the unfit.

"Before he embarked on his shooting spree," writes Sewell, "Auvinen posted a lengthy apologia on the internet. Styling himself a 'social Darwinist,' he said that natural selection appeared not to be working any more... the 'stupid, weak-minded people reproduce faster than intelligent, strong-minded ones'. The gene pool was sure to deteriorate if society continued to guarantee the survival of the second-rate... He understood that life was just a meaningless... outcome of a long series of random mutations, so there might not be much point in doing anything at all. But eventually he had decided he would do his bit by becoming a natural selector, aping the pitiless indifference of nature."

In his postings, Auvinen pleaded that people not write him off as a psychopath, or attribute his actions to the effects of violent movies or computer games. He said his concerns about natural selection should be taken seriously: "No mercy for the scum of the Earth! Humanity is overrated. It's time to put natural selection and survival of the fittest back on track."

No Underlying Moral Authority

Despite his 20 years as a BBC investigative reporter, Sewell was nonetheless surprised by his findings, and ends his report with this:

> "Darwin also taught that morality has no essential authority, but is something that itself evolved – a set of sentiments or intuitions that developed from adaptive responses to environmental pressures tens of thousands of years ago. Whether an individual opts to obey a particular ethical precept, or to regard it as a redundant evolutionary carry-over [is] thus a matter of personal choice... wackos such as Harris and Auvinen can just as readily interpret it as a licence to kill... Today's evolutionary scientists... describe ethics as merely an illusion produced by genes. From a Darwinian perspective, there is nothing objectively wrong with shooting your classmates."[31]

Today, the "Natural Selection's Army" web site is shut down. But a computer game called "Natural Selection" (in which teams compete to wipe-out the unfit) remains popular, as do "Natural Selection" t-shirts, a sign, says Sewell "that Darwin's term is still associated by many teenagers with sudden and extreme violence." The Columbine killers have also "spawned a gruesome personality cult - there is even a computer game in which players adopt the roles of Harris and Klebold, which features original TV footage of the killings."[31]

Chapter Thirteen

DARWINISM'S WAR AGAINST THE NEW EVIDENCE

Living in the bloodstream of tropical snails is a one cell animal we might call "Casi" (*Capsaspora owczarzaki*). Casi is so microscopically tiny that the world didn't know it existed until 2002.[1]

Casi startled scientists. Inside they found all the genetic equipment for making modern animals – leopards, elephants, whales. Casi's species has been around for well over a billion years, and has changed little in all that time. This meant that the genes for making complex animals existed a billion years before life moved beyond the stage of single cells.[2]

This presented a profound mystery. The Darwinist story was that the random appearance of new genes led to new traits, and eventually to new species. But what if new traits don't arise from new genes? What if *evolution* is not driven by the random appearance of new genes? If true, this would require more than just tinkering around the edges of old ideas. Casi pointed to a *fundamentally different* concept of how evolution works.

Genes Are Ancient

For centuries, fossils were our only window into the past, providing a picture of primeval life's size, shape and form, but revealing little else.

In the 1990s, new gene decoding technology changed this, shedding light on ancient biology. Genes became tiny time machines, carrying scientists deep into the past, allowing them to see things which don't fossilize, like biology, behavior, and color. The primeval landscape came alive in rich textures, vivid colors, and vibrant activity. Neanderthals with pale skin and red hair hiked the Alps.[3] Giant mammoths with curly auburn fur lumbered across frozen tundra.[4]

Dinosaurs were a particular surprise, emerging with colorful feathers. A 160 million year-old dinosaur called Microraptor (like the velociraptors that raced through *Jurassic Park*) sported splendid iridescent black plumage. *Anchiornis* ("near bird") had black and white "racing stripe" feathers.[5]

Microraptor
Courtesy Xu Xing and Institute of Vertebrate Paleontology China

Anchiornis
By Michael DiGiorgio, Courtesy Yale Univ.

The ability to decode the genome put a spotlight on the New Darwinist claims about genes. They said that random change meant each species would have its own unique set of genes. If the same trait appeared in different species, each had evolved independently. Eyes, for example, were said to have evolved separately as many as 60 times.

Deciphering animal genomes allowed scientists to test these claims. What did they find? Each species does *not* have its own unique genes. Quite the opposite. Animals, for example, all share the *same* genes – like those in Casi. "The whole spring of animal diversity has been fed by essentially this same set of genes," said University of Wisconsin geneticist Sean Carroll.[6] "Natural selection has not forged many eyes completely from scratch, [rather] there is a common genetic ingredient to making each eye type."[7]

This meant the genes that made Casi also made sharks, frogs, hamsters and humans. They also built the dinosaurs. "It's not an overstatement to say that this blew biologists minds," says Carroll.[8]

The human genome proved a particular shock.

The Darwinists said each gene made a unique protein which in turn formed a unique trait. Eye genes make eye proteins which form eyes. Heart, skin, and other genes do the same. The more complex a creature, they said, the more protein-making genes it will have.

When the human genome was decoded in 2003, it contained the same number of protein-making genes as a mouse. In fact, all animals, from earthworms to elephants, have roughly the same number of these genes. The entire Darwinist theory of genes instantly collapsed.

New Traits Come from Rearranging Old Genes

These discoveries held two important insights.

First, it turned upside down the Darwinist claim that life began in simple form, and became more complex as new genes emerged. All the genes needed for complex life had existed from the beginning.

Second, new traits don't come from the random appearance of new genes, but from an astonishing process Darwinists never imagined – *rearranging* old genes in new ways. "We first believed [diversity] would involve all sorts of novel creations, starting from scratch, again and again and again," says Sean Carroll. "We now understand that, no, evolution works with packets of information and uses them...in new and different combinations, without necessarily having to invent anything fundamentally new."[9]

Differences between species come not from different genes, but from differences in how genes are *organized*. This, of course, was Barbara McClintock's 1951 idea of evolution, re-emerging half a century later.

Nature's many different body plans, for example, are all built using the same genes. The *Hox* genes that determine where a caterpillar's legs and antennae go, also determine where a butterfly's wings appear. These same genes guide the placement of arms, legs, eyes, ears, heads and toes in every type of animal, from flies and frogs to crabs and camels.

This process goes beyond genes to all of biology. Nature uses the same parts over and over. Harvard's Marc Kirschner and his University of California colleague, John Gerhart, compare evolution to rearranging Legos:

> "The components of living things are like Lego blocks. Size and shape...or an anatomical part of the organism can be varied by reusing common components in new combinations and amounts. The blocks do not change but the arrangement does. Linkages are readily made and broken."[10]

Beyond body parts, all life shares the same basic processes. "A big surprise of modern biology," say Gerhart and Kirschner, is that "the details of metabolism are the same in bacteria and humans; basic cell organization and functions are similar between yeast and humans...developmental strategies in fruit flies are strikingly similar to those in humans."[11]

In short, new traits evolve by *rearranging* old genes, parts, and processes in new ways.

FROG'S TEETH AND DINOSAUR TAILS

This explained one of nature's most puzzling mysteries, the reappearance of ancient traits long thought to be extinct, called "atavisms." Occasionally snakes grow pelvic bones, whales develop primitive legs and chickens grow teeth. (This is where the saying "rare as hens teeth" comes from.)

Nature is full of such strange reappearances. Frogs lost their teeth 200 million years ago. But 170 million years later they reappeared in one species. Darwinists said the teeth had "re-evolved" from scratch, but couldn't explain why. Oddities like this, inexplicable with Darwinism, suddenly make sense if traits emerge from changing the relationships between genes.

This got people thinking about another old question: did birds evolve from dinosaurs? Fossil hunter Jack Horner, who advised on the movie *Jurassic Park*, believes they did. He points to the many traits birds and dinosaurs share: unique dagger-like teeth, retractable claws, hollow bones, body scales, feathers and oblong eggs.

A bird is a dinosaur with a slightly different gene arrangement, says Horner, "so all we have to do is tweak it a little bit and we should have a dinosaur."[12] McGill University paleontologist Hans Larsson did just that.

Birds are born with seven vertebrae in their tails. But early in embryo development, they form a tail with 15 vertebrae, like dinosaurs. Before birth, these extra bones disappear. "For 150 million years this kind of tail has not existed in birds, but they have always carried it deep in their embryology," says Larsson.[12] By changing a few gene signals, Larsson stopped the extra bones from disappearing, and the birds were born with dinosaur-like tails.

"Even more exciting," says Larsson, "is that all the scaffolding [around the tail] is enlarging too." Changing just a few genes had triggered a cascade of other changes to support the longer tail. "I see no reason why we could not develop...a completely dinosaurian looking tail in a modern bird," he says, "the system for developing that is in place."

At the University of Wisconsin, Matt Harris and John Fallon tried the same thing with teeth. Altering a few gene switches led to the birth of birds with the distinctive dagger-shaped teeth of dinosaurs, like the Tyrannosaurus Rex. Harris found that birds have the genes needed to make teeth, they are just "turned off." Harris also found he could turn the scales on a bird's leg into feathers. Presumably the reverse is true, bird feathers can be turned into dinosaur scales.

Today we know that birds didn't evolve from dinosaurs. Birds *are* dinosaurs, direct descendants of a line of dinosaurs that didn't become extinct.

Retaining Old Genes: The Transition to Complex Life

The ability to decode an entire genome was the zenith of a devastating period for Darwinists. While new discoveries were systematically disproving their claims, a series of older discoveries, long buried by Darwinists, reemerged to answer important questions about evolution.

The first life was a single cell creature with an outer membrane, but everything inside swirled around in one giant soup.

The first great evolutionary transition was the appearance of a new type of cell which contained separate sections inside. Each section was enclosed in its own membrane, and specialized on one job. This turned the cell into a network of robust little factories.

This new type of cell appeared abruptly, seemingly from nowhere, and swept the earth in an instant. Remarkably, each of the internal sections were complex from the start – fully elaborate chemical systems. There was no Darwinian gradual build up, no incremental change over time.

How cells made the leap from chaotic soup to self-enclosed complex factories was one of evolution's biggest puzzles. In 1966, 28-year-old biologist Lynn Margulis solved the mystery with a stunning insight. The transition occurred when one cell swallowed another cell, but instead of turning it into lunch, the inside cell was harnessed to do specialized work. Both cells had the same genetic equipment, but the inner cell was now free to specialize on one task.[13]

This process, called symbio-genesis, made complex life possible. For example, early cells got their energy by absorbing light and turning it into food. Margulis showed that when one cell swallowed another, the inside cell could now specialize in turning light into food (forming an area called a chloroplast). When a second cell was absorbed, it could focus on the tasks involved in reproduction (forming an area called a mitochondrion).

Evolution's First Leap in Complexity

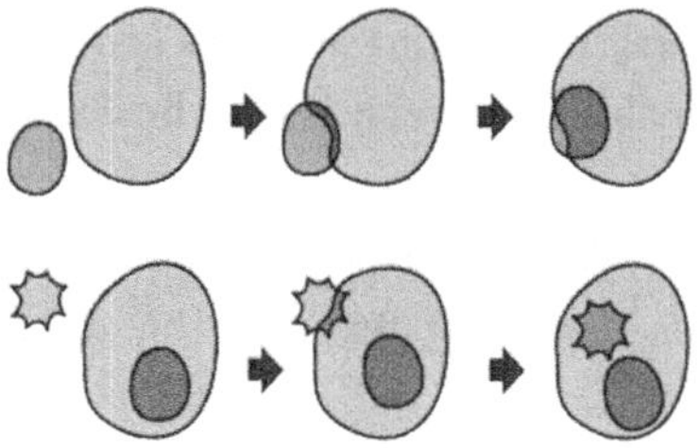

One cell can absorb others.

Most of life's specialized functions originated this way, said Margulis. "Random mutation does not lead to new species or even to new organs or new tissues," she said. Rather, "the major source of...evolutionary novelty comes from the acquisition of...entire sets of genes."[14]

Acquiring more genes, even if they are identical to what you already have, allowed creatures to form new genetic relations, creating more complex traits.

Today we know this process occurred throughout the history of life. It is now recognized as one of evolution's most important processes.[15]

Margulis's discovery resolved other evolutionary mysteries that Darwinism never could. Everyone recognizes a sea star, the small star shaped ocean animal often seen on beaches. In the middle of their development, sea stars abruptly swim off and become a completely different type of animal. A few months later they change again and return to being a sea star. University of Liverpool marine biologist Donald Williamson says, "I can't see how one animal with one genome could do that."[16] But, he says, absorbing all the genes of another animal does make sense of it.

This insight also explained other mysteries. Butterflies start as eggs, hatch as caterpillars, then wrap themselves in a cocoon and emerge as flying insects. Moths, beetles and bees go through a similar process. In every case, the young have entirely different bodies, behavior and biology from the adults. The idea that such a complex process evolved many times, each one from piling up millions of random changes, defies mathematical probability. However, if one animal can absorb all the genes of another, as Margulis proved they did, then the mystery becomes easy to explain.

Margulis was an outspoken critic of Darwinism ("it will look ridiculous in retrospect, because it is ridiculous"[17]), and for decades her discovery was marginalized. By the 1990s, however, Margulis's work was recognized as a major breakthrough. "Everybody is waking up" to the importance of her work, said University of Wisconsin geneticist Margaret McFall-Ngai in 2009. "Basically Lynn was way, way ahead of her time." But pick up any biology textbook today and "she'll always be in there."[17]

LYNN MARGULIS

COOPERATIVE EVOLUTION: EXCHANGING GENES

Another key discovery that reemerged in the the1990s also had its roots in the past.

In the1970s, geneticist Carl Woese and his colleagues at the University of Illinois discovered a third form of life, *Archaea*, (in addition to bacteria and the cells that later evolved into plants and animals). This overturned conventional ideas about the origin of life.

Woese and his group also discovered exchanging genes was not limited to bacteria. *All* early life did it, a process he called "gene swapping" (or "horizontal gene transfer").[18] This meant that if a beneficial new gene arrangement appeared in one cell, it would quickly spread to others. "It is the community as a whole which evolves," said Woese.[19]

This had powerful implications. Cooperatively sharing adaptive traits is the *opposite* of what should happen with natural selection. This meant that for the first two billion years, evolution was *not Darwinian.*

On November 3, 1977, after years of careful confirmation, the National Science Foundation issued a press release on the breakthrough. A front page story appeared that morning in *The New York Times*: "Scientists Discover a New Form of Life."

Woese's colleague, Ralph Wolfe, remembers receiving many calls that morning, most liberally laced with obscenities. One of the few "free of four letter words," recalls Wolfe, came from biologist Salvador Luria who said, "Ralph, you must dissociate yourself from this nonsense."[20]

"But Lu," pleaded Wolfe, "the data are solid and support the conclusions. They are in the current issue of PNAS [*Proceedings of the National Academy of Sciences*]." Luria, however, was adamant: "You're going to ruin your career!" Similar calls came in all morning, says Wolf. "I wanted to crawl under something and hide."

The discovery of widespread gene exchange eventually overturned old classifications and led to major revisions in textbooks. Gene exchange also led to revolutionary new techniques for fighting disease, curing cancer, and making antibiotics more effective. Woese went on to win many of science's highest awards, including the National Medal of Science.

But all that would came later. In 1997, on the 20th anniversary of the announcement, the journal *Science* ran an article about Woese called "Microbiology's Scarred Revolutionary," the story of a man who "started a scientific revolution - and paid a price."[21]

Woese described the time after his discovery as the "Dark Age." Evolution, he said, "became the private domain of a quasi-scientific movement, who secreted it away... effectively disguising the fact that their primary concern with it was ideological, not scientific."[22]

But Woese would not be intimidated. Facts were facts, and he stood by them. The first 50% of life's evolution simply did not work as the Darwinists claimed. "We can't expect to explain... evolution if we stay locked in the classical Darwinian mode of thinking," he said. "The time has come for biology to go beyond."[23]

Cooperative Evolution: Uprooting Darwin's 'Tree of Life'

Woese's discoveries pointed to something else important about evolution. Darwin said the history of life looks like a tree, each new species neatly branching from old, dubbed the "Tree of Life."

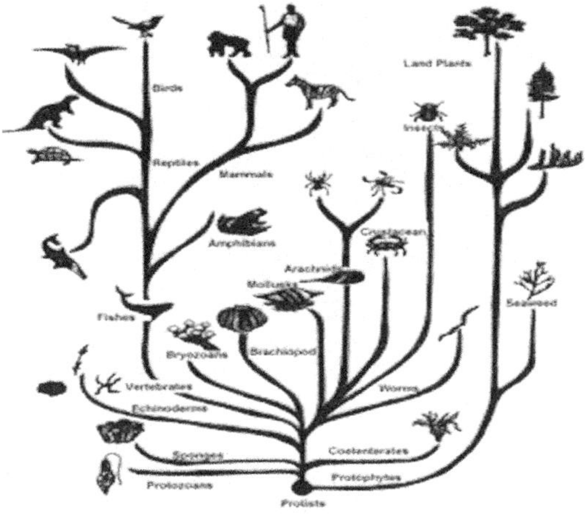

Tree of Life

The idea of life as a branching system was old. In ancient Greece,Theophrastus ("the father of botany") arranged plants in a branching system. In the 1730s, Carl Linnaeus organized all plants and animals in a branching structure. And by the mid-1800s, Richard Owen and others described the evolution of species as a branching system.

To this old idea, Darwin added his own idiosyncratic claim. After an evolutionary split, natural selection will lead each species to develop its own unique set of traits (in modern terms, they will have unique genes).

But it was clear to Linnaeus, and to Victorian anatomists like Owen, that after a split, species developed strikingly similar traits. The discoveries of Woese and Margulis explained some of this, showing that species continue to swap genes after a split, meaning their traits and genes will not be unique. Who was right, Linnaeus, Owen, Margulis and Woese, or Darwin?

"DARWIN WAS WRONG," declared the cover of *New Scientist* in 2009. Britain's *Guardian* newspaper headlined, "Evolutionary biologists say crossbreeding between species is far more common than previously thought, making nonsense of the idea of discrete evolutionary branches."[24]

The headline in another leading British newspaper, *The Telegraph*, was "Darwin's Tree of Life 'Wrong and Misleading' Claim Scientists."[25] *The Telegraph* reported that scientists "believe the concept misleads us because his theory limits and even obscures the study of organisms and their ancestries...Evolution is far too complex to be explained by a few roots and branches, they claim."

In "Uprooting Darwin's Tree," *New Scientist* editor Roger Highfield said "most biologists now accept that [Darwin's] tree is not a fact of nature."[26] *New Scientist* explained the problem this way:

> "The tree-of-life concept was absolutely central to Darwin's thinking, equal in importance to natural selection...The tree has been the unifying principle for [a Darwinian] understanding the history of life...For much of the past 150 years, biology [tried] filling in the details of the tree.. but today the project lies in tatters, torn to pieces by an onslaught of negative evidence."[27]

What was the "negative evidence"? In a nutshell, gene swapping. The true extent of gene exchanges is now firmly established and "surprisingly, it turns out to be the rule rather than the exception," said *New Scientist*.[27]

Let's be clear about what happens. Plants and animals do evolve from earlier species in a branching system. But it doesn't work as Darwin claimed. After a split, natural selection does not take over and hone a unique set of genes for each species. Instead, they continue swapping genes, even with totally unrelated species. This means all creatures are full of genes picked up from others. "We are walking communities," said Lynn Margulis.[28]

Gene exchange shredded Darwin's claim that evolution is isolated branches, each with unique traits individually formed by natural selection. In the largest study of its kind ever undertaken, geneticist Michael Sanderson compared the genes of 500,000 plants to understand how they're related. What did he find? Not isolated branches with unique genes, but a "blurry, inscrutable thicket" of exchanged genes.[29] Biologist Tal Dagan estimates that 80% of the genes in microbes were acquired from other species.[30]

As life became more complex, plants and animals continued to swap genes, even between different species. The breadth of exchange is extraordinary – snake genes in cows, insect genes in fish.[31]

Today, it's clear that gene exchange was, and remains, a major force in evolution. "It's part of a revolutionary change," says University of Exeter biologist John Dupré.[27] "Our standard model of evolution is under enormous pressure. We're clearly going to see evolution as much more about mergers and collaboration than change within isolated lineages."

COOPERATIVE EVOLUTION: DARWINISTS ATTACK THE EVIDENCE

What was the Darwinist response to these discoveries?

Unfortunately (but as usual) they attacked the evidence. *New Scientist* was a special target. Paul Meyers called the magazine "déclassé...yet another rag pandering to the gullible."[32] He ridiculed the magazine's highly respected editor, Roger Highfield. "Pity Roger Highfield, he has been target[ed] by a number of big name scientists who have been hammering him in a small typhoon of outraged private correspondence."

The "big name" scientists Meyers referred to were the same Darwinist who had for decades attacked anyone who criticized Darwinism – Jerry Coyne, Daniel Dennett and Richard Dawkins. "What on earth were you thinking when you produced a garish cover proclaiming that 'Darwin was wrong,'" the trio wrote.[33] "As you surely know, many readers will interpret the cover...as being...about evolution."

Well, yes, the story *was* about evolution: one of Darwin's most important claims is untrue. "We have no evidence at all that [Darwin's Tree of Life is] a reality," said biologist Eric Baptiste at France's Pierre and Marie Curie University. "Now that we know more about evolution, it's time to move on."[27]

Nowhere did the attackers cite any incorrect facts. But facts weren't the issue, religion was. "Your cover," wrote the Darwinists, is "handing the creationists a golden opportunity to mislead school boards, students and the general public about the status of evolutionary biology."[33] The cover will be "abused by creationists," said Paul Meyers. "I expect we're going to have to face the headache of many school board meetings where that cover is flaunted as evidence that students ought to be taught about how weak Darwinism is... [it will] bring nothing but joy to the ignorant."[32]

But it was the Darwinists who were misleading people. The whole point of the story was that Darwin's idea of evolution as *isolated* branches, with *unique genes* and traits created through natural selection is wrong. "[Darwin's] tree of life is being politely buried, we all know that," says University of California biologist Michael Rose.[27] "Our whole fundamental view of biology needs to change. Biology is vastly more complex than we thought."

The idea that genes can be exchanged alters medical thinking about everything from childhood diseases to how we might cure cancer. Should society be kept in the dark about this discovery just to protect Darwinism? The Darwinists seem to think so.

Darwinists Distort Evidence to Oppose Religion

Ironically, these discoveries made a stronger case for evolution than Darwinism ever did. If the Darwinists were interested in advancing the idea of evolution, they would have embraced them. But their cause was not evolution, but religion. It was simply not possible to *scientifically* challenge Darwin without being attacked as a "creationist helper."

Following his article "Why Pigs Don't Have Wings," which outlined Darwinism's many failures, well-known evolutionary theorist Jerry Fodor joked that he'd been forced to enter the "witness protection program."[34] Like everyone who attempted to focus on evidence rather than ideology, Fodor was loudly shouted down as a creationist helper. "In my opinion," blasted Michael Ruse, "Fodor's piece is grotesquely and immorally irresponsible." Not because his facts are wrong. But rather, said Ruse, because "every creationist in North America is salivating today."[34]

The *scientific evidence*, Fodor calmly responded, makes it clear we need a new theory of evolution. "All I wanted to argue is that whatever the [new] story turns out to be, it's not going to be [Darwin's] selectionist story."

Other scientists agreed. "Fodor's piece seems pretty reasonable to me, in fact kind of obvious," said evolutionary anatomist Stuart Newman. It "has nothing whatsoever to do with creationism."[34]

For Darwinists, however, evolution was never just about the evidence. It always involved their battle with religion. This "has *everything* to do with creationism," said Ruse. "To write a piece [criticizing] natural selection in that way, is to give a piece of candy to the creationists."

Darwinism Fails: The Lack of Evidence

A major problem for Darwinists is that millions of examples of natural selection should exist. But they don't. Few, if any, proven examples exist. Robert Cecil made this point at the 1894 British Association for the Advancement of Science meeting, and it was still being made a hundred years later.

In 1987, Janis Antonovics, president of the American Society of Naturalists, told the Society that as a graduate student in the 1960s "I remember the excitement and eagerness with which we studied natural selection. I am now also appalled by it. Why on earth, a hundred years after Darwin... were we still excited by demonstrating natural selection?"[35] The answer, he said, is that despite decades of worldwide effort by thousands of scientists, only a dozen to twenty *possible* examples of natural selection have ever emerged.

Antonovics went on to say that his most pressing concern as president of the Society was the damage being done to research by modern Darwinism:

> "The [Darwinian] Evolutionary Synthesis failed in many serious and insidious ways...We [must] earnestly work to dismantle it...only [then] can we free ourselves of many of the methodological, conceptual, and even socio-religious difficulties that plague evolutionary biology."

You may wonder if other evidence exists that supports Darwinism, and accounts for its acceptance in the popular media. It does not. In 2009, for the 150th anniversary of the *Origin of Species*, the journal *Nature* published "15 Evolutionary Gems" in support of Darwinism.[36]

Two things were remarkable about the article. First, what it contained: a few minor examples of evolution, only five of which related directly to natural selection, and all five were better explained by other evolutionary processes. After 150 years, Darwin's evolution had precious little evidence of any substance.

More striking is what "15 Gems" (and similar articles that year) did *not* contain. Despite the grand claims made for Darwinism, it created nothing like the broad integration of evidence that emerged in geology from Plate Tectonics, nor the deep understanding of nature's basic processes that emerged in physics from Quantum Mechanics. There was nothing like the comprehensive explanation for the evolution of the eye we will see in chapter 21. Nor did it lead to new discoveries, like those which flowed from Pasture's germs theory of disease. (Not a single important 20th century discovery was predicted by Darwinism.)

That is what good theory does – it explains, integrates and predicts. Darwinism did none of that. The problem, however, lays not with evolution, but with the attempt to support Darwinism. *Nature* was founded to promote Darwinism and they gave it their best shot. The few minor examples offered in "15 Evolutionary Gems" was the best that could be mustered.

INVENTING FALSE FACTS MAKES EVOLUTION EASY TO DISCREDIT

Darwinism is presented as enlightened science, a theory built on well-established facts. Those who dispute it are painted as uneducated luddites, clinging to unscientific superstitions. The social conflict that embroils Darwinism is cast as the inevitable result of the challenge evolution presents to religious beliefs.

The reality is that the Darwinists created and sustained the controversy over evolution. Their lack of evidence, denial of established facts and imaginary processes made evolution easy to discredit. Think of all the ideas put forward by Darwinists that proved entirely fictitious – blending, gemmules, the Weismann Barrier, the Central Dogma, kin selection, selfish genes, the Handicap Principle and many others. Even natural selection has little, if any, supporting evidence.

Despite this, *any* scientific criticism of Darwin was branded "anti-evolution" and "creationist." Defending Darwinism became synonymous with defending science. "Neo-Darwinists...treat any criticism of the current paradigm as a breach of the scientific worldview that will admit the fundamentalist hoards," says Canadian biologist Robert Reid. "Consequently, questions about how selection theory can claim to be the all sufficient explanation for evolution go unanswered or ignored."[37]

These unanswered problems mean that opponents of Darwinism are able to point to solid *scientific* evidence that contradicts Darwinism. Since they are told Darwinism and evolution are the same thing, they (mistakenly) believe these fact also disprove evolution.

Many biology teachers find themselves in a somewhat similar bind. Research from Pennsylvania State University shows that in high schools there is a reluctance to teach Darwinism, not just in the South as often thought, but across the nation.[38] There is "a pervasive reluctance of teachers to forthrightly explain evolutionary biology," say the researchers. Among high school biology teachers *nationally*:

- 12% advocate creationism,
- 28% teach straight Darwinism, and
- 60% refuse to endorse either Darwin or creationism.

Are these teachers poorly trained, or religiously biased? On the contrary, most are well-educated and believe in evolution. They also understand that the scientific evidence does not support Darwinism. High-profile New York University scholar Thomas Nagel recently made the same point.

In *Mind and Cosmos: Why the Neo-Darwinian Conception of Nature Is Almost Certainly False* (published by Oxford University), Nagel argues that Darwinism has little supporting evidence. Meanwhile, he says, the evidence used by advocates of Intelligent Design to criticize Darwinism is grounded in solid science.[39] Because of this, says Nagel, these critics "do not deserve the scorn with which they are commonly met. It is manifestly unfair." Rather, he says, "I believe the defenders of intelligent design deserve our gratitude for challenging a...view that is ripe for displacement."

"My skepticism is not based on religious belief," says Nagle, an atheist. Rather, "it is just a belief that the available scientific evidence... does not require us to subordinate common sense."

"Even if one is not drawn to [their] explanation," he says, the scientific problems they point to "should be taken seriously." For example, he asks,

> "In the available...time since the first life forms appeared on earth, what is the likelihood that, as a result of physical accidents, a sequence of viable genetic mutations... occurred that was sufficient to permit natural selection to produce the organisms that actually exist?" His answer: it is "highly implausible that life as we know it is the result of a sequence of physical accidents together with the mechanism of natural selection."[40]

Facts matter, says Nagel, and the facts demand a response. "It is no longer legitimate simply to imagine a sequence of gradually evolving [forms], as if their appearance through mutations in the DNA were unproblematic

- as Richard Dawkins does."[41]

It is a measure of the sweeping change in thinking underway that dozens of reviewers cheered Nagel's criticisms, purely on scientific grounds:

- "Fascinating... a call for revolution." *National Public Radio*
- "A model of carefulness, sobriety and reason." *The Guardian*
- "[Nagel has] performed an important service with his withering critical examination of some of the most common and oppressive dogmas of our age." *The New Republic*

The rigid adherence to ideology over evidence led many others to conclude that Darwinism was indeed an "oppressive dogma." "It's an unfortunate fact that Neo-Darwinian ideas have evolved into almost a religion," said geneticist Hugo Hoenigsberg.[43] Lynn Margulis said history will ultimately judge Darwinism to be "a minor twentieth-century religious sect."[42] Stephen Jay Gould called modern Darwinism a "cult."

This "cult" mentality has compounded the confusion over evolution. Geneticists James Shapiro recalls an evolution conference at which Richard Dawkins said, "I will not only explain that Darwin had the right answer, but I will show that he had the *only possible* right answer." Knowing that in science, better explanations always emerge, Shapiro remembers thinking "the Creationists have a point. They are dealing with a form of religious belief on the 'evolution' side. Dawkins' transformation into an aggressive proselytizer for... atheism confirms this conclusion."[44] Shapiro says that,

> "One of the Creationists' main... arguments [is] that evolutionists are simply militant atheists in drag, who care more about dissing religion than about understanding evolution. Dawkins' ill-considered crusade just bolsters their position."

The effort to defend Darwinism against contradictory evidence is the core of evolution's many scientific problems. Stuart Newman, the biologist who discovered key aspects of how the skeleton evolved, offers a prime example. "If you look at the [2004] Pennsylvania legal case on the teaching of evolution," he says, "there was a very solid identification of evolution with Darwin's theory of evolution."

> "This was very reinforced in the public mind – if you believe in evolution, you believe in Darwin's theory because it's supposedly the same thing. And if you don't believe in Darwin's theory, you must believe in something supernatural. This is not at all valid and...a big mistake because we know there are other [evolution] mechanisms involving self-assembly... it's

> an unfortunate error that some advocates of evolution are making by adhering so closely to... Darwinian dogma."[45]

Newman (like Nagel and others) believes that defending Darwinism undermines the credibility of science. For example, says Newman, the skeleton did not evolve in tiny increments, as Darwinists claim. "Thus, everybody who's been assuring [a] skeptic that [evolution is] all incremental turns out to be wrong. It really undermines confidence in science if people are always being subjected to what we call 'hand-waving arguments' that all complexity had an incremental origin."[45]

Calls for a New Theory of Evolution

By 2000, the unscientific "hand-waving arguments" of Darwinism were leading to rebellion. "The accumulating evidence calls for a radical rethink of how evolution works," said geneticist Russell Bonduriansky.[46]

This wasn't coming from the fringe, but from leading scientists. Darwinism "is totally wrong," said Lynn Margulis. "It's wrong like infectious medicine was wrong before Pasteur. It's wrong like phrenology was wrong. Every major tenet of it is wrong."[47] (Phrenology is the pseudo-scientific 19th century idea that the shape of the skull reflected a person's mental abilities and personality traits.)

One of the "strongest messages" in current scientific conferences, says eminent Canadian biologist Brian Hall, is the large number of key questions that remain entirely unsettled about evolution.[48]

In 2000, Niles Eldredge, one of the world's leading paleontologists, made the case to the scientific community that "ultra-Darwinists" had turned evolution into a second-class science. "The group I label 'ultra-Darwinists,' and with whom I have been in active disagreement for over a quarter century, have created a demonstrably shallow picture of the evolution process."[49] Eldredge argued that evolution could recover only by returning to the "patterns and predictability" central to all science:

> "Patterns cannot forever be denied," he said. "The physical world changes in a regular and intelligible manner. And those changes have profound effects on the evolution of life. These evolutionary effects are regular and law like. And they... demand incorporation into evolutionary theory."[50]

This was the heart of the problem: the denial that evolution is predictable, and thus has direction. As Jerry Coyne said, "the distinguishing feature of [Darwin's] evolution is the complete absence of 'laws' or 'forces' that push it in a single direction."

Michael Ruse said that "to put direction into evolution is to be a supporter of the non-scientific theory of Intelligent Design."[51] Does this mean that cosmologists who predict the evolutionary future of stars, planets and the universe itself are advocating Intelligent Design? Or that geologists who predict the evolutionary direction of continents and mountains are advocating creationism? The idea is absurd. But that, in effect, is what Darwinists are claiming.

(Coyne, Ruse and others confuse two different issues: that natural laws make change predictable and give it direction, versus the idea that this predictably and direction mean nature was "divinely designed" to be this way. They confuse the scientific reality of natural laws with a philosophical question about the source or purpose of those laws and predictability.)

Darwinists had trapped evolution in their strange world partly by separating it from the other sciences, and claiming that the predictable evolution seen in the other sciences comes to a screeching halt with biology.

The challenge for Darwinists is this: if evolution in physics, chemistry, cosmology, geology and every other physical science is predictable, why would evolution in biology be unpredictable? If the evolution of atoms, molecules, minerals, stars, planets and continents has direction, why would the evolution of plants and animals suddenly become directionless?

Would nature really have two types of evolution – one for biology that is unpredictable, and another for everything else in the universe that is predictable? Darwinists insist the answer is "Yes." But most scientists call this claim unscientific and counter to the facts.

In the past, the Darwinists picked off their opponents one at a time, marginalizing or destroying their careers. By 2000, however, a transformation began to emerge. The most respected figures in evolution began meeting to discuss Darwinism's many problems.

One group that met for several years included David Jablonski, head of Evolutionary Biology at the University of Chicago, Marc Kirschner, head of Harvard's Systems Biology Department, pioneering geneticist Eva Jablonka, the Vienna University's Gerd Muller, one of Europe's leading evolution theorists, and a dozen other equally prominent scientists. In inviting others to their meetings, the group wrote,

> "The challenge seems clear to us. How do we make sense... of the astounding advances in biology since the 1940s, when [neo-Darwinism] was taking shape? Not only have we witnessed the molecular revolution, from the discovery of DNA to the genomic era... we just don't have the *theoretical tools* necessary to make sense of the... diversity and complexity of living organisms."[52]
>
> More simply put, science needs a new theory of evolution.

Modern Darwinism had failed. New ideas were needed to make sense of life's complexity. In 2007, Douglas Erwin, a senior scientist at the Smithsonian Natural History Museum, took the problem to the pages of *The New York Times* in a story headlined, "Darwin Still Rules but Some Biologists Dream of a Paradigm Shift":

> "In the past few years every element of [the Darwinian] paradigm has been attacked... Development, ecology, genetics and paleontology all provide new perspectives on how evolution operates."[53]

Around the same time, Carl Woese wrote, "the study of evolution is poised to cast off a century of dogma and become a true science... Biology can no longer afford to keep the study of evolution within the narrow confines of the so-called modern [Darwinian] synthesis."[54]

Woese, James Shapiro and other leading scientists began actively calling for a "21st Century view of evolution."[55]

In 2009, *New York Times* science reporter Carl Zimmer wrote, "Today a number of biologists argue that it's time for a new understanding of evolution... Some researchers argue that many patterns of nature...cannot be reduced to the effect of natural selection...They may be following rules of their own."[56]

A hint of the emerging new concept of evolution was in Zimmer's words: "many patterns of nature may follow rules of their own." Patterns that follow their own rules. Yes, that was the key to understanding evolution.

PART 4

HOW EVOLUTION *REALLY* WORKS

Four Major Discoveries

Evolution is a fact. So if Darwin is wrong, what is right?

The answer has emerged in just the last decade. A new concept of evolution that applies not just to biology, but to every level of nature.

The Central Mystery of Evolution

The mystery of evolution begins with a single fact. All living things are made from the same material - 96% water, with a bit of calcium, carbon and minerals.

Arranged one way, this mineral water forms a two-ounce tulip that can make oxygen, but does little else. Arranged another way, the same mineral water became a Tyrannosaurus Rex, seven tons of mobile terror that could see, hear, smell, hunt down and devour its prey. Both the tulip and T-Rex are made of exactly the same stuff. The only difference between them is how that stuff is *organized.*

This also applies to behavior. Termites and tigers are made from the same mineral water, yet they have entirely different instincts and social behavior.

How can a new pattern of mineral water create such different behavior? How can rearranging mineral water *create behavior* of any kind?

This mystery about evolution is part of a larger mystery about life. How can an *organization pattern* turn lifeless molecules into breathing, feeling creatures?

This is part of an even larger mystery, about the nature of existence itself. At every level of the natural world, there is no new matter, only the same old matter organized in new ways. The universe began with tiny particles called quarks, which possess only a few simple properties. Everything that followed came from organizing quarks in different patterns.

How can a "pattern" turn quarks into radiant stars, rain forests, elephants and people? Stars and snowflakes are made from the same quarks, yet stars are hot and snow is cold. How can both emerge from quarks? Water is "wet," but quarks are not. Where does that come from?

This mystery can be distilled to three questions:

1. **How** can something *immaterial*, like an organization pattern, turn quarks into stars, planets and continents?
2. **Why** does nature always make the same things, that always evolve the same way? Throughout the universe, nature makes the same types of atoms, stars and galaxies, which always evolve the same ways. This is how scientists can predict the evolutionary future of every star in the universe.
3. **Where** is the "organization plan" stored? When nature makes things (like atoms or stars) it always uses the same plan. Quarks don't contain the plans for making these things, so where do they come from?

The same three questions apply to biology and life:

1. How can an organization pattern turn quarks into toadstools and flamingos? How can an *organization pattern* create life at all?
2. Why does nature make only a few basic types of life? For all its seeming diversity, everything alive is built from just 15 basic body plans. When a new species appears today, it is a variation on one of these plans.
3. Where is the "organization plan of life" stored? Every day nature makes billions of living things, and uses the same plan each time. This means there must be a common plan for them. Where does nature store these plans?

In the Middle Ages, people believed that seeds and eggs held a miniature version of each life form (called a "homunculus"), that simply grew larger. In the 20th century we believed that genes held the plans, serving as blueprints for making life. Today we know that all animals *share the same genes*. So the plan can't be stored in genes. But if not there, where does the organization plan of life come from?

Over the last decade, science has made a series of remarkable breakthroughs and can now answer these questions.

Since the beginning of time, nature has created no new matter, only new relationships. The secret to the existence of everything in the universe is found in these relationships – *how* they form, *why* they form and what happens *after* they form. Science has solved (in part) the mystery of how relationships, and the patterns they create, form the natural world.

These discoveries about relationships and patterns led to a new concept of evolution. It is a comprehensive theory, that unites all of evolution under a single process. It integrates evolution at *every level* of nature – from atoms and stars to plants and people – and at *every stage* of natural history – from the origin of the universe, to the appearance of life and emergence of the human mind.

This new theory also transforms our basic concept of matter and reality. It explains how something *non-physical* like a pattern, can create something *physical*, like a tulip. It also explains how a pattern can create new behavior and new abilities, like our senses - sight, smell, hearing. In short, it transforms our basic ideas about nature and reality.

Chapter Fourteen

Science Discovers

HOW NATURE MAKES *EVERYTHING*

The Hidden Order in Nature

> **If we look at nature's work, we seem to catch a glance of a system. Things are so strictly related that from any one object the parts and properties of any other may be predicted.**
>
> *Ralph Waldo Emerson, "Nature"*

The new theory of evolution emerged from four transforming discoveries. The first is that everything in nature is made through naturally forming relationships in a process called "self-assembly."

Discovery 1: Self-Assembly: Nature's Hidden Creation Process

If you think, as the Victorians did, that nature is like a clock, you look for parts and forces that push the parts around. This was Newton's universe.

The 20th century revolution in physics revealed a profoundly different picture of nature in which time and space vary, and the universe emerged from particles with strange, often bizarre properties, such as the ability to be in two places at the same time.

This triggered a series of revolutions in the sciences, which overturned our basic concepts of reality and transformed our understanding of nature.

In the 1970s, as this revolution unfolded, scientists began detecting unexpected patterns that spanned the entire natural world. In seemingly unrelated areas – the birth of a galaxy, an ant colony, brain activity – the same patterns reappeared.

Scientists surprisingly discovered that even "chaos" has patterns. Edward Lorenz at MIT discovered that shapes which appear random or chaotic – clouds in the sky, smoke billowing from a fire, water swirling down a drain – have predictable patterns.[1] And the patterns in these "chaotic" forms matched those that appear elsewhere in nature.

A powerful set of laws appeared to govern these underlying patterns. This pointed to a deep, previously hidden layer of nature. In 1984, a group of scientists at Los Alamos National Laboratory (famous for cracking the secrets of atomic fission), left to establish the Santa Fe Institute, dedicated to penetrating this hidden world, traversing its mysterious landscape and returning with answers.

They succeeded.

In one of the most important scientific discoveries ever, they found that *everything* in the universe was created using the same simple rules. All the creations of nature emerged from this process – stars, oceans, orchids, antelopes and the human brain. This simple process made the universe complex, and thus was dubbed "Complexity Theory."

Despite its name, Complexity Theory is based on a simple process called "self-assembly." Scientists describe self-assembly as much like building with Legos,® where a few simple parts can be combined to form anything, from a simple hut to a complex city.

The "Lego" parts that nature uses have properties, which attract or repel each other, much like magnets. Attraction causes the parts to pull themselves together – to "self-assemble" – into naturally forming relationships.

Each new set of relationships becomes the platform for building the next level of relationships, pushing nature up the ladder of complexity.

Self-assembly

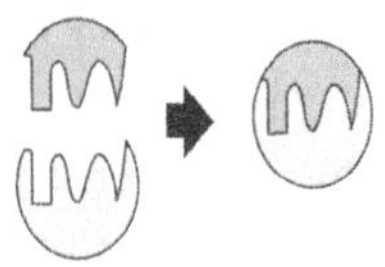

Properties naturally attract each other.

Creating Patterns

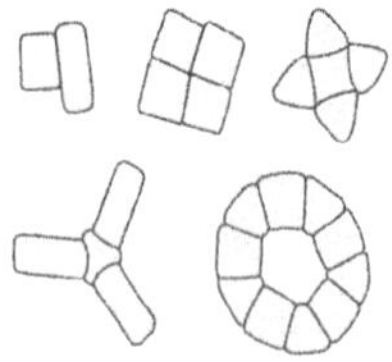

A few properties can combine in many ways.

Creating Complex Forms

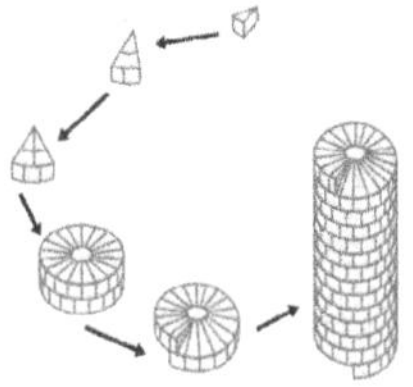

Each new pattern becomes the platform for creating the next, higher level of complexity.

Everywhere scientists looked they found self-assembly at work. Harvard's George Whitesides says self-assembling processes are common throughout nature, from the level of atoms up to entire galaxies.[2]

Physics, chemistry and biology are all filled with examples of self-assembly. It is nature's universal process for making and organizing everything. This simple process turns particles into atoms, atoms into massive stars, and organizes stars into vast galaxies. The same process also created living things, turning atoms into plants, animals and people.

Research into Complexity grew explosively in the 1990s as every area of science began using it to plumb nature's deepest secrets. In 1999, Charles Shank, then director of the Lawrence Berkeley National Laboratory, called Complexity the science of the "new millennium."[3] The BBC hailed it as "one of the three great theories which revolutionized science" in the 20th century (along with Quantum Physics and Einstein's Relativity Theory).[4]

Complexity Theory is so central to understanding nature that many scientists believe it will define the scientific agenda in the 21st century.

Biologists Neil Pearce and Franco Merletti write, "it is difficult, nowadays to open a popular science magazine or leading science journal, without reading about complexity."[5]

Today, most universities have a Complexity program, and many have entire institutes devoted to Complexity research. It's a major factor in every aspect of life – physics, chemistry, cosmology, geology, biology, zoology, medicine, neuroscience, social psychology, anthropology, economics and business.[6-18] Even evolution.[19] *Especially evolution.*

New Relationships Bring New Things into Existence

Self-assembly is more than a new idea about how nature *works*. It is a new concept of what nature *is* – a new idea about *existence* itself.

We tend to think different things are made of different matter (the tulip and T-Rex, for example). But this is not true. Everything is made of the same stuff. It is *new relationships* which bring *new things* into existence.

One of the great truths Einstein revealed is that matter and energy are the same thing, just *organized differently*. Change the way energy is organized and it becomes matter. Change the way matter is organized and it becomes energy. This is how batteries power your flashlight, and lumps of uranium are turned into electricity.

The idea that new things emerge, not from new matter, but from organizing old matter in new ways applies at *every level of nature*.

The universe began as a mass of particles. Separately they were just a chaotic swirling mess. The universe we see today emerged only when these particles began to interact and form relationships, creating new forms.

Nature became more complex by forming new relationships between those old particles. The differences between our Sun, a sunflower, and a sunfish come entirely from organizing those old particles in new ways. Physicist Eric Chaisson says the 14 billion year history of the Universe can be summed up as "hydrogen is a light odorless gas which, given enough time, changes into people."[20]

New relationships are also the basis of life. The ability to capture light and transform it into matter (called photosynthesis) is what allows life to exist. The cell later turns this matter back into energy. The cell doesn't create anything new, it merely transforms light from one state to another. And it does this by *changing the relationship* between light and the elements around it.

Nature Became *More* Complex by Forming *More* Complex Relationships

In short, scientists discovered that nature became more complex by forming more complex relationships through self-assembly. This explained many old mysteries, including the origin of nature's basic elements and where the sun gets its energy.

In the Victorian era, when factories were fueled by coal, scientists imagined the sun as a giant coal burning furnace. This idea disappeared in the 1900s when an American astronomer pointed out that while Pennsylvania held enough coal to power the United States for centuries, it wasn't enough to light the sun for even one 1,000th of a second.

In the 1950s, physicist Fred Hoyle eventually solved the mystery.

The universe began as massive clouds of hydrogen. Gravity began to pull these clouds into balls. As gravity's crushing pull increased, the hydrogen atoms broke into pieces. The pieces formed a new relationship which could withstand the pressure of gravity, creating a new type of atom called helium. The electrons that were left out of this new relationship shot out from the ball in vast waves of heat and light – a star was born!

This process, called fusion (forcing atoms together), explained the source of the sun's energy and revealed the origin of helium, nature's second basic element. Hoyle realized that all of nature's basic elements – the 118 atoms in the periodic table – came to exist this way.[21] Inside a star, gravity's relentless pull kept forcing atoms to break up. The parts would then form a new, more complex relationship, creating a new element – carbon, oxygen, iron and all the rest.

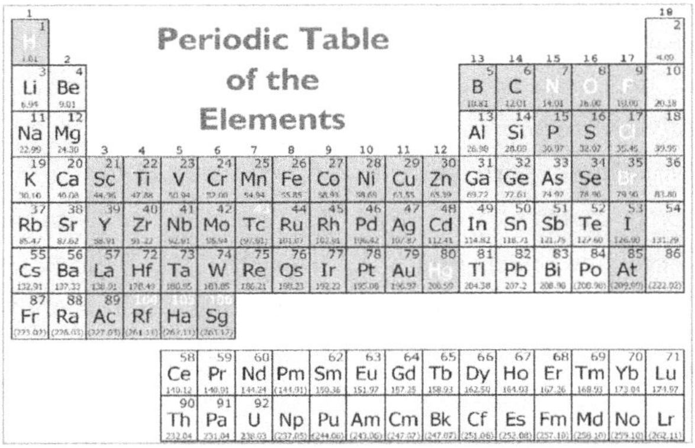

Hoyle said that eventually gravity's pressure builds until the atoms are so compressed the star explodes, spewing the new elements into space where they eventually form planets.

In one fell swoop, self-assembly allowed Hoyle to explain how the universe we see came to exist, integrating the origin of stars with the creation of the elements, the explosion of supernovas, and the formation of planets. The self-assembly process also allowed Hoyle to correctly predict that certain cosmic bodies (at that time undiscovered) must exist – dwarf stars, red giants, quasars and others – all of which astronomers later found.

At the level of life, self-assembly governs the entire biological world. "Self-assembly...is centrally important in life," says George Whitesides.[22] He says this is how the complex molecules of life – proteins and genes – are formed, and how they combine to create the basic processes of biology. [23]

Whitesides' colleague, Mila Boncheva, says that the cell itself – the basis of all living things – is formed through the self-assembly process:

> "The cell contains an astonishing range of complex structures such as lipid membranes, folded proteins, structured nucleic acids, protein aggregates, molecular machines, and many others that form by self-assembly."[22]

A landmark 2008 study made a powerful case that even the most complex parts of the cell, the nucleus, which holds and organizes genes, is made through self-assembly.[24]

Discovery 2: The Emergence of New Properties

The second major discovery about evolution is that, as nature becomes more complex, new properties appear in a process called "emergence."

Scientists had long assumed that all the properties in nature existed when the universe began. They did not.

The universe began with a handful of properties. But as it became more complex, new properties – new physical aspects of nature – came into existence. This confirmed Niels Bohr's discovery that while physics, chemistry and biology obey the same laws, as nature becomes more complex, unique properties emerge which can not be reduced to any of the properties at the lower level.

What makes emergent properties remarkable is not that they are new, but that they are *entirely different* from anything that precedes them. When elements unite, their properties don't blend the way red and blue create purple. Nor do they add together, the way 2+2 = 4. Instead, with each change, the old properties *completely disappear* and *entirely different* properties appear. In nature, 2+2 sometimes equals 7 or 9, or something previously unimagined, like yellow.

Nobel physicist Murray Gell-Mann described "emergence" this way: over time, the smallest elements in nature (quarks) combine to form atoms, molecules, stars, planets, flowers and jaguars, each with previously unimaginable new properties utterly unlike anything possessed by quarks.[25]

Emergent Properties

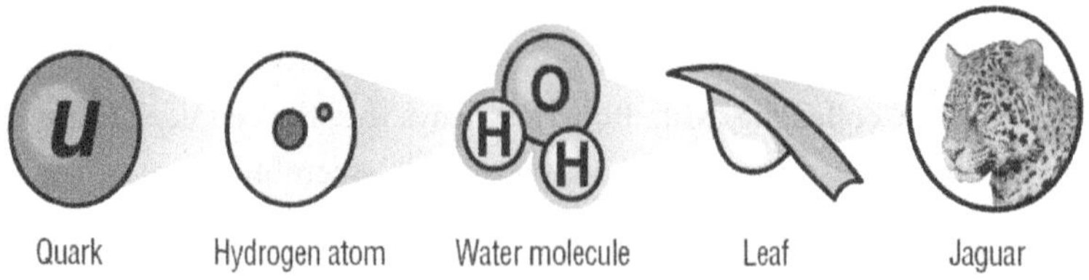

Each new level of complexity leads to the emergence of new properties.

Emergent properties means that in nature, the whole is not just more than the sum of its parts, it is *fundamentally different* from the parts. This shattered old ideas about what can exist in nature. In this new reality, objects are made from parts, but *cannot be reduced* to the sum of those parts. Basic elements like iron and oxygen, for example, each have physical properties unimaginable in quarks.

As these elements combine with each other, still more new properties emerge. Oxygen and hydrogen, for example, form water, which starts as a gas, floating unseen in the air. When cooled, water molecules link up, forming a liquid. This is what happens when the cool night air causes water to form the dew drops you find on the morning grass.

As a liquid, water molecules have new properties, like being "sticky," linking with each other in long lines. This "stickiness" is so powerful it can defy gravity, pulling water molecules uphill, or to the top of a 300-foot redwood tree.[26]

When many water molecules stick together they become the raging force of rivers, carving huge canyons from the hardest rock.

Water's "stickiness" also serves as biology's basic transport system. As water moves from one part of the body to another it pulls along nutrients, food, and other items the body needs.

At exactly 32° the water molecules abruptly stop moving, and instantly form a crystal (ice), with a new set of physical traits. Same water molecules, yet entirely new properties.

Emergent Properties Appear Abruptly and Fully Formed

Important to our understanding of evolution, emergent properties do not gradually appear. There are no in-between states, no gradual shift from one state to another, no mix of old and new properties. Instead, the new properties appear instantly. They either exist or they don't. This is how atoms instantly jump from one state to another (the famous "quantum leap"). On one side are the properties of the old atom, and on the other side the new properties of the new atom appear instantly, with all their abilities in full force .

Discovery 3: Nature Limits What Can Exist

Self-assembly does more than explain *how* nature makes things. In a third major discovery, scientists found that self-assembly determines *what* nature can make. It determines *what can exist.*

In theory, the vast number of particles in the universe could form billions of different relationships, making billions of different things. But that doesn't happen. In nature, to form a relationship, the properties of both parts must *precisely match* each other, the way a key must precisely fit a lock. (Chemists actually call these connections the "lock and key" process.)

Here is the problem in nature. Very few parts match each other. Thus, only a few things can exist. This explains why nature makes only 118 different types of atoms. And why, after 14 billion years of evolution, nature still makes *only* those 118 atoms.

This limit on atoms in turn restricts the types of stars and galaxies that can form. Out of billions of possibilities, nature makes only seven types of stars and four types of galaxies.

These restrictions on what can exist hold true as nature moves up the ladder of complexity. For two molecules to combine (creating what we call chemistry), their patterns must exactly match each other, like the lock and key. This limits the chemical combinations nature can form.

The same "lock and key" process applies to biology.

Because few patterns match, the chemical combinations that can form are limited, which restricts what can exist in biology. This begin with the basic "building blocks of life," called amino acids. Only 20 exist, and they combine in a limited number of ways. This restricts the type of life that can exist. What appears to be a cornucopia of diversity in nature is, underneath, the same things rearranged in different ways.

Something similar happens to genes, which are made from molecules called nucleic acids. Only five nucleic acids exist. Every gene is made by rearranging these five elements. This limits which genes can exist, and thus what can be inherited.

Nature Limits What Can Exist

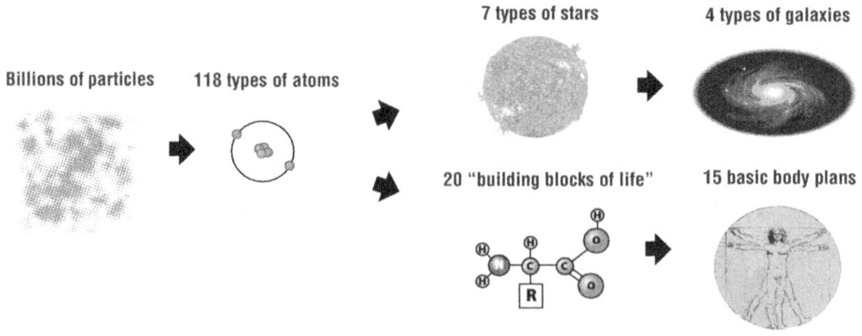

You don't have to be a rocket scientist to see that all these restrictions have profound consequences for evolution. Despite the seeming diversity of nature, all plants and animals are made from just 15 basic body plans.

What appears to us as nature's great diversity comes from the ability to use a few elements in remarkably different ways. For example, bird feathers, dinosaur scales and alligator claws are all made from the same material, called keratin. Packed together, keratin cells form a tough scale. Thinly organized they form a light feather. A feather is just a hollow scale. A scale is just a very dense feather (explaining, as we will see, why dinosaur scales could evolve into bird feathers).

Organized in other ways keratin cells become spider silk, animal fur, deer antlers, the hooves of cattle, sheep and goats, and human fingernails, hair and skin. In other arrangements, keratin forms bird beaks, porcupine quills, the baleen plates whales use for feeding, and the shells of tortoises, turtles and terrapins. All of this from just one type of cell.

What Nature Can Make is Predictable and Universal

Two more remarkable developments came out of these discoveries.

First, while emergent properties are not predictable, the *type* of things nature can make are predictable. Because nature's most basic properties combine in predictable ways, the things they will form *are also predictable.* This is how scientists could look at the Periodic Table, see that some gaps existed and predict the missing elements that would fill those gaps.

Second, because nature's basic properties are universal, the relationships they will form are universal. This means that what nature makes will be the same *everywhere in the universe.* Thus, the same 118 elements are created throughout nature.

This predictable, universal creation process holds true as nature moves up the ladder of complexity. While trillions of stars exist, only seven types of stars exist. This is why cosmologists can look at a cloud of cosmic gas and know the type of star it will form, and the evolution cycle it will go through.

The same is true for solar systems and galaxies.[27] The same types exist everywhere in the universe. This is how astronomers can predict the future of galaxies billions of light years away. Slight differences in the original conditions mean that the *number* of stars, and *their places*, will be different in each galaxy. But the *type* of stars created in each will be identical.

NATURE'S STRICT SEQUENCE

Another aspect of the self-assembly process, crucial for understanding evolution, is that everything emerges in a strict sequence. The orderly appearance of the elements is a snapshot of the underlying progression that self-assembly creates in nature. The basic elements all form in a precise order, each building on the last. Oxygen is central to life, but can't appear until after nature first makes half a dozen other elements, each in sequence.

Stars, galaxies and planets also form in a sequence. Before a planet can exist, a massive star, a Red Giant or a Super Giant (a star one million times bigger than the Sun), must form, burn itself down (making elements like carbon, oxygen and iron), then explode, spewing those elements into space.

These elements then form a new type of star which burns down (forming more new elements) and then explodes, sending these new elements into space. Only after stars have gone through this two step evolutionary cycle, creating all the elements, can planets like the Earth come into existence

Everything nature makes emerges in a similar sequence.

Only after the Earth cooled, creating water and minerals, could life emerge. Early life forms created new types of minerals. Only after these minerals existed could more complex life, like plants, appear. And only after plants existed could the first animals evolve.

THE HIDDEN ORDER OF NATURE

The predictable relationships of self-assembly determine both *what* can exist, and the *order* in which it will come into existence. This comprises what scientists now call a "hidden order" in nature.[28]

This has profound consequences for understanding evolution. Just as the few types of atoms and stars that can exist are predictable, universal and evolve in a strict sequence, the same holds true for living things. The few types of life that can exist are predictable, universal and emerge only in a strict evolutionary sequence.

This "hidden order" in nature also means that if there is life on other planets it will almost certainly be similar, if not identical, to life on Earth (as Richard Owen said 160 years ago).

A NEW CONCEPT OF NATURE: FROM PARTS TO PATTERNS

The discovery of this previously hidden order in nature united the sciences in a new and unexpected way. Not through shared parts, like particles, but though a shared *organization process*, one that operates at every level of nature and leads to formation of everything from atoms to alligators.

This was the culmination of a process that began with the revolution in physics 100 years ago in which Newton's mechanical cosmos of "parts and forces" was replaced by a universe made of "properties and relationships."

The great Karl Popper said this was one of the most profound shifts in ideas in all of human history. In this new view of nature, matter is irrelevant. Relationships and patterns bring things into existence. "The world is no longer a machine," wrote Popper. "It now can be seen... as an unfolding process of realizing possibilities"– poetic language from such a hard nosed scientist.[29]

The discovery of this underlying organization process, and the "emergent properties" which flow from it, led to three concepts which transformed our understanding of nature, evolution, and reality itself.

First, *relationships* bring *things* into existence. *New relationships* bring *new things* into existence. This is the single most important insight of the Evolution Revolution. All that is awe-inspiring in the universe exists not in things, but comes from the relationships *between* things.

Nature evolves by creating new relationships. New types of life, new behavior and new abilities, all emerge solely from the creation of new relationships between old stuff. This extends even to the human mind. Eric Kandel won a Nobel Prize for his discovery that human learning and memory are stored, not inside neurons, but in the space *between* neurons.

Second, emergent properties mean that the *reality* found at one level of nature can't be reduced to the *reality* of a lower level. Atoms can't learn or remember, but animals do both. Learning is real, but does not exist anywhere in the atoms that make up animals.

Emergent properties explain why emotions (fear or feeling safe) are real, but can't be reduced to biochemistry. They explain why leading neuroscientists today say the human mind is both entirely natural and yet entirely unique, possessing abilities not found in animals.[30]

Emergent properties also have profound implications for our approach to solving medical problems and our concept of human nature. For decades, Darwinists reduced problems like cancer to the activity of genes and reduced the mind to brain chemistry. But University of California biologist Richard Strohman said it is not possible to reduce cancer to the rules that govern genes, nor human intelligence to the activity of neurons. Genes are involved in cancer and neurons in intelligence, said Strohman, "but the cause of both lies elsewhere, at *higher levels of organization*."[31]

The Immaterial Creates the Material: A New Type of Reality

Emergence is one manifestation of an even more transforming concept about nature: *immaterial* things bring *material* things into existence. (Pause for a moment to let that sink in: *immaterial* things can bring *material* things into existence.)

This is the third key to the new understanding of evolution: much of nature and evolution emerges from immaterial things, like relationships, and patterns.

The Quantum Revolution in physics revealed that many material aspects of nature (like gravity and the other basic forces) have no physical dimensions. In other words, everything "material" in nature is not necessarily physical.

The Complexity Revolution now reveals that immaterial things, like organization patterns, bring material things into existence. In fact, most of what we think of as matter is a product of organization patterns.

For example, the physical properties of iron exist solely as a result of a set of relationships - a particular pattern - which quarks form. Take the pattern away and the physical properties of iron disappear. Bring the pattern back and the physical properties reappear. The particles remain the same. All that changes is the pattern they form. These material properties do not exist in the particles, but in the immaterial patterns the particles form.

"Central to this concept," says biologist Paul Grobstein, "is that there are aspects of reality which need to be characterized not in terms of matter or energy, but rather in terms of the particular organized form taken."[32]

Patterns can also bring important *non-physical* things into existence. Albert Einstein pioneered this discovery, showing that neither time nor space exist on their own, but are created as the universe expands. Relationships bring other non-physical dimensions of nature into existence, like feelings. Atoms don't feel pain or excitement, but animals feel both.

Importantly, these non-physical aspects of nature effect physical things. For example, while time has no physical dimensions, it effects every part of nature.

Sometimes the effect of new patterns is so far reaching they bring new *laws of nature* into existence. For example, scientists long believed that all the laws of nature existed at the beginning of the universe. But they did not. Robert Laughlin, who won the 1998 Nobel Prize in Physics for work involving emergent behavior, explains:

> "Early in the twentieth century it was discovered that atoms, molecules and subatomic particles are described by... rules so different from Newton's that scientists struggled to find proper words to describe them."[33]

Newton's legendary laws turn out to be emergent properties, appearing *only* after nature reached a high enough level of complexity. Laughlin says that because we live in a world created by Newton's laws, our everyday lives are based on a reality that did not exist at the beginning of the universe.

How can an immaterial pattern create something material? Where do the physical aspects of these properties exist *before* the pattern forms? Where do they go *after* the pattern changes?

Science is still trying to understand this deep mystery about nature. But it is nonetheless a discovery central to our understanding of evolution.

Scientists call the ability of patterns to create a new reality "higher organizing principles." Laughlin and his colleague, physicist David Pines, say that, "higher organizing principles and nothing else" determine emergent properties.[34] This is what Nobel physicist Erwin Schrödinger was referring to when he wrote, "living matter...cannot be reduced to the usual physical laws... because its *structures* differ from all that we have ever studied."[35]

Self-assembly answers the first two questions we asked about life and evolution: how does nature makes things, and why does it always makes the same things.

But where does nature store the plan of life for making these things? And what are the "higher organizing principles" that scientists like Pines and Laughlin refer to? And how do they relate to evolution?

Chapter Fifteen

Science Discovers

WHERE NATURE STORES INFORMATION

The Hidden Plan of Life

The gap between a collection of molecules and even the most primitive cell remains enormous.[1]

Chris Wills, biologist, University of California, 2009

Discovering the hidden order of nature explained much about the universe, evolution and life. But it came with a mysterious information gap. Where is the information – the "plans"– for making things?

There is no information inside atoms for making stars, planets, galaxies or anything else.

Genes were even more disappointing. We were told they held the plans for making life. Different genes made different creatures. But that proved untrue. All animals share the same genes. Ditto for plants.

Despite this, every day nature makes billions of trees, bees and other creatures, and does it the same way each time. So there must be a plan for making each one. But where does nature store this development information? For many, this was biology's biggest mystery.

Genes Aren't Templates for Development

Three discoveries deepened the mystery of development and how nature makes living things. The first shattered old ideas about genes.

The problem began in the 1960s when Francois Jacob and Jacques Monod at France's Pasteur Institute found that genes have "on-off" switches, which determine when a process starts and stops.[2]

In the 1970s, geneticists Phillip Sharp and Richard Robert rediscovered McClintock's mobile genes, and confirmed they regulate the body by moving around and flipping the switches of other genes on and off.

The importance of McClintock's "mobile regulator" genes grew rapidly in the 1980s as geneticists uncovered entire networks of them, working in close coordination to manage the body's most complex processes.[3]

But these discoveries presented a mystery. What was managing the "regulator" genes, telling them where to go and what to do?

The situation became more confused in the 1990s when it was discovered that genes don't exist in the simple way we long thought. They often share overlapping DNA with other genes, or include DNA from other genes that combine in odd new ways, including "small genes entirely contained within larger genes."[4]

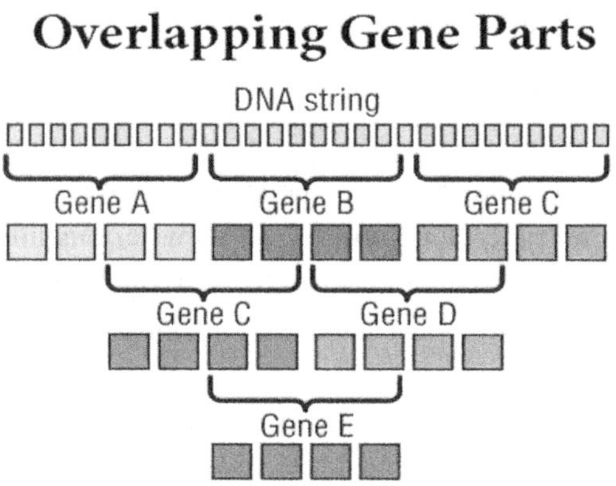

The same pieces of DNA often work in multiple genes.

Geneticist Peter Portin says that "discoveries such as overlapping genes, split genes, and alternative splicing [forced] biologists to rethink their understanding of what actually made a gene... a gene."[5]

The problem became so acute that Richard Dawkins was reduced to saying a gene is like a "recipe," made with different ingredients.[6] But this only deepened the mystery. If genes are assembled from a "recipe," what is organizing the "recipe"?

GENES AREN'T SUFFICIENT TO EXPLAIN DEVELOPMENT

In another surprise, science discovered that there are far too few genes to supply the vast information needed for development. As Richard Lewontin put it, "a complete description of genes and their actions does not come close to producing the program of life."

The vast gap between the number of genes and the information needed for development is seen in the human brain. By week eight of development, when the baby weighs just a few ounces, the brain is already making 1,700 neurons *every second* –100,000 a minute. By month nine, the baby's brain creates a staggering 20,000 neurons per second – *1 million every minute.*

To make each neuron, millions of molecules must arrive at just the right place in the brain, then organize into a cell. All new cells have identical material. To become part of the brain (instead of a tooth or toenail), a cell must flip thousands of chemical "switches" at precisely the right time, and in the right order to become a neuron.

And that is the easy part of making a brain.

After this is done, the newly formed neuron must find "home." While the brain is making neurons, it is also rapidly building veins, arteries and many other things. The neuron must navigate this complex environment, and insert itself in exactly the right place among billions of cells.

Once there, the neuron must reach out and link up with as many as 9,000 other neurons, and get each connection precisely right. By month nine of development, the baby's brain forms a staggering 200 million connections *every second.* Imagine linking 200 million telephone lines every second, over and over again, and getting every connection exactly right.

The potential for error is so high that if neurons were computer chips, the baby's brain would malfunction 300 times a minute, 18,000 times an hour. But that doesn't happen. Despite the unimaginable complexity, millions of babies are born every day with brains that work perfectly.

The question biologists asked is how does nature organize this colossal project, and get it right each time? An instruction manual for organizing the brain's 300 billion cells, and trillions of connections, would require vastly more information than provided by all the genes in the body, let alone the limited number dedicated to the brain.

A similar problem applied to making everything – the body's complex nervous system, its elaborate network of blood vessels, the intricately balanced hormone system, stomachs, eyes, arms and thousands of other things. Even making a pinky finger is the end result of a process that involves trillions of precisely executed instructions.

Far too few genes exist to account for all this. Biologists were stumped. Where was all this information coming from? More importantly, how could it be the same every time nature made something?

What is Organizing the Genes?

Another mystery emerged when scientists found that the genome can completely shatter, scattering bits and pieces of genes all over, then correctly re-assemble everything.[7] Craig Venter, whose group was central to decoding the human genome, points to an organism that can be hit with three million rads of radiation and not die. "Its chromosome gets blown apart," says Venter, "but it stitches everything back together and just starts replicating again."[8] Each time the scrambled pieces come back together in precisely the right order.

We think of genes as the instructions that organize everything. So how can genes get all scrambled up and then correctly reassemble? What is organizing the genes?

Part of the mystery was solved by Princeton geneticist Laura Landweber who discovered a "pathway for gene assembly."[9] In the cell, molecules create what she called "an elaborate cascading assembly line of clipping, coiffing, and adorning" that builds each gene. Reassembling scrambled genes is a variation on this "assembly line" process that kicks in when a gene needs to be fixed or reassembled.

This led biologists to ask what creates the "assembly line" that makes genes? And how does a cell know when a gene needs to be fixed? In other words, what is organizing the genes and telling them what to do? This pointed to the existence of an organization process *above* genes, one that had the information for making genes, and guiding what they do.

Discovery 4: Relationships Contain the "Plan of Life"

Biologists eventually found the organization process "above genes," and realized it also held the missing information needed for development. Where did they find the organization plan? "The short answer," said biologist Richard Strohman, "is that the program is in no one place. It is distributed at *many levels*."[10]

In the fourth major discovery about evolution, biologists found that the information needed for development is stored in the body's naturally forming relationships. In an unfolding process, the relationships which form at each step in development create a unique pattern that serves as the instruction for the next step.

The relationship at each level becomes the platform and information used in making the next level in development. Genes are only part of this larger process, explained Strohman, "they cannot, in themselves, specify the details of life." Genes are one element in a vast network of molecules, cells and tissue which *together* hold the "organization plan of life."

Creating Complex Forms

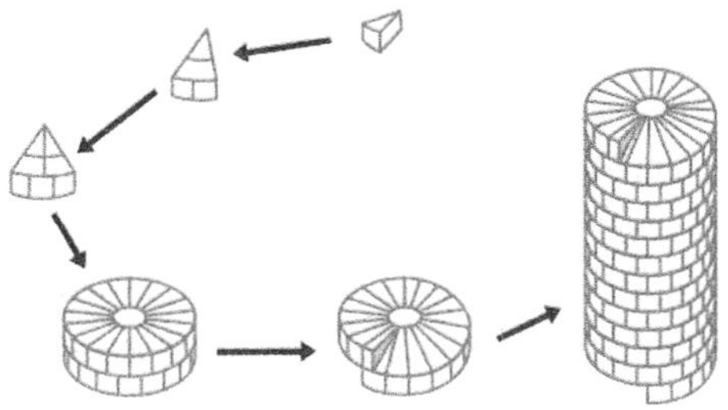

Each new pattern becomes the platform for creating the next, higher level of complexity.

The "plan of life" is embedded in these relations, emerging as they unfold. At each location, these elements combine to create a unique pattern – a unique chemical signal – that is the information for what should happen there. That change in turn creates a new pattern – a new chemical signal – which serves as the information for the next step in development.

The idea that each stage of development creates a unique pattern that provides the signals for building the next stage is today widely accepted. MIT geneticist Evelyn Fox Keller says "current research obliges use to acknowledge that these instructions are not written into the DNA itself but are distributed throughout."[11]

This utterly crushes the old idea that genes are blueprints for traits. It was also a blow to the idea that genes control life, and we are "lumbering robots" built to serve genes. The body's "higher organization process" controls genes and how they act, not the other way around.

Patterns Are Information

Discovering that an unfolding set of relationships creates patterns that are the information for development makes the process both more complex and far simpler than previously imagined.

Embedding information in trillions of rapidly changing relationships makes development complex in ways we can barely grasp. It far outstrips the analytic ability of our most powerful computers.

At the same time it makes development surprisingly simple. The complex, ever changing landscape of development emerges from a few simple rules (such as "opposites attract" and "likes repel").

The mathematical genius Alan Turing, creator of the modern computer, showed how this works in 1952. Turing demonstrated that with a few common shapes (like triangles), and just two rules, he could quickly create very complex patterns. Turing focused this process on biology and showed that despite the body's great complexity, it could be formed the same way, using a few simple rules and parts.[12]

Turing showed that everything in nature could develop this way. The "organization plan" for everything is stored in its own naturally forming relationships, and the patterns they create as they unfold.

Turing's "pattern formation" work later became central to modern physics, chemistry, and eventually biology.[13] The genius of this system is that as life evolves, the information needed to build the higher levels of complexity is easily stored in the next layer of relationships. No genes are needed.

Turing's "pattern formation" work showed something else important for understanding evolution. While the rules and parts are simple, complex new patterns emerge with each change. This explains how nature can abruptly shift to a new state which looks and acts completely different from the old – how a fluffy cloud can suddenly become a lashing rainstorm or a cascade of glittering snowflakes. (The ability of a new pattern to radically alter things is also the key to the great mystery of the origin of new species. See Chapter 18).

Relationships Contain the Information for Making Cells

Self-assembly and naturally forming relationships are the same thing. Biologist Stuart Kauffman famously called self-assembly and its naturally forming relationships "order for free," and said they explain why everything in nature evolves as it does.[14]

In an unfolding sequence, quarks, guided by nothing more than the relationships between themselves, form atoms, which in time organize themselves into stars, solar systems and entire galaxies. These naturally developing relationships formed increasingly complex patterns which held the information for making stars, planets and eventually entire galaxies. These patterns contained the information for making the universe complex.

The same process explains how evolution works at every level of nature.

Cells, for example, are made from proteins, molecules so complex that how they had evolved was a deep mystery. But in 2008, in the largest study of its kind ever, a group at Cambridge University discovered that proteins evolved through nature's basic self-assembly process. "Our model of self-assembly allows the reliable prediction of evolution and assembly of a [protein] complex," they wrote.[15] Proteins, it turns out, form and evolve much the same way as atoms, based on their internal properties and naturally forming relations.[15]

The same self-assembly process operates all the way up the ladder of life.

Starting at the smallest level, the properties of molecules cause them to unite, forming amino acids, which naturally unite in ways that form specific proteins.[16] At a higher level "the innate properties of molecules lead to their self-assembly" into cells.[17] For instance, the protein that forms the cell wall creates a unique signal that causes the surrounding molecules to literally lift it up and insert it into the wall.

The entire cell is built the same way.

Relationships Contain the Information to Make the Body

Self-assembly and naturally forming relationships also explain how cells then make an entire body. Take, for example, those handy things we call arms.

When the embryo is little more than a wrinkled blob, two microscopic bumps called "limb buds" form where the arms will eventually grow. In the 1940s, biologist John Saunders discovered that the outer ridge of these limb buds had what he called "mysterious powers" over arm development. Working with animals, Saunders found that if he clipped off this ridge, the arm never developed. If he grafted a second ridge onto the bud, a second arm began to grow.

Exactly why this happened was a mystery until it was discovered that the *physical shape* of the ridge creates a unique signal which triggers the production of "growth molecules." They then begin building a line of cells that eventually form the arm.

When the arm reaches the right length it abruptly stops growing. But how does the arm know which length is right? In the past this would have been explained by a gene, an "arm stop growing gene," or something similar. But no such gene exists. So how does the arm know when to stop?

In the 1990s, Harvard biologist Clifford Tabin discovered that as the arm grows, the end gets further away from the "growth molecules" produced by the ridge. Eventually the effect of the growth molecules becomes so weak that new cells stop forming and the arm stops growing.[18]

In other words, it is the *physical relationships* between the parts of the developing arm that determine its length. In simple terms, big animals have big arm ridges, which lead to more "growth molecules," more cells and longer arms. Small animals have smaller arm ridges, fewer growth molecules, fewer cells, and shorter arms.

It is this sequence of physical signals that makes an arm. Genes are involved, helping turn some processes on and off. But they are simply responding to near by physical signals. Their activity is guided by the larger organization process unfolding around them.

The same process creates every part in the body. "The lessons learned in the limb give you insights into how you build a face, or how you build a heart," said Tabin.

Each cell has the equipment to be any part of the body. What leads them to become a blood vessel, bone or something else, are the thousands of signals they receive from all around them.

This can be seen in the human brain. "The most complex biological structures [are] derivable purely from simple physical processes," says neuroscientist Chris Cherniak.[19] He explains that the vast information needed to organize the brain's billions of cells and trillions of connections is supplied largely by naturally forming physical relationships (what he calls the "harmony of neuroanatomy and physics").[20] Ordinary cells, using nothing more than the surrounding signals to guide them, create the brain's complex systems.[21]

Brain Development

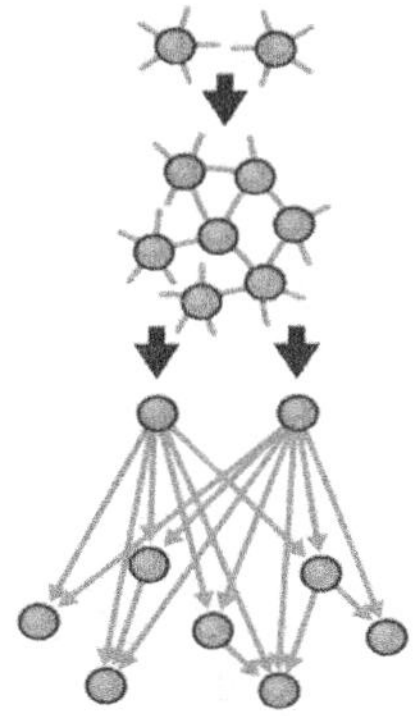

The physical properties of neurons lead them to naturally "self-assemble" into a brain.

In just the last few years, a remarkable set of discoveries has confirmed the idea that the brain self-assembles itself.

Sitting a lab dish surrounded by the right chemicals, human stem cells will begin to turn themselves into neurons. In 2008, researchers showed that when these lab developed neurons are put together in groups, they naturally "self-organize" into layers, like those in the brain's cerebral cortex. On their own, these neurons naturally create the brain's basic architecture.[22]

Even more startling, research showed that the neurons are transformed in a series of "waves," each one creating the *unique type* of neuron needed to form the next layer of the cortex, building it up layer by layer.[23] Neuroscientist Oliver Brüstle, says these "self-organizing" processes create "miniatures of the developing human brain... recreating the early steps in the formation of the human brain's cerebral cortex, which is evolutionarily the most complex tissue in the animal kingdom"[24]

Later work showed that identical neurons, placed in a different context, will "self organize" into primitive eye shapes (called "optic cups").[25] In a third context the same neurons naturally form the brain structure that creates spinal fluid.[25]

Then, in 2013, researchers provided what they called "unprecedented detail on how closely these self-organizing [structures] match human brain development." This included neurons that "segregate into distinct areas [matching the] different brain regions."[26]

These discoveries show the human brain making itself, much as Chris Cherniak describes.

But don't get the idea we're close to recreating the brain. "Despite these compelling data," says Oliver Brüstle, "the realization of a 'brain in a dish' remains out of reach."[24] The reason is that these lab generate brain regions are primitive, are not connected to each other (an unimaginably complex process by itself), and because they lack blood vessels, the inner parts of these groups become "a dead zone of cells starved of oxygen and nutrients." Creating a fully integrated, living brain is vastly more complex than these experiments might indicate. What they do show, however, is that underlying all the brain's complexity is nature's basic self-organization process.

The same self-organization process applies all across biology. A group that includes University of Bristol biologist Nigel Franks says that "recent research... reveals that even the most sophisticated structures are self-organized [using] surprisingly simple behaviors... that rely only on local information."[27]

In short, living things create them, using the same self-organization process that created the cosmos. To be sure, the patterns are far more complex, but they form using the same simple rules that govern organization throughout all of nature.

RELATIONSHIP PATTERNS EXPLAIN MANY OF EVOLUTION'S MYSTERIES

Storing the development plan in a series of unfolding relationships solved some of development's biggest mysteries.

One key puzzle was how genes can do many different jobs. "Within an embryo the same molecules are used over and over again," says Clifford Tabin. "No one would have expected there would have been so few signals used to form an embryo. If you have a signal that says make a heart, you wouldn't expect [the same signal also] makes a limb. But that's exactly what you find."[18]

To understand how this works, think of genes and molecules as similar to words. It is a whole sentence that determines the meaning of a word. Depending on the context, the word "cool" might mean a low temperature, or stylish or simply "ok."

Just as context changes the meaning of a word, it changes a gene's effect. Because of this a gene can perform different jobs just by showing up in a new place.[19] For example, genes used in jellyfish metabolism show up millions of years later involved in human hearing.

(This discovery is also transforming medical research. "We now understand that genes act in concert," writes biologist Jan Jensen, "thus, we may need to test these genes in proper context, before they reveal their true capabilities."[28] See Chapter 22 for more on how this is changing medicine)

Storing the development plan in relationships also solved the mystery of how the size arms and legs, hands and feet, and everything else in the body, always grow in perfect proportion to each other. As they develop, each part is responding to signals from all the other parts, balancing with the others.

This explains why, if an embryo gets more food and grows bigger, the arms, legs and other parts always grow bigger to match. And why, if there is less food, everything becomes proportionally smaller.

Storing the plan in a set of unfolding relationships makes development remarkably flexible, a self-balancing system that creates the same thing each time, while perfectly adjusting to changes.

This also resolved the old Nature versus Nurture conflict: how the body creates the same things every time, yet flexibly adapts to new experiences.

The problem is easily seen in the brain. At the core of the conflict, says NYU neuroscientist Gary Marcus, is "the evidence that the brain – like the body – can assemble itself without much help from the outside world," At the same time "the evidence shows that little about the brain's initial structure is rigidly cast in stone."[29] He says:

> "To an earlier generation the evidence for innateness and... flexibility seemed almost irreconcilable... The brain is capable of awesome feats of self-organization – and equally impressive feats of experience driven reorganization."

But the conflict is more apparent than real, says Marcus. "Self-organization and re-organization are two sides of the same coin," he says, each the product of the same basic process. The Nature side is "right that significant parts of the brain are organized even without experience." And the Nurture side is "right to emphasize that the structure of the brain is exquisitely sensitive to experience," such as learning and memory.[29]

How Relationships Create Social Behavior

The simple physical relationships of self-assembly and self-organization also explain a surprising amount of the social behavior in nature. (Scientists refer to self-assembly when they talk about how nature makes "things," and self-organization for how it creates social behavior.)

Take the tiniest creatures in most backyards, a harmless single-cell fungus called "slime-mold." When food is easy to find, each cell wanders off and takes care of itself. But when the food dries up, thousands of solitary slime-mold cells suddenly pull together and unite as one.

These mass combinations are complex, but always work well, with all the cells coordinating their behavior, and taking on different tasks

The question biologists asked was how can simple cells pull off this organizational feat? How do they coordinate such complex behavior? And how do they "decide" who will do each job?

Biologists assumed a leader must orchestrate the process. We grow up with the idea of leaders – the brain directs the body, parents guide children, mayors manage cities. Biologists assumed that a leader emerged to guide the slime-mold colony. How else could thousands of individual cells possibly coordinate with each other, and know what job each should do?

Decades of research, however, failed to turn up a leader cell. Nor did they find anything resembling a central process. To the contrary, slime-mold are all identical. None has any ability that would allow it to become a leader.

As biologists looked more closely they found the same problem throughout nature. Bee colonies, for example, were thought to be directed by a queen. But the queen is nothing but a giant egg-laying machine, helplessly spewing out larva all day, and not involved in guiding the hive. In a surprise, biologists found that *no one* was in charge. No "leader" bee directs or coordinates anything. The colony just organizes and manages itself.

This was baffling. Such a decentralized process defied common sense. Each individual has only a fraction of the information needed to coordinate group activity. Without centralized management, how could the behavior of thousands of bees be synchronized? Without a leader, how would each bee know which job to do? How could a smoothly functioning system possibly emerge if decisions are made by individuals with no knowledge of what is happening elsewhere in the hive?

To answer those questions, watch a flock of flying geese. "Collectively they form the image of a giant delta-shaped bird that moves as purposefully as if it were a single organism," says Cornell scientist Michael Macy.[30] "Yet the flock has no 'group mind' nor is there a 'leader bird' choreographing the formation." How do they manage this magical synchrony?

Although the birds can't tell the effect of their behavior on the group, the basic principles of aerodynamics, plus their individual actions, lead the birds to naturally form the "optimal V-shape."[31] "Each bird reacts to the movement of its immediate neighbors who in turn react to it," says Macy.

"The result is the graceful dance-like movement of the flock whose hypnotic rhythm is clearly patterned, yet" it has no leader or central organization.

Patterns Provided the Information for What to Do

The flying geese reveal the two remarkable aspects about social behavior.

First, much of it is the *inevitable outcome* of naturally forming *physical relations*, and the patterns they create. These physical patterns, for example, explain how slime-mold manage their remarkable synchronization process.

When a slime-mold cell gets hungry it's behavior changes. If a whole group is hungry, the chemical effect of their new behavior *physically pulls them together*. Once joined, the *physical structure* they form causes the cells on the inside to receive different *physical signals* from those on the outside, automatically leading the two groups to specialize on different tasks.

Here is the crucial point: much of the social behavior we see in nature is like this. It is not the result of genes, or natural selection, or any similar process. It emerges solely from the properties of physical relationships, like those seen in slime-mold and the flying geese.

The ability of physical processes to create social behavior starts at the most basic level of nature. If electrons are grouped together they automatically begin to coordinate with each other, leading to group behavior.[32] Other types of particles, influenced only by the other particles in their neighborhood, also synchronize their behavior.

At a higher level, a cloud of atoms, each equipped with "nothing more than knowledge of their immediate neighborhood," will organize themselves to form complex shapes, like crystals.[33]

Moving up the ladder to life, physical patterns provide the information that turns millions of molecules into a single cell, billions of cells into a single animal, and thousands of animals into a single social system. This is how neurons organize themselves into a brain, and embryonic cells organize themselves into guppies and puppies. It is how ants, bees, birds, bison herds and wolf packs coordinate their behavior.

(This also means that much of the social behavior we see today was *embedded in the initial laws of nature* at the start of the universe, a fact that evolutionary scientists are still grappling with.)

Complex Behavior from Individual Choices

The second remarkable thing about complex social behavior is that it emerges *solely* from individuals responding to what is directly around them. These individual acts, taken all together, give rise to organized group behavior.

In an ant colony, for example, where no one is in charge, the information for what each ant should do emerges from the patterns they create together. "What matters to an ant," says Stanford biologist Deborah Gordon, "is not the particular message at each interaction," but rather "the *pattern* of interactions the ant experiences."[34]

The combined behavior of all the ants creates a pattern that determines the job each one decides to do (look for food, clean the nest, feed the queen). When their individual decisions are combined, says Gordon, the effect is to coordinate behavior across the entire colony.

Individuals play the same role in creating social behavior as properties do in creating things. At each place, the combined behavior of the individuals creates a unique pattern. This pattern provides the information, or signal, for what they should each do next. This means the group doesn't need a leader, who can see everything, for social behavior to synchronize.

This is "self-organizing" social behavior, and it is seen throughout nature. "Self-organization of social systems can be seen at all levels of biology," says Dutch biologist Charlotte Hemelrijk, "from cells to organisms and communities."[35] We see it in schools of fish, cattle herds, wolf packs, and groups of bats, rats and penguins. Hemelrijk says that individual interactions, plus a few rules, can create a wide range of complex social behavior including group coordination, specialization and elaborate communications systems:

> "Although individuals are governed by simple rules," she says, "their interactions with each other and their environment leads to complex patterns."[35]

This cooperative behavior is what gives every biological system its adaptive power. From cells to large social groups, cooperation allows them to do things no single individual could achieve on their own. This may seem obvious. But given that Darwinism says conflict leads to the evolution of complex social behavior, it is important to note that the natural world shows exactly the opposite is true.

Relationships are Nature's Hidden Information System

"Organisms can be proud to have been their own designers," says biologist Kalevi Kull. This, however, requires us to think differently, "to look at the organism as a self-organizing and communicating system," says Kull.[36]

Biology's communication system works much like human language.

When we speak, a random string of words has no meaning. Meaning emerges *only* when the words are organized into a *particular pattern*. Rearranging the pattern gives the words a new meaning.

The same holds true in biology. The patterns formed by biological relationships serve to convey information. New relationships convey new or different information.

Relationships serve as nature's basic communication system. They are, in effect, the universe's hidden language. The patterns they form create a communication system that works at every level of nature.

Atoms exchange information, as do molecules. Water, for instance, is an important part of life's communication system, able to send information about light, temperature, food and minerals from the top of a 300-foot Sequoia tree to the bottom and then back up again. With water, "all the molecules are able to feel the structure of all the other molecules that are present, so they can work as a whole rather than as individuals," says applied sciences professor Martin Chaplin. "Water acts as the contact between biological molecules...imparting information among them."[37]

The ability of relationship patterns to convey information applies throughout the elements of life. In his Nobel lecture, Jean-Marie Lehn said the molecules of life have sophisticated abilities, including "recognition, communication and information processing."[38]

The Hidden Order of Life

The patterns created by relationships show nature's ingenuity. It doesn't need physical things, like genes, to hold information. It embeds information in the unfolding patterns of development and evolution, allowing them to store an infinite amount of information without using anything physical.

We saw that the constraints on the patterns nature can make represents a previously "hidden order" in nature. In biology, similar constraints on the patterns that can form reveals a previously "hidden order to life." And this, in turn, has profound implications for understanding evolution.

Chapter Sixteen

Science Discovers

HOW EVOLUTION WORKS

The Hidden Code of Evolution

> **Major changes in our understanding of evolution have gone virtually unnoticed. Surprising discoveries have radically altered our views of evolution, contrary to the previously accepted neo-Darwinian view.**[1]
>
> *Edward Ziff, biologist, New York University, 2006*

Complexity Theory, at its heart, is a theory of evolution. The rules for making things are also the rules for changing them. Evolution is simply change over time. Thus self-assembly explains how things are *made*, how they *change* and how they *evolve*.

Evolution is simply the story of how self-assembly made nature complex.

How does the Complexity Theory of Evolution compare with Darwin's?

Distilled to its essence, Darwinism made six claims about evolution. He said change occurs in tiny increments, which makes evolution extremely slow as the tiny changes gradually build into complex traits. Which traits will prove helpful is determined by the environment (retaining heat is good in the arctic, but death in the desert). In a competitive, "survival of the fittest" struggle those with helpful traits survive, while others die. This is natural selection.

Darwin said arbitrary changes in the environment mean that which traits will survive is unpredictable, and thus evolution has no direction. (As Jerry Coyne said "the *distinguishing* feature of [Darwin's] evolution is the *complete absence* of 'laws' or 'forces' that push it in a single direction.")

In short, Darwin said evolution is a slow, gradual, competitive process, driven by unpredictable external forces in which the losers are eliminated.

These claims have now been scientifically tested. The results show that every major claim made by Darwin has evolution not just wrong, but *backwards*. Darwin's goal was to eliminate predictability in evolution. Because evolution is predictable, Darwin had to create a theory that is the exact opposite of how nature actually works.

How Does Evolution Work?

	Darwin said evolution is	Science shows evolution is
1.	**Slow and gradual**	**Abrupt and Rapid**
2.	**Tiny increments**	**Big leaps**
3.	**Competitive** "survival of fittest"	**Cooperative**: via relationships
4.	**Eliminates** losing genes & traits	**Preserves** genes & traits
5.	**External** issues decide survival (e.g. the environment)	**Internal** issues decide survival (self-assembly process)
6.	**Random** (no law predicting change)	**Predictable** (laws predict change)

Complexity Theory, on the other hand, matches what science has discovered. It shows that evolution (at every level of nature) is rapid, with big jumps in complexity based on the laws of self-assembly. Nothing is eliminated, only rearranged. Most importantly, it shows that change is predictable. Evolution has direction.

Complexity science turns every major Darwinist claim upside down. Either all the world's Complexity Institutes have nature wrong, or Darwinism has evolution wrong. Look at the evidence and decide for yourself.

RANDOM CHANGE OR PREDICTABLE CHANGE?

In its 150 year history, nothing has been more contentious about Darwinism than its claim evolution is directionless. In Darwin's words, evolution is no more predictable than which way "the wind blows."

In the 1960s, Darwinists said random gene mutations made the origin of new traits unpredictable and this made evolution directionless.

Darwinists knew how improbable the idea of random mutation was to other scientists. "The most serious objection to the [Darwinian] theory of evolution" wrote Theodosius Dobzhansky, "is that since [gene] mutations occur by chance and are undirected, it is difficult to see how mutation and selection can add up to the formation of such beautifully balanced organs as, for example, the human eye."[2]

Today, says geneticist Barbara Wright, most scientists "share Dobzhansky's intuitive conviction that the marvelous intricacies of living organisms could not have arisen by... random mutations."[3]

Calling "random mutation theory one of the great wrong turns in the history of science," Lynn Margulis said "the importance of random mutation in evolution is wildly overemphasized."[4] An accumulation of random changes, she said, "does not lead to new species or even to new organs or new tissues."[5] The evidence shows she was right.

Self-Repair Makes Random Change Impossible

For starters, nature is extremely efficient at eliminating the type of random changes on which modern Darwinism is built.[6]

Self-assembly provides nature with a built in self-repair mechanism. The same mechanisms that assemble genes also repair them.[7] Processes, like the "SOS" repair mechanism (shown below), use the surrounding genetic material as a template to repair damaged genes.[8] Elements that don't match the template are rejected and disposed of. Then, much like a magnet, the template attracts the right parts to properly fix the gene.

DNA Damage Repair

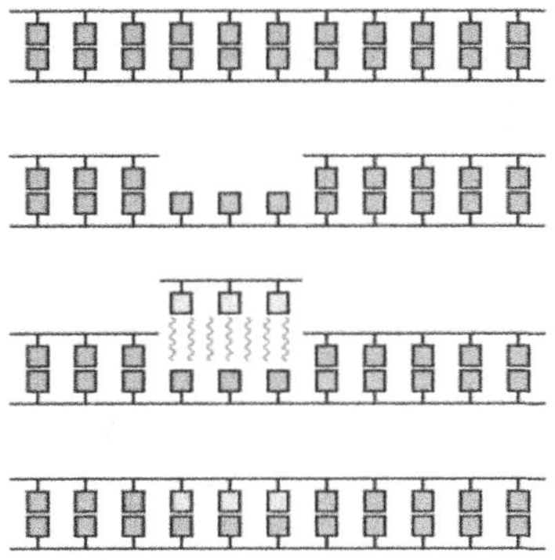

One side of DNA naturally attracts the correct molecules to repair itself.

Using what is called the "gap filling" process, other repair mechanisms can replace large sections of damaged DNA anywhere in the body.[9] Entire networks of genes repair themselves this same way in response to damage.[10]

At multiple points the cell also performs what are called "proof-reading" checks which ensure that incorrect genes are identified and disposed of.[11] As evolution progressed, and cells became more complex, the sophistication of these repair mechanisms also increased.[12]

Over the course of evolution, cells also created thousands of specialized molecules called "chaperones" which, as their name implies, ensure that genes form properly and are guided to the right places, eliminating errors along the way.

In short, the gene "code is highly resistant to errors relative to random change," says University of Colorado geneticist Rob Knight.[13]

Random Change Leads to Death or Chaos

Random change never made much sense. For example, most genes work in pairs, switching each other on and off (a process called "reciprocal regulation"). This means they must *change in pairs*, a process called "co-evolution," making random change highly unlikely.

Also, genes are made by a complex assembly line of molecules, and can't evolve unless this "large cast of supporting molecules evolves with it."[14] Mathematician Richard Bernhard calculated that:

> "If every mutation were really random and had to be tested against the environment for selection or rejection, there would not have been enough time to evolve the extremely complex biochemical networks and regulatory mechanisms found in organisms today."[15]

In Bernhard's estimate, random change would required life to have begun eight billion years ago, twice the age of the earth.

Many genes share overlapping DNA with each other, and a surprising number of genes are nested entirely inside other genes. Changing the DNA of any of these genes would make a mess of many other genes.[16]

Making the idea of random change even more unlikely is the fact that many genes perform multiple jobs. For example, cancer tumors die if they're cut off from their blood supply. Looking for a way to do this, University of Texas geneticist Edward Marcotte isolated ten genes essential for building blood vessels. He also discovered these same genes do many other jobs central to life, like build cell walls.[17]

And while some genes have only one job, they often do it in many places. For example, during development, a *Hox* gene tells the body where to place the head, tail, arms, legs, wings and other parts. In one place it might signal where to put an eye, in another where to put a finger or a toe.

It would be impossible to randomly change any of these genes without creating utter chaos inside the body.

And then there is the "Fatal Change Paradox."

A hundred years of research going back to Mendel, shows that genes are linked in complex networks. A random change in one gene would create chaos for hundreds of others. In the1960s, Ralph Hinegardner at Columbia University showed such changes "would almost certainly" prove fatal. On the contrary, he said, the evidence shows that "the genetic code, once established, would...remain invariant."[18]

Today we know he was right. University of California biologist John Gerhart and Marc Kirschner at Harvard, say that central to modern evolution is the realization life used the same genes "for hundreds of millions or even billions of years, because changing them would be lethal."[19]

A New Concept of Evolution

Duplication and Specialization

These discoveries confirmed that random gene changes are not the source of evolution. So if they don't create evolution, what does?

Part of the answer emerged in 1970, when Japanese geneticist Susumu Ohno showed that genes frequently make a copy of themselves, and the duplicate gene then begins to specialize on one part of its former job.[20]

Imagine a carpenter who makes chairs. When his two sons grow up they also become carpenters but they specialize, one making kitchen chairs, the other desk chairs. Over time, they modify their tools a bit to match the chairs they make. Both still do the same job (making chairs), and use the same basic tools. But because they've specialized, they can make more complex chairs, better suited to different needs.

Called "duplication and specialization" or "duplication and divergence," this is one of evolution's key processes. Ohno said it explains both the great simplicity of living things (all sharing the same genes) and their great diversity (gene copies specializing on some task).[21]

The duplication and specialization process occurred mostly with the "mobile regulator" genes discovered by Barbara McClintock. These duplications led to the emergence of entire networks of regulator genes which made complex life possible "The creation of additional regulatory systems," Ohno said, created the "big leaps in evolution."[19]

Later research proved Ohno correct.

At the California Institute of Technology, Edward Lewis discovered the *Hox* family of "regulator" genes, which guide the placement of body parts in all animals (for which he won the Nobel Prize).[22] Lewis found that during evolution, the repeated duplication and specialization of a single gene led to the evolution of a "family" of closely related genes, all performing variations on the same job.

In the case of the *Hox* family, Lewis found that variations of this gene help organize the bodies of ants, angelfish, antelopes and every other animal on Earth. These genes signal where to place eyes, legs, fins, antennae and other body parts.[23] Rearranging when and where *Hox* genes send their signals changes where a body part will appear. This explained how the legs on ants and antelopes can be guided by the same gene, yet appear in different places.

In the decades that followed, entire networks of duplicated genes were found to control processes at every level of biology. Path-breaking geneticists like Eric Davidson and Uri Alon showed that large families of duplicated genes organize themselves into intricate systems that manage biology's most complex processes.[24] Davidson, widely acknowledge as "the worlds leading expert" in this area, says regulatory genes are now recognized as "a major force in evolution."[25]

In the 1990s, a second system of regulator genes was uncovered by John Mattick. Made from RNA (another type of genetic material), this system is so vast that some geneticists believe it is as important to evolution as DNA. Among other things, RNA makes the molecules which form DNA, and thus controls how DNA changes. "We have fundamentally misunderstood the nature of genetic programming... for the past 50 years," said Mattick.[26] It appears that complex life involves "a vast and hitherto hidden layer of regulatory RNAs."

Nobel Prize winner François Jacob earlier summed up the new thinking, calling it "nature's tinkering. After duplication, the cell acts like an engineer at his work bench, tinkering with the new gene's switches (its "regulatory circuits") to determine when and where it will be active. This in turn alters traits and evolution.[27]

Regulator Genes Explained Crucial Facts about Evolution

These findings vindicated Ohno and made sense of crucial facts about evolution.

The discovery that evolution occurs when genes make duplicate copies of themselves explained why genes are both ancient and yet differ over time.

It also explained how plants and animals can change without killing themselves. Using old genes in new ways greatly reduced the possibility of deadly errors.

These discoveries also resolved the "junk genes" mystery.

The Modern Darwinists story of evolution was that genes make proteins, which then make traits. Life became complex when more of these genes appeared. Humans were said to have up to 100,000, versus a fruit fly's 20,000.

This was never more than another just-so-story, but it soon became textbook dogma. As Richard Lewontin put it, "from *Science* to *Science Digest* to the *Reader's Digest*, the word has spread that DNA makes proteins and proteins make us."[28]

But this story also proved untrue.

When fully decoded in 2003, the human genome turned out to have a whopping 2 million genes, but only 1% of them (22,000) made proteins, roughly the same number as an earthworm.

Biologists asked about the other 99% of genes. What do they do? Darwinists quickly invented a just-so-story, claiming they did nothing. They were "junk DNA," the remnants of ancient evolutionary events, abandoned as life became more complex.

The genome told a quite different story. What Darwinists called "junk DNA" were duplicate genes that had gone on to specialize on some task. The "*clock*" gene, for example, started out regulating simple things, like daily feeding. Over time it duplicated and specialized to regulate everything, from monthly menstrual cycles, to annual migration and hibernation cycles.

More than debunking old Darwinist myths, these discoveries revealed how evolution works. They showed that while the number of protein genes remained the same, throughout evolution the regulator genes kept duplicating, allowing life to become more complex.[29] "The irony," said John Mattick, "is that what was dismissed as junk...may well [be the genes] that underpinned the emergence...of complex life."[26]

They showed that regulator genes are part of a vast instructional process, and *new relationships* between them can create evolutionary change. *Hox* genes, for example, which direct the placement of body parts, are like mobile traffic cops, telling a cell to "stop, and turn left." Altering where they send these signals changes where traits will appear.

In 2012, the human genome was found to hold an astonishing four million gene "switches," regulating every aspect of the body. Eric Green, director of the National Human Genome Research Institute, said, "What we learned is how complicated the human genome is [and] the immense number of switches that are choreographing how genes are used."[30]

Today, scientists say "hardly any aspect of genome evolution or function is not somehow linked to gene duplications... The duplication of genes may form a cornerstone in the evolution of biological complexity."[31]

McClintock and Ohno had gotten that part of evolution right.

EVOLUTION INFORMATION IS EMBEDDED IN RELATIONSHIPS

Although these discoveries explained many mysteries, they left a big gap. Just as genes alone were inadequate for explaining development, they proved unable to explain evolution. Duplicating and rearranging genes, even millions of them, didn't contain enough information to account for all the complexity in nature. The problem is seen in the brain.

A simple worm has 300 neurons, each with a thousand connections.[32] These links can be organized in 40 million different ways. Yet nature produces close to the most optimal wiring of these connections.[33]

The human brain, with trillions of connections, could be organized in hundreds of trillions of different ways. Yet neuroscientist Chris Cherniak shows it also has close to optimal wiring. Only one plan in *100 million* would be superior to the human brain's current organization.[34]

How is this extraordinary optimization explained? Certainly not by random changes and natural selection. Randomly testing *one combination per second* would take billions of years longer than the universe has existed. Clearly an impossible task.

Even re-arranging millions of genes couldn't explain this level of optimization.

Many other aspects of biology revealed the same problems. They are far too complex, and optimal, to be explained by natural selection or by re-arranging genes.

RELATIONSHIPS CONTAIN EVOLUTION'S 'MISSING INFORMATION'

So if genes alone don't explain these aspects of evolution, what does? Where is nature storing all the additional information needed to make the brain and other complex systems, like hormones and the heart?

Scientists eventually found evolution's missing information. The same naturally forming relationships which guide development also guided evolution. Just as the "plan of life" is contained in the relationships of development, the "code of evolution" is embedded in a set of relationships.

Evolution and development are mirror images of the same process.

In development, each level provides the information for building the next. The same is true for evolution. The relationships that form at one level of evolution serve as both the platform and the information (the "code") for creating the next level of evolution. Genes are just one part of this process.

For example, Chris Cherniak says these physical relationships (what he calls the "harmony of neuroanatomy and physics") led to the brain's extraordinary organization and optimization. The brain is "self-organizing," he says, "with no need of [other] evolutionary mechanisms for its creation."[35]

"Most complex biological structures [are] derivable" this way, says Cherniak, "purely from simple physical processes."

The idea that the "code of evolution" is contained in a series of physical relationships, which emerge in a specific sequence, explains one of evolution's deepest mysteries. During development, an embryo *repeats every stage* of its evolutionary history - a process called "recapitulation."

This process, which occurs in all animals, was famously summed up in the 1880s by biologist Ernst Haeckel as "ontogeny recapitulates phylogeny"– the development process repeats the evolution process. During development, humans for example, develop fish gills, and later primate tails, repeating parts of our evolutionary history. Other animals go through a similar sequence.

Recapitulation was one of evolution's great mysteries. If genes hold the information for making a creature, why not just make it straight away? Why create the potential for thousands of errors by starting deep in the evolutionary past and building up through *every stage* of one's evolutionary history? What purpose could this possibly serve?

Self-assembly provided the answer.

Much of the information for development does not come from genes, but from physical relationships that unfold in a specific sequence. To get this sequence right, plants and animals must start with their earliest form, and then progress through each stage of their evolutionary development.

These ancient structures serve as development's crucial building blocks.

They provide both the platform and the instructions for creating the next level of development.

If you want to make a human, and get it right, you must build up through stages that include worms, fishes and then early primates. If you want to make a modern bird, you must first make an ancient dinosaur. This is why, if you stop human embryo development midway through, you find fish gills. And why, if you stop a bird's development midway, you find a dinosaur tail.

This need to build up by recreating the stages of evolution appears to be a universal aspect of biology. For example, life's basic building blocks, proteins, are made through a self-assembly process that repeats the steps of their evolution history. "The assembly pathway mimics the evolutionary pathway that has taken place over...millions of years," say the geneticists at Cambridge University who made the discovery.[36]

"This represents a molecular analogy to Haeckel's paradigm of embryonic development," they say, where each step in making a protein "represents a form that appeared in its evolutionary history." These evolutionary steps provide both the platform and the instructions for building the protein's next layer.

There is evidence that genes are built much the same way. Repeating each evolutionary step may be how nature creates complex forms in general, with each level providing the critical framework for properly building the next level.

The 'Code of Evolution' is in Relationships

These discoveries have profound implications for understanding evolution. We already know that self-assembly's unfolding relationships store the information for the evolution of everything in the cosmos – atoms, molecules, minerals, stars, planets, and the universe itself. It appears they also store the information for the evolution of everything in biology.

Look what happens when a cell reproduces. It creates a duplicate copy, not just of genes, but of millions of molecules, each a signal in a vast instruction network. Biologist Michal Locke says these relationships "imply that a pattern...has also been inherited...not just genetic material."[37]

These relationships, and the patterns they form, amount to a hidden "code of evolution." Just as the "plan of life" is stored in self-assembly relations, the "code of evolution" is stored in the same relationships, emerging as they unfold over time. In biology, each stage of evolution serves as the platform and the instructions –"the code"– for creating the next stage of evolution.

This discovery integrates biological evolution with evolution in the other sciences. The same process guides evolution at every level of nature.

It also makes biological evolution predictable, based on the same properties that make evolution in physics, cosmology and chemistry predictable. And just as the evolution of stars and planets has direction, so too the evolution of proteins, genes and life itself has direction.

Chapter Seventeen

Adaptive Evolution

STRESS ALTERS THE BODY, BRAIN AND BEHAVIOR

The genome is a highly sensitive organ that in times of stress initiates its own restructuring and renovation.

Barbara McClintock, Nobel lecture, 1983

In broad strokes we now know *how* evolution works. But we don't know *why* it happens. Why didn't life stay the same? What led the first cells to eventually transform into toadstools and terriers?

Despite the rise of Darwinism, this question remained entirely unresolved. In 1967, Nobel laureate Peter Medawar put it this way: "the main weakness of modern evolutionary theory is its lack of an [explanation for] variation." We don't know *why* things change, he said. "We therefore have no convincing account of evolutionary progress – of the inexplicable tendency of organisms to adopt ever more complicated solutions to the problem of remaining alive."[1]

Today, a big piece of that mystery is solved.

The answer, in a word, is stress.

Nature makes things by forming relationships. Evolution occurs when those relationships change. Stress does that. Stress alters nature's basic relationships, leading to change and adaptive evolution. Here's how it works.

Nature's Inherent 'Plasticity'

Nature creates things by forming stable relationships between parts. That is the heart of self-assembly. Stress alters that basic stability (called equilibrium or homeostasis), leading plants and animals to adjust somewhat, without altering their basic biology.[2]

The ability to change while remaining fundamentally the same is called "plasticity." We see this when we bend a straw, twist a wire, or gain weight. The shape changes, but the essence stays the same.

Everything in nature is "plastic" to some extent. Atoms can absorb energy without fundamentally changing. Stars absorb asteroids and comets, but remain essentially the same.

All living thing are naturally plastic.[3] Trees on a wind swept cliff bend and grow at an angle. Snails reshape their shells to match the flow of water around them.[4] The body's inherent plasticity allowed whales to reshape their legs to act as fins. Even the human brain rewires itself in response to the stress of learning and other experiences.[5]

The idea the body is plastic, and can change shape without altering its core biology, is old. But in the 1980s scientists, like pioneering geneticist Mary Jane West-Eberhard, showed that plasticity is central to explaining adaptive evolution. Plasticity shows that everything in biology is not predetermined by genes. In real life, creatures are always changing.

This "compels us to reexamine the major themes of evolutionary biology in a new light," says West-Eberhard.[6] Most biologists agree, believing the adjustments the body makes in response to stress explain one of nature's oldest mysteries: adaptive evolution.

STRESS CHANGES THE BODY

We usually think of stress as bad – the strain of personal problems, anxiety over money and the like. But stress can also be good. Exercise makes our muscles stronger and our body more resilient. The thrill of a roller-coaster ride even creates stress some people find fun.

Sometimes we stress our bodies without realizing it. The mental effort of doing a crossword puzzle or learning a new skill stresses the brain, causing it to rewire itself. Even changing our diet is a form of stress, forcing the body to adjust to new foods. Shifting from Asian to American or Mexican foods can throw the body into disarray for months.

STRESS CHANGES CAN BE INHERITED

It was old news that stress alters the body. The question was, could these changes be inherited? Darwinists said emphatically "No," declaring it was biologically impossible. And for decades this was accepted as dogma. But in 2000, Duke University geneticist Randy Jirtle did something that startled the scientific world.

Taking a group of mice prone to diabetes, cancer, obesity and short lives, Jirtle put them on a special diet that transformed them into slim, healthy creatures. Jirtle already knew the diet would have this effect. What surprised him was finding that the *next generation* of mice were also slim and healthy even though they had *never gone on the special diet.*[7]

This was remarkable. The changes occurred without altering any DNA, and were inherited in a single generation. "It was a little eerie and a little scary," says Jirtle, "to see how something as subtle as a nutritional change... could have such a dramatic impact."[8]

Similar results were soon uncovered by other labs. Researchers had long known that nutrition boosts the immune system, anxiety depletes the body, and learning rewires the brain. These were all part of the body's natural plasticity. What surprised them was finding that these changes occurred without altering any DNA, and that some were immediately inherited by the next generation.

How was this possible? And what did it mean for evolution?

Lamarck's Direct Inheritance of Changes

The idea that changes in the body can be inherited is Jean-Baptiste Lamarck's legendary concept of evolution through "acquired traits": changes in a parent's biology can be directly "acquired" by the next generation.

First presented in 1809 and later backed by a strong body of evidence, Lamarck's "acquired traits" became the dominant explanation for adaptive evolution throughout the Victorian era and well into the 20th century.

Lamarck's theory "had great breadth of appeal across both national and disciplinary boundaries," says historian Richard Burkhardt.[9] It integrated well with physiology and biology, and explained the rapid inheritance of changes seen by bacteriologists.

In the 1860s, Lamarck's theory was validated when biologist Louis Pasteur discovered that people attacked by a virus became more susceptible to future attacks from the virus, and this increased vulnerability *was inherited by their children.*

By the 1920s, the rapidly developing field of immunology showed that children can also inherit *increased resistance* to a disease from their parents, making immunologists almost uniformly pro-Lamarck and anti-Darwin.[10]

Around the same time, geneticists like Michael Guyer at the University of Wisconsin, showed that even changes in behavior can be passed directly from parent to child, further confirming Lamarck's concept.[11]

Creating Imaginary Barriers to Direct Inheritance

Lamarck's acquired traits also showed that evolutionary change could be rapid and highly targeted, not slow and directionless as Darwin claimed. Of all the ideas Darwinists needed to destroy, Lamarck's theory topped the list. After their 1940s takeover of evolution, they began an aggressive effort to discredit direct inheritance by inventing imaginary barriers to the process.

In the 1880s, biologist August Weismann cut the tails off of five generations of mice without seeing any changes. Although the experiment was later discredited (even Weismann called it flawed), it became a pillar of modern Darwinism. Naming it the "Weismann Barrier," Darwinists held it up as proof that Lamarck's direct inheritance was impossible.

In 1957, Francis Crick updated the Weismann Barrier for the genetic era. The genes in eggs and sperm (the raw material of inheritance) can't be altered by any changes in a parent, said Crick. Thus changes in the parents biology can't be inherited by their young. Calling this the "Central Dogma of Biology," Crick said it conclusively proved Lamarck wrong.

Crick later acknowledge he had no evidence for the Central Dogma. "All we had to work on," he wrote, "were certain fragmentary experimental results, themselves often rather uncertain and confused."[12] In other words, Crick had nothing we would call scientific evidence. Despite this, the "Central Dogma" soon entered textbooks as an established fact.

The Weismann Barrier and Crick's Central Dogma were later used to discredit anyone who sought to study direct inheritance. Lamarck's ideas faded in the 1950s, and then vanished entirely from science. Or so it seemed.

Secretly, Lamarck's theory lived. Harvard's Stephen Jay Gould said that for decades the "open secret" among paleontologists was the *lack of any evidence* for Darwin's gradual change, and the abundance of evidence for Lamarck's rapid evolution.

Darwinists denied this evidence, and attempted to browbeat paleontologists into submission. "Paleontologists have paid an exorbitant price for Darwin's argument," wrote Gould:

> "We fancy ourselves...students of life's history, yet to preserve [Darwin's] account of evolution by natural selection we view our data as so bad that we almost never see the very process we profess to study."[13]

Geneticists also discovered widespread evidence of direct inheritance. Fearing attack they describe their findings in other terms, but privately they acknowledged it was Lamarck's evolution in vivid action.

REMOVING THE IMAGINARY BARRIERS TO DIRECT INHERITANCE

While geneticists could see direct inheritance, they couldn't explain how it works. This changed in 1970 when Howard Termin and Richard Baltimore found that RNA can change DNA in a way that is directly inherited (a process called "reverse transcription"), garnering them a Nobel Prize.[14]

This discovery proved the Weismann Barrier and Central Dogma were wrong. "CENTRAL DOGMA REVERSED" ran the headline in *Nature* that year.[15] Although never more than a figment of Crick's imagination, the Central Dogma was so well accepted that many scientists were shocked when it was shown to be untrue.

Today, hundreds of studies show that changes in a parent's biology can be directly inherited.[16] The process discovered by Termin and Baltimore is one way. (This is also how the HIV virus inserts its DNA into humans, causing AIDS.)

Another way is by altering the "on-off" switches that surround genes (a process called methylation).[17] Diet, exercise and other activities can flip these switches, altering how genes behave and the job they do. This is what transformed Randy Jirtle's mice.

In another process, mobile genes carry changes directly into the DNA of the eggs and sperm used in inheritance.[16] In 2008, an international team led by Antonin Bukovsky at the University of Tennessee Medical School found that, contrary to conventional thinking, women regularly make new eggs, which can incorporate new genetic changes.[18]

MIT biologist Susan Lindquist discovered that a special type of protein (called a "prion") can also carry changes from one generation to the next.[19]

EPIGENETICS: AN INHERITANCE SYSTEM ABOVE GENES

These discoveries revolutionized our ideas about inheritance, showing genes are only one part of a much larger process called "epigenetics." The *BBC*, in *The Ghost In Your Genes* (2005), said that today "biology stands on the brink of a shift in the understanding of inheritance. The discovery of epigenetics, a whole new frontier in biology."[17]

'Epi' means 'above.' Epigenetics is a system of inheritance which sits above genes, controlling what they do. It's "a revolutionary new science," says Adrian Bird at Edinburgh University, "an antidote to the idea that we are hard-wired by our genes."[21] It's as though genes "learn" says Bird, and this "learning" is then passed to the next generation.

University of Michigan geneticist David Nanney coined the term "epigenetic" in 1958 to describe all the inheritance events not explained by conventional genetics.[22] But the idea was rejected by Darwinists and lay dormant for decades. "It was controversial in 1958," says Nanny. "It took 40 years for epigenetics to become a major word in the vocabulary of modern biology."[23]

Today it is a very major word. Over 2,500 research studies and dozens of conferences are conducted every year on epigenetics.

A pioneer in these discoveries, geneticist Eva Jablonka showed that epigenetic processes allow the body's 'plastic' changes to pass from one generation to the next, as Lamarck said. This has profound implications. "The whole discourse about heredity and evolution will change," said Jablonka.[24]

Epigenetics shows that the direct inheritance of adaptive changes is an everyday occurrence. "Lamarckian evolution is a fact in plants," says Geneticist Hugo Hoenigsberg, from microscopic fungus to the most complex flowers.[25] Animals of every type – worms, mice, humans – also show it.[26]

Direct inheritance is seen in every type of trait, from the simple (adjusting to new levels of sea salt), to the complex (changing the shape of bones), to behavior (altering how a mother treats her young), to the purely mental (reading rewires the brain).[27]

Two things are important about epigenetics. First, it shows changes in the body, brain and even behavior can be inherited in a single generation.

Second, it shows that many of these changes come from everyday experiences. What we eat, drink, see and hear, how we behave, even what we think and feel, can create changes that are passed on to future generations.

Not long ago these ideas would have been called heretical. But no longer. From maize to mice to men, Lamarck's epigenetic inheritance is found everywhere.[28] "A new world is opening up," says Canadian geneticist Moshe Szyf, "one that is so much more complex than the genomic world."[29]

Direct inheritance of changes explains how the immune system stays robust, and why humans lost their ability to make some essential chemicals during evolution (see below). Direct inheritance also holds the potential for medical breakthroughs, revealing how viruses evolve, cancers spread, and how we might harness the body's own immune system to fight both (see Chapter 22 for the full implications).

Direct inheritance even alters our core ideas about the cause of disease, and the importance of our experiences to health and wellness. The life our parents and grandparents lived – what they ate, breathed, thought and felt – influenced the body and brain we inherited. Our behavior, in turn, effects what our children and grandchildren will inherit.

Adaptive Evolution: Rapid, Targeted and Helpful

The rediscovery of Lamarck's "acquired traits" did more than revolutionize our ideas about inheritance. It transformed our understanding of adaptive evolution.

In a famous 1988 experiment, Harvard geneticist John Cairns showed that when bacteria are stressed (by lack of food or changing temperatures), they begin to duplicate the genes related to the stress. They engineer the duplicate genes with slight changes that help counter the stress. This confirmed Susumu Ohno's idea that the duplication and specialization of genes was central to evolution.

Cairns said this could also be the "mechanism for [Lamarck's] inheritance of acquired characteristics," allowing life "to evolve new capabilities."[30] This caused an uproar at the time, but Cairns' results were later confirmed. Barry Hall, director of the Bellingham Research Institute, for example, showed such changes: [30]

- Occur primarily to genes related to the stress,
- Are mostly adaptive and helpful,
- Occur 100 *million* times faster than possible with random mutations.

The idea that these changes are random, said Hall, represents "the improbable stacked on top of the highly unlikely."[31]

A decade of additional research showed that the changes produced by stress were "almost exclusively beneficial."[32] A wide range of other evidence showed that "*most* of these evolutionary changes are probably adaptive."[33]

Life, it seems, can create exactly the changes it needs to adapt. University of Chicago geneticist James Shapiro calls this ability "natural genetic engineering." "The capability of cells has gone far beyond what we had imagined," he says, with "evolutionary capabilities unimaginable in the first six decades of the twentieth century."[34]

Life appears able to engineer helpful adaptations to even the most extreme events. In April 1986, at the Chernobyl power plant in Ukraine, a nuclear reactor exploded, spewing radioactive debris onto the soil for miles.

Despite this, nearby plants and flowers adapted, not just surviving but flourishing in the contaminated soil. "If you visit the area, you'd never think anything bad had happened there," said Martin Hajduch, one of a group of geneticists who studied the area.[35] "Somehow plants were able to adapt to the radioactivity," said Hajduch, and "we wanted to understand what kind of molecule changes were going on." They concluded that the plants had rapidly evolved a sort of "protective shield" against radioactivity.[36]

Adaptive Evolution: The Body's Natural Plasticity

The body's natural plasticity leads to a surprisingly wide range of adaptive changes, allowing animals to inhabit any environment while remaining basically the same. This made it a powerful force "in driving evolution."[37]

Spiders, for example, evolved the ability to spin silk 400 million years ago, and the basic process changed little afterward. What *did* change were the *different uses* spiders developed for silk, allowing these formerly shy creatures living deep in burrows, to leave these hideaways and master every ecology on Earth.[38]

The small lemurs that live on the islands off Africa made similar adaptive changes, without basically changing. On the Comoro Islands, lemurs are "active at night, stay in trees, and feed by hanging upside down from their hind feet to suck nectar out of flowers."[39] On Madagascar, they are active during the day, eat leaves and travel on the ground. "This kind of shift is quite common," says noted anthropologist Jeffrey Schwartz. "These lemurs have the same teeth, the same feet, the same eyes, but if the environment changes, they change their activity and their diet, not their anatomy."[39]

And don't forget those penguins, living on both hot tropical isles and frozen arctic ice while remaining the same.

Adaptive Evolution: 'Plastic' Changes are Easily Reversed

Because the body is plastic, adaptive changes are easily reversed. This allows animals to switch back and forth between traits as needed, and accounts for the many zig-zags in evolution that are impossible to explain with Darwinism.

Beetles, for instance, emerged 300 million years ago and today make up a quarter of all animals. Their extraordinary success is explained, not by random gene changes, but by their ability to shift back and forth between a set of traits which allowed them to live on land or water. Beetles made this switch at least 10 times over the course of their 300 million years.[40]

Another example of reversible changes are Darwin's Galapagos Island finches. They were long held up as a classic example of Darwinian evolution, but like other "classic" examples it fell apart on close inspection.

Peter and Rosemary Grant, who studied the birds for decades, showed that as the amount of rain varied, the seeds on each island changed – more rain, bigger seeds, less rain, smaller seeds. The finch beaks then changed to match the size of the seeds.[41]

Darwinists said these changes came from random gene mutations. Natural selection then determined which new genes and beaks survived, eliminating birds with the less adaptive genes and beaks.[42] In 1999, Ernst Mayr summed up the Darwinian explanation this way:

> "The development of the heavy [finch] bill was a slow process, probably involving dozens of small mutational steps, each one surviving only if proving its usefulness in the actual test of selection. It must be admitted, however, that it is a considerable strain on one's credulity to assume that finely balanced systems [can] be improved by random mutations."

The idea these changes came from the random appearance of new genes does more than "strain one's credulity," it is simply wrong. Beaks always use *the same genes.*

Beaks change when bones growing from the head change their shape.[43] These changes are guided by many genes, none of which are unique to finches, or even birds. All animals, from agama lizards to zebra fish, use the same genes to make their jaws.[43] The difference in the size and shape of each jaw (or beak) depends on when and where these shared genes are active.[44]

Beak changes also don't involve Darwin's slow, gradual natural selection. The finches, it was discovered, switched between small and large beaks many times.[45] Sometimes they fully reversed size in just a few years, far too quickly to be the result of natural selection.

Rather than Darwinism, this rapid reversibility is an example of adaptive evolution through Lamarck's "acquired traits." As the seeds become larger the finches used more force to crack them, causing their beaks to get bigger (just as any bone gets bigger when stressed). Smaller seeds required less stress, leading to smaller beaks. These differences in size alter when and where genes are active. Those change are then directly inherited by the next generation through epigenetics. (For the details of how this works, see "Defining and Proving Darwinian Evolution" at the back of the book.)

Adaptive Evolution: The 'Behavior First' idea of Evolution

Connecting stress with changes in plasticity and the direct inheritance of changes radically rewrites our concept of evolution.

First, it reverses the focus from external to internal processes. These adaptive changes look nothing like the evolution Darwin envisioned, with life passively dying off from the harsh force of natural selection. Instead, life *actively adapts* to challenges. "We need to develop a new [model of change] based on a view of the cell as an active entity," says geneticist James Shapiro. "The emphasis has to be on what the cell does to its genes, not what the genes direct the cell to execute."[34]

The idea that the cell's behavior determines what genes do, and how they change, is part of a new concept called the "Behavior First" model of evolution. It was originally developed by anthropologists who faced a problem. Darwinism says random mutations alter genes, which then change the body's biology, and this later leads to new behavior. But the evidence in human evolution shows the opposite sequence. Consistently, new behavior emerged *before changes in the body* appeared. And those changes always came *before changes in genes*.[46]

To explain this, leading anthropologists reversed the old Darwinian paradigm and developed the Behavior First model of human evolution.[47] New behavior changed the body, and then altered our genes. Although initially controversial, the Behavior First model is now well accepted because, like all good theories, it matches the evidence.

It also resolves many old mysteries about human evolution. Take Vitamin C, which is essential to survival. We humans don't make it. "We should be able to make it since we have all the genes," said French geneticist Rebecca Stevens.[48] But we don't. How could we lose the ability to make something so crucial to life?

The Behavior First idea of evolution explained this. It appears that in the ancient past our ancestors ate so much fruit (packed with Vitamin C) that the genes for making it "turned-off" from lack of use. Many animals that made fruit a staple of their diet – bats, guinea pigs, certain birds – also lost the ability to make Vitamin C. "If you're surrounded by a vitamin for a long period of time, then you don't need to use the gene," says Cambridge geneticists Katherine Helliwell.[48] Lack of use leads to a loss, just as Lamarck said.

The Behavior First concept of evolution also meshes with new discoveries which show changes in biology *precede* gene changes, not just in human evolution, but throughout nature.[49] Evolutionary biologist Mary Jane West-Eberhard summed up these discoveries saying, "I consider genes followers, not leaders, in adaptive evolution."[50]

This means adaptive evolution is the opposite of the Darwinian idea change originates with random gene changes. But it matches the process Lamarck described. Some adaptive changes are driven by new behavior, others by changes in the environment (the stress of cold, for example, can lead animals to grow more hair). But in both cases, the body adaptively responds by altering its biology, which then leads to genetic change.

ADAPTIVE EVOLUTION: LIBERATION FROM THE ENVIRONMENT

The second big change in our understanding of adaptive evolution is that plasticity and direct inheritance *free life from the effects of the environment.* This is geneticist Ivan Schmalhausen's famous idea. The opposite of what Darwin envisioned, Schmalhausen saw evolution as a *progressive liberation* from the environment.[51] He said evolution is emergence of adaptive traits which counter external stress and maintain the body's internal balance.

The ability to counter the environment has existed from the beginning of life. For example, "bacteria are 'smart' about how they evolve," says Baylor College of Medicine geneticist Susan Rosenberg.[52] In response to bad conditions they activate internal processes which help them, creating for example, "cold shock genes" which make protective proteins when the temperature drops, and "heat shock genes" that do the same when the temperature gets too high.[53]

Plants also actively counter the environment. "Plants constantly monitor their light environment in order to grow and develop optimally," says Cornell geneticist Haiyang Wang.[54] "Plants can 'see' their surroundings using a battery of photo-receptors to determine the color, intensity, direction, and duration of light around them," and then they adjust to maintain internal balance.[55]

As life grew more complex, it became increasingly insulated from the environment. Many trees, for example, have an internal 'thermostat' that maintains their leaves at a constant 70° temperature. Animals use fur, fat, shells, and other adaptations to offset outside changes. Sometimes animals even create their own "mini-ecosystems" – burrows, nests and hives – that entirely shelter them from the environment.

Contrary to what is commonly thought, throughout evolution life has changed the environment more than the environment has changed life.

When the Earth initially formed, for example, it was enveloped in a thick cloud of carbon-dioxide. The first cells, in making energy, absorbed the carbon-dioxide, broke it up and spit out the oxygen. This transformed the Earth into an oxygen-rich world, allowing life to move onto land. These cells were "responsible for the oxygen in the atmosphere...that made plants grow and let us breathe air and allowed us to walk on the Earth," says UCLA biochemist James Lake.[56]

As plants became more complex, they produced new minerals, which further altered the landscape, and provided the raw materials for even more complex plants, which in turn created hundreds of new types of minerals.

These new minerals allowed complex animals to emerge, which further changed the environment. "Animals have engineered the biosphere over evolutionary time," explains Cambridge geologist Nicholas Butterfield.[57] "Animals don't just occupy the modern biosphere, they permeate its structure and define how it works."

Adaptive Evolution: The Immune System

The idea of adaptive evolution as a "liberation" from outside changes is nowhere better seen than in the immune system, that remarkable invention which destroys invading infections that threaten internal stability.

As we know, self-assembly has its own built-in self-repair mechanism. Biologist Gerald Edelman won the Nobel Prize for showing that the immune system is a variation on this basic self-repair process.

Plants, animals and even bacteria all have what is called an "innate" immune system based on the self-assembly process. Viruses and other invaders send off a signal that doesn't fit in with body's other processes, making them appear like a damaged part, which the body disposes of. (This response also explains why the body rejects transplanted organs, viewing them as "invaders.")

More complex animals have a second, "adaptive" immune system, which can craft a unique identifier and defense against invaders, and then remember both. This ability to "remember" both the invader and the defense allows the adaptive immune system to effectively fight a huge array of deadly parasites, bacteria and viruses.

Today we know this ability emerged from the evolutionary union of McClintock's mobile genes and Lamarck's acquired traits.

In 1998, Evan Eichler at Case Western Reserve showed the genome is filled with thousands of DNA fragments which can be "cut out" and re-combined.[58] That same year, geneticists David Schatz and Martin Gellert showed how these "cut out" fragments are used to fight disease.[59]

What happens is that the immune system instructs mobile genes to carry the bacteria or virus to a special area, where it finds gene fragments that match the invader's shape.[60] It then assembles the matching fragments into what is called an "antigen"– an infection fighter. Because the antigen matches the shape of the invader, it can attach itself to the invader and use biology's "lock and key" process to break it into harmless pieces.

Once the immune system has found an effective antigen, it begins to rapidly duplicate it, eventually flooding the body with these infection fighters. This explains why a fever appears when we're fighting an infection (the body is working hard to make antigens), and why the fever builds gradually, hits a peak and then abruptly "breaks," as the antigens finally win the battle.

Afterwards, the immune system retains a copy of both the antigen *and* the invader, allowing it to "remember" the invader and rapidly respond to future infections. The response is often so fast that the person is said to be "immune" to the disease.

A wide range of research shows that some new antibodies are inherited by the next generation, revealing what biologists like Hans Lange call "a Lamarckian dimension of the immune system."[61]

The ability of mobile genes to recombine pieces of DNA accounts for the extraordinary capabilities of the "adaptive" immune system. "No other genes in mammals are split up like this and then recombined at the DNA level," says David Schatz.[62]

Using a few simple parts, the immune system can produce an unlimited number of antigens, each tailored to match and breakdown a specific invader. From this perspective, the immune system is an infinitely adaptive repair mechanism that evolved from nature's basic self-assembly processes.

ADAPTIVE EVOLUTION VIA MOBILE GENES AND DIRECT INHERITANCE

The ability of genes to recombine revealed that they, too, have an adaptive plasticity. "The genome is now known to have a flexibility referred to as plasticity," says Pierre Capy at France's Genetics and Evolution Laboratory. "The genome is like a puzzle made of parts that can move from one position to another and, through exchange, deletion, insertion or amplification, generate new combinations...with different functions."[63]

This makes McClintock's mobile genes central to creating and inheriting many adaptive changes.[64] "I see more and more people looking at [evolution] this way," says Nobel laureate Werner Arber. "Whether by radically rearranging themselves, making use of mobile elements to generate variation, or causing stretches of DNA to mutate at high rates, genomes are showing that they can help themselves cope with a changing environment."[65]

While adaptive evolution has many complex dimensions, at its essence the process is simple. In response to stress (especially from new behavior) creatures adaptively alter their body and biology, which leads to genetic changes, that are inherited by later generations. That's how evolution works.

This means that living things are not as Darwin envisioned, passively waiting for natural selection to determine their fate. On the contrary, they create exactly the changes they need to cope. They *actively adapt,* guiding and molding their own evolutionary future.

Chapter Eighteen

The Origin of Species

NEW RELATIONSHIPS CREATE NEW FORMS

The origin of species occurred in stressed environments.[1]
Roger Lewin, 1983

Acquired traits and epigenetics explained many evolution mysteries. But they didn't answer the big question: what causes new species to appear?

No amount of adaptive change leads to a new species. Plants alter the shape of their leaves, animals change in size and shape. Yet their core traits remain unyieldingly the same. Trees stay trees, they don't slowly evolve into flowers. Insects don't slowly evolve into mammals.

Whales adapted to life in the sea. Yet, after millions of years in the water, they didn't evolve into fish, but remained mammals – breathing with lungs (never developing gills), giving birth to live young (never laying eggs), and nursing their young with milk (never abandoning them to feed themselves). And while their legs were reshaped to work like fins, the bones inside remained the same as any animal walking on land.

The big question was not how did trees change their leaves, or whales reshape their bones, but how did trees and whales come to exist in the first place?

The short answer, again, is stress.

Nature makes things by forming relationships. Stress alters those relations. In adaptive evolution, stress slightly *modifies* them. However, on a few occasions, stress *completely rearranges* the relationships. It rewires all the connections. This creates *new types* of things. At the simplest level, the stress of gravity forces particles to rearrange themselves, creating new types of atoms. At a higher level, stress forces plants and animals to rearrange their biology, creating new types of life – new species. Here's how it works.

Darwin's Slow, Divergent Evolution

If Darwin's most contentious claim was that evolution is directionless, his second most disputed idea involves the speed of evolution.

Central to Darwin's theory was the claim that new species emerged from random changes. The only way complex life could develop from random changes was through the slow accumulation of many tiny modifications. Thus, it was crucial to Darwin that evolution be slow and gradual.

But all the evidence contradicted Darwin. Life has existed for four billion years, blanketing every corner of the planet. Yet in all that time, and across all those varied landscapes, the evidence showed new species appeared only a few times, and each time was abrupt and rapid, emerging with forms astonishingly different from anything prior.

Worse for Darwin, when new species appeared they didn't begin in some simple form and become complex over time. Instead, they were fully complex *right from the start.*

Darwin said this picture of evolution was wrong. Although the evidence was all against him, Darwin said future discoveries would vindicate his claim that new species had evolved slowly and gradually.

Nature's Rapid, Convergent Evolution

In the 100 years after *The Origin of Species*, the evidence was greatly expanded, and the picture it presented was unequivocal: new species appear in short, abrupt bursts. Darwin had this part of evolution exactly backwards.

In 1909, the world's greatest trove of fossils was discovered in the Canadian Rockies, near Burgess Shale. In the decades that followed, tens of thousands of magnificently preserved specimens filled out the history of life in great detail. Harry Whittington, unofficial dean of the digs at Burgess Shale, and the other fossil hunters saw a clear pattern. New types of life emerged abruptly, in short, sharp bursts. This was especially true of the Cambrian Explosion, 540 million years ago, when modern plants and animals first appeared.[2]

The findings confirmed what had been known for a hundred years, and cemented the failure of Darwin's claim that evolution is slow and gradual.

However, a second, unexpected discovery rocked fossil hunters. During these bursts of change, nearly *identical species* arose multiple times. Similar "animals arose independently in different areas," wrote Harry Whittington.[3] (Modern genetics now confirms this.[4])

The implications were huge. It meant new species don't emerge in isolated branches, each individually honed by natural selection. Whittington, reflecting a widely held view, wrote that because of this "I look skeptically upon [Darwin's] branching diversity...based on a single kind of animal."[3]

Paleontologists also realized that if similar species emerge multiple times, without sharing a common ancestor, it meant evolution was pushing life along a narrow pathway. Evolution has direction.

Although the evidence eventually became overwhelming, the Darwinists continued to deny it, voting at a 1959 conference to "reject the idea that the fossil record shows signs of directed evolution."[5]

No Intermediate Species Exist

In another blow to Darwin, no "intermediate" species were found at Burgess Shale, or anywhere else. Darwinists continued to hold up a few creatures as intermediate forms, but later genetic analysis revealed they are not.[6] Ocean-dwelling flatworms, for example, long thought to be the bridge between primitive and complex animals, were found to have the same genes as complex animals.[7] "It was nice to have this intermediate group," said British geneticist Max Telford, "now we've lost that."[8]

Other intermediate species disappeared, leaving the gaps between each level of evolution greater than ever. Stephen Jay Gould wrote that "the absence of... intermediate stages between major transitions, remains a persistent and nagging problem."[9]

In each evolutionary leap, the differences between old and new species were so great that paleontologists found it impossible to even *imagine* what an intermediate form might look like. "All of a sudden, you realize that you haven't got the faintest idea [how] to sketch a... common ancestor," says University of Bristol paleobiologist Philip Donoghue.[10]

Looking at the evidence, Darwinist Ernst Mayr despaired. "Intermediates between species are not observed," he wrote, instead "all species are separated from each other by bridgeless gaps."[11]

These huge gaps were called "evolution's deepest paradox." They were certainly impossible to explain with Darwin's ideas. "A large number of scientists...have unfortunately gotten the idea that the fossil record is far more Darwinian than it is," says University of Chicago geologist David Raup, famous for his work charting evolutions largest cycles.[12]

"There is probably some wishful thinking involved," said Raup. "In the years after Darwin, his advocates hoped to find predictable progressions. In general, these have not been found – yet the optimism has died hard and some pure fantasy has crept into textbooks."[12]

By the 1970s, the "pure fantasy" of Darwin's gradual evolution of new species reached a breaking point. Two of the world's leading paleontologists, Niles Eldredge and Ian Tattersall, wrote that "the fossil record flatly fails to substantiate... finely graded change."[13] New species appear "with a bang,...often with no evidence of prior evolution," they said. For decades the Darwinists claimed gradual evolution had occurred "somewhere not yet discovered." But an exasperated Eldredge and Tattersall said "evolution cannot *forever* be going on somewhere else. The record jumps, and all the evidence shows that the record is real: the gaps we see reflect real events in life's history."[14]

A New Theory: Punctuated Change

In 1970, Eldredge teamed up with Stephen Jay Gould in an effort to explain the "real events in life's history" seen by paleontologists.

Building on the work of Richard Goldschmidt (which showed new species emerge in "decisive steps...without transition"), Gould and Eldredge said that all species display the same two features.[15]

First, they appear abruptly. "A species does not arise gradually by the steady transformation of its ancestors; it appears all at once and fully formed." Second, species change little afterwards. Despite many adaptive changes, most animals look the same at extinction as they did when they first appeared millions of years earlier.[16]

When this model was applied to the origin of species going back a billion years, it matched the evidence perfectly. Gould and Eldredge formalized these ideas in a famous 1972 paper which dubbed the process "Punctuated Equilibrium."

Punctuated Equilibrium was well received. For paleontologists it perfectly matched the evidence they saw, and it was widely viewed as an important advance in understanding evolution.

Darwinists however, saw it as a threat to their "directionless" change. They launched a vicious attack on Gould and Eldredge that went on for decades – at conferences, in journals and the popular press, and against them personally. John Maynard Smith completely dismissed the idea of Punctuated Equilibrium, and ridiculed Gould as "a man whose ideas are so confused as to be hardly worth bothering with."[17]

Repeating their familiar pattern, in addition to the attacks, Darwinists created new imaginary explanations for the facts. To explain the long periods without change Sewell Wright said animals hit "adaptive peaks" and could no longer evolve. John Haldane said evolution had "speed limits." These ideas were always proven with "statistics." Precisely how they worked at the level of real biology, no one could say.

When these imaginary processes proved empty, Darwinists shifted to their third tactic, calling the evidence irrelevant. Mark Ridley, for example, made the astonishing argument that fossils have "never been part of the evidence for evolution":

> "A lot of people just do not know what evidence the theory of evolution stands upon. They think the main evidence is... in the fossil record... However the gradual change of fossil species has never been part of the evidence for evolution."[18]

In true Darwinist form, Ridley sought to distract people from the evidence against Darwinism by dragging "creationism" into the discussion. "In any case," he continued, "no real evolutionist...uses the fossil record as evidence in favour of the theory of evolution as opposed to special creation."

However the issue was not creationism versus evolution, but Darwinism versus the facts: new species appear rapidly, not gradually.

For his part, Gould understood that ideology, not science, drove the attacks on him and Punctuated Equilibrium:

> "The field of paleontology has strongly challenged the Darwinian premise that life's major transformations can be explained by... tiny changes produced generation after generation by natural selection. Why should Darwinian fundamentalism be expressing itself so stridently when most evolutionary biologists have become more pluralistic in the light of these new discoveries? I am no psychologist, but I suppose that the devotees of any... cult must dig in when a general threat arises."[19]

Adaptive Changes Don't Create New Species

Despite the attacks, Punctuated Equilibrium was excellent science. It integrated the evidence from different areas of evolution under a single process.

David Jablonski, head of Evolutionary Biology at the University of Chicago, says it represents "a core of solid analyses that are very convincing, with an impressive array of examples...from snails to horses," all of which evolved in abrupt jumps.[20]

Well-known dinosaur hunter Jack Horner discovered that even Tyrannosaurus Rex, the terror of *Jurassic Park,* evolved in three abrupt jumps, each followed by millions of years with little change.[21]

Looking at major evolutionary events, like the emergence of mammals, Oxford University's Curator of Zoology, Tom Kemp, says "the concept of punctuated equilibria, evolutionary change occurring in a series of jumps, with relatively little change in between...accounts for this rather well."[22]

Punctuated Equilibrium was discovered to be widespread in nature, seen in the evolution of lakes, forests, deserts, oceans, continents and the Earth itself, all of which go through abrupt transitions.

It was later found that this is how *most* major changes occur in nature, starting with the emergence of the universe in the "Big Bang," through the evolution of atoms, stars and everything else – even human evolution. The first human stone tools appeared abruptly, as did the first symbols and art, the first cities 5000 years ago, and the abrupt rise of the Industrial Revolution 250 years ago. Albert-László Barabási, an authority on such transitions, says most of human history is "characterized by bursts of rapidly occurring events separated by long periods of inactivity."[23]

At the World Academy of Arts and Science meeting in 1999, the group's president, anthropologist Jeffrey Schwartz, summed up the evidence:

> "Most major groups of organisms...appear in the fossil record as Athena did from the head of Zeus, full-blown and raring to go, in contradiction to Darwin's depiction of evolution as resulting from the gradual accumulation of countless infinitesimally minute variations."[24]

The evidence was now definitive. The Darwinists were simply wrong. No future discoveries would rescue their claim that new species emerge slowly and gradually. They had this aspect of evolution *exactly backwards.*

ARE THERE TWO TYPES OF EVOLUTION?

So how do new species evolve?

Going back to the Victorians, naturalists believed that two different evolution processes must exist. One for small adaptive change (explaining how a whale leg is reshaped to function like a fin), and a second for the emergence of new species (explaining how whales came to exist in the first place).

Modern paleontologists made the same argument: there is one process for adaptive changes (micro-evolution) and a separate process that creates new species (called macro-evolution). In 2000, Douglas Irwin at the Smithsonian Natural History Museum, put it this way:

> "One of the oldest problems in evolution... remains largely unsolved. Historically, the neo-Darwinists... stressed the predominance of micro-mutations in evolution. But that doesn't explain new species. Macro-evolution is more than repeated rounds of micro-evolution."[25]

But if small adaptive changes don't lead to new species, what does? The answer to this mystery began with two important Victorian discoveries.

Decades before Darwin, the great French biologist Georges Cuvier, said "natural history has a rational principle," which he called the "correlation of parts." Animals are not a loose collection of parts, said Cuvier, but a well-integrated whole. "The different parts of each being must be coordinated in such a way as to make possible the total being."[26] Each part must balance with everything else.

The Darwinists dismissed Cuvier's insight, but modern science shows he was on to something important.

So was Richard Owen, who argued that nature has basic "organizing forces" and "geometric laws."[27] Owen said these laws determine the form and structure of animals, and how they will evolve.

Although on opposite sides of some issues, Cuvier and Owen agreed that all the elements of biology are tightly integrated, and thus must perfectly balance with each other for life to exist. This insight emerged again 150 years later, in one of the most important new discoveries about evolution: Danish scientist Per Bak's discovery of *how* and *why* things in nature abruptly change from one form to something completely different.

Bak's Big Breakthrough: The "Goldilocks Effect"

For years, Bak studied everything from galaxies to clouds, ant hills, avalanches and earthquakes to understand how and why they change.

In 1987, while working with sand piles, he and his colleagues Chao Tang and Kurt Wiesenfeld found that adding more and more sand didn't change things for a long time. The pile just got bigger. But at a critical point, adding just a single grain – one microscopic bit of sand – led to a huge avalanche.

They asked why the pile didn't change gradually, as more sand was added? And why did a single tiny grain suddenly trigger a full-scale collapse?

What Bak discovered is that stress had been building-up in the sand pile all along, but the pile had managed to absorb the stress without changing.

But at some point, adding a single grain created more stress than the pile could handle. At that moment the finely balanced connections that hold the grains together instantly broke, creating a chain reaction. When one connection breaks it creates too much stress on the next, which breaks, and so on, creating the avalanche.

As they fall, the grains reorganize themselves, seeking a new, more stable shape. Once the new shape is found the avalanche stops, and the pile can again absorb more sand (and more stress) without changing.

Bak knew that everything in nature seeks stability. He grasped that when too much stress upsets this balance, things reorganize their parts to create a new form that is stable. He called this process "self-organizing equilibrium."[28] Everything in nature seeks stability (equilibrium). When stress becomes too great, the parts rearrange themselves to create a more stable form.

Physicists were already familiar with "self-organizing equilibrium," seen in nature's most basic processes. For example, the increasing stress of gravity causes the bonds between the parts of a hydrogen atom to break. The parts then form a new, more stable set of relations, creating a new type of atom (helium). More stress later leads those bonds to break, forcing the parts to form new relationships and a new type of atom.

These changes don't occur gradually. Like Baks' avalanche, the parts instantly leap from one type of atom to another (the famous "quantum leap").

This process is also how liquids turn into gases, and how elements like iron become magnetic, not gradually, but in sudden, instantaneous shifts.

As scientists began to look, they found the same process everywhere. "Bak's idea of a 'self-organizing state of equilibrium' spread rapidly through the scientific community" says Mogens Jensen at Denmark's Niels Bohr Institute. It became "an established part of many scientific disciplines and remains one of the most cited works in" the history of science.[29]

Bak's Breakthrough Shows up Everywhere

In self-organizing equilibrium Bak had discovered two universal truths about nature – *why* things change from one state to another, and *how* they make those changes.

Why things change: Everything in nature, from atoms to eco-systems, seeks balance. When stress becomes too great, pushing relationships out of balance, new, more stable connections must be found. Think of this as the

"Goldilocks effect"– nature wants things to be not too loose, not too tight, but balanced "just right."

How they change: Change is abrupt and rapid. It has to be. When parts are out of balance a new form must quickly be found, otherwise the parts collapse into chaos.

Scientists discovered this is how major changes occur at every level of nature. The universe, for example, began in the "Big Bang," an explosive event in which a microscopic bit of matter began expanding at an unimaginable speed, covering billions of miles in mere seconds, changing rapidly as it went. Then things abruptly calmed and the universe changed little for a billion years.

Later, in another explosive burst, the first galaxies appeared. They too then changed little for billions of years. Stars act much the same way, forming quickly, burning steadily for millions of years, then abruptly exploding and collapsing into a cinder over a period of a few days.

For planets, like the Earth, geological change shows a similar pattern, with a buildup of stress leading to explosive volcanic eruptions and violent earthquakes.

Climate follows a similar cycle. Over the last 500 million years, a series of abrupt changes occurred, especially sudden coolings which led to the emergence of ice ages (sometimes in just a few decades). These ice ages last for millions of years, followed by periods of rapid warming. Drought, floods and other weather changes follow the same patterns of long term stability, followed by abrupt changes, often affecting the entire globe.[30]

The question was, did a similar process apply to life?

In the 1990s, Spanish scientists Ricard Solé and Susanne Manrubia applied the idea of "self-organizing equilibrium" to a tropical rain forest and found it explained the balance of the life there. They concluded the rain forest had *evolved* through a process in which the elements constantly reorganize themselves into a state of harmony with each other.[31]

Solé later teamed with Mexican scientist Octavio Miramontes to look at ant colonies.[32] They found that the number of ants in a colony increases until they achieve an optimal level of information exchange. At that point, the colony abruptly stops growing. If for some reason, the number of ants goes above or below the optimal level, the ants adjust the colony boundaries in or out to reestablish just the right density for perfectly balanced interaction.

Other research showed that "self-organizing equilibrium" applied throughout biology.

Self-organizing equilibrium eventually explained how most complexity arises in nature.[33] The breadth of its effect was astonishing, seen in fields as diverse as physics, chemistry, cosmology, geology, biology, ecology, neurology, economics, sociology and all types of evolution.[34]

"Bak was delighted his theory could be applied to virtually every branch of science," says Mogens Jensen, "and that it stands as one of the most successful interdisciplinary contributions to science in recent decades."

Four Predictions about Evolution and the Origin of Species

Self-organizing equilibrium, which became central to modern Complexity Theory and the Complexity Theory of Evolution, makes four predictions about the origin of new species.

First, it says big changes will be rare. Stable forms resist change. Like trying to separate two magnets, only extreme force can pry stable relationships apart.

Second, when major changes occur they will be abrupt, fast and far reaching. They will look much like one of Bak's landslides.

In an Alpine avalanche, for example, snow can fall on a mountain for months. Tension builds but little changes. Then a single snowflake hits, and stress passes a tipping point. The relationships between all the snowflakes on the mountain suddenly break. A landslide begins. Billions of snowflakes frantically race to find a new "equilibrium."

In the rush and crush of change, the avalanche reshapes everything in its path. The landscape ends looking entirely different from its start. There is little to connect the new landscape with the old, even though one emerged from the other.

In life, if stress becomes more than a creature can manage, all the old connections break. A biological avalanche races through the body, desperately seeking a new set of relationships – a new pattern – that will prove stable. The process goes from start to finish in mere moments (in geologic time), and ends with a biological landscape that looks so different from its start that there is little to connect new species to old, even though one emerged from the other.

This explains why Darwin's intermediate life forms (his "missing links") are never found. When species change they travel so fast and far that no intermediate states exist. Bak's self-organizing equilibrium makes scientific sense of their absence, and eliminates this as a "problem" for evolution.

(The rapid change in Bak's "self-organizing equilibrium" is similar to Alan Turing's idea of "pattern transformation shifts," which explained how nature abruptly moves from one state to a second state, that looks and acts nothing like the first - how billowy clouds can suddenly become a lashing rain storm or a glittering cascade of snowflakes.)

Third, self-organizing equilibrium predicts that large scale change leads inexorably to death or greater complexity.

The body is a web of well-balanced parts. Imagine the mobiles that people put above a baby's crib. The body is like that mobile, only with *millions* of parts and processes held in precise balance – the heart must pump just the right amount of blood, which must carry just the right nutrients to millions of spots, then return to pick up just the right amount of oxygen, and so on. Millions of finely balanced events occur every second.

When stress forces the body to rearrange itself, all those processes must find a new set of balanced relations. As life becomes more complex, with more pieces to re-arrange, stable patterns are harder to find. Four parts are easy to balance. Four million are nearly impossible.[35]

Rebalancing Everything

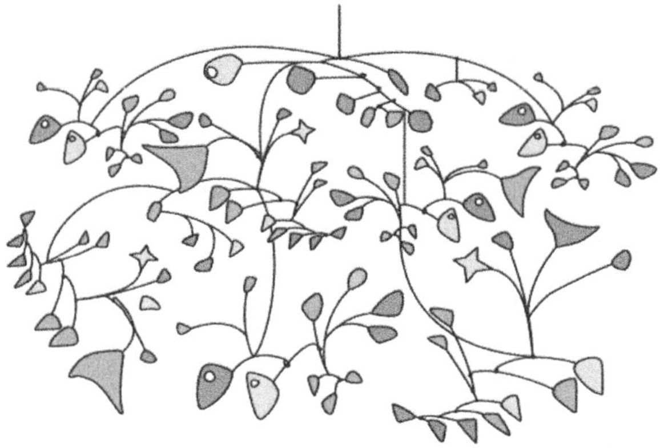

A new species requires millions of parts to perfectly re-balance with each other.

This means that as life progresses, with more pieces to re-arrange, *fewer new forms will survive*, and those that do will look *increasingly different* from their ancestors – far more complex, with previously unimagined new forms and abilities. "Basic plans are rare," says University of Helsinki biologist Jukka Jernvall. "The majority of them go belly up, as they did after the Cambrian, never to be replaced."[36] For example, he says, life churns out millions of different types of insects, each with "more glorious modifications," but it makes "no more new kinds of things such as insects."

Lastly, self-organizing equilibrium predicts that these far-reaching readjustments will be based, not on external forces, like natural selection, but on *internal forces*. A group that includes France's Jean-Louis Deneubourg put it this way: "in self-organizing systems, pattern formation occurs through interactions *internal to the system*, without intervention by external directing influences."[37]

This focus on internal forces starts at the most basic level, where the innate properties of particles lead them to form atoms, and continues up the ladder of complexity to molecules, minerals, and the basic body plans of life.

In short, Complexity Theory predicts that major shifts in evolution will be rare, rapid, far-reaching and determined by internal forces – what Richard Owen called the "laws of structure," and modern scientists like Stuart Kaufman and Chris Cherniak call the "laws of physics."

Complexity Theory predicts new species will appear much the same way as new atoms – in an abrupt leap with no intermediate forms, emerging fully complex from the start, with never before seen bodies, abilities and properties.

And this is exactly what the evidence reveals.

EVOLUTION'S MAJOR TRANSITIONS

Life began abruptly, four billion years ago, with single-cell creatures that instantly blanketed the Earth. The biology and genetics of these creatures was remarkably complex from the start. "Everything needed for evolution existed in bacteria billions of years ago," says Cambridge University evolutionary scientist Simon Conway Morris.[38] These creatures blanketed the Earth for almost two billion years.

Then two billion years ago, seemingly from nowhere, the first multi-cell creatures appeared, including the earliest forms of plants and animals. Stefan Bengtson, a Swedish paleo-zoologist involved in key discoveries, says they possessed remarkable new abilities, including sophisticated cell-to-cell signals and specialized responses that allowed them to form complex structures and do new things.[39]

Seaweed began to float on the ocean surface, small worms burrowed into the ocean floor.

This new type of life dominated the Earth for another billion years.

Then, 540 million years ago almost everything on earth abruptly died in a mass extinction. A burst of new life forms followed immediately, in what is called the "Cambrian Explosion." In a single leap, sea life went from spongy worms to complex animals, like trilobites, and opabinia with its long vacuum cleaner nose that sucked up food and placed it in its mouth.

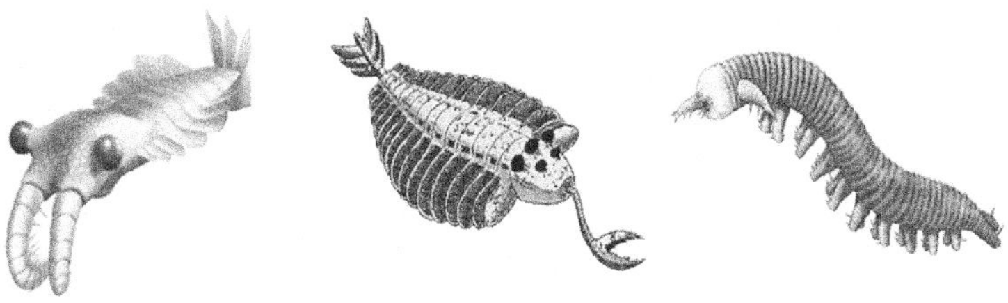

CAMBRIAN ANIMALS

University of California biologist James Valentine says recent discoveries show the shift was "even more abrupt and extensive than previously envisioned."[40] What emerged were not old species with new features, but entirely new creatures with never before seen abilities – eyes, noses, gills, guts, legs and fins. Life could now see, smell, and walk the sea floor, or surf the ocean currents in search of prey.

A similar mass extinction and explosive wave of new species occurred 250 million years ago at the Triassic era. It began with the "Great Dying" – 96% of ocean life and 70% of land animals disappeared in the (geological) blink of an eye.[41] This was followed by a wave of new creatures. Marine life abruptly leapt to more complex forms. New types of animals appeared, including the first dinosaurs, with never before imagined features and abilities, like feathers and flying.

The same cycle repeated itself 200 million years ago with the appearance of mammals and most modern animals. The first mammals were about the size of a mouse and emerged in a single leap, with an abundance of new features including a new type of heart and jaw, the ability to hear, grow hair, nurse their young, and a host of others.

These abilities allowed mammals to occupy every environment on earth – the frigid arctic, steamy tropics, arid deserts, stormy seas, forest tree tops and underground burrows.

Within a few million years, mammals exploded into thousands of variations on their basic form, including the ancestors to modern animals like dolphins, horses, elephants and humans.[22]

The Major Leaps in Evolution

(mya=million years ago)

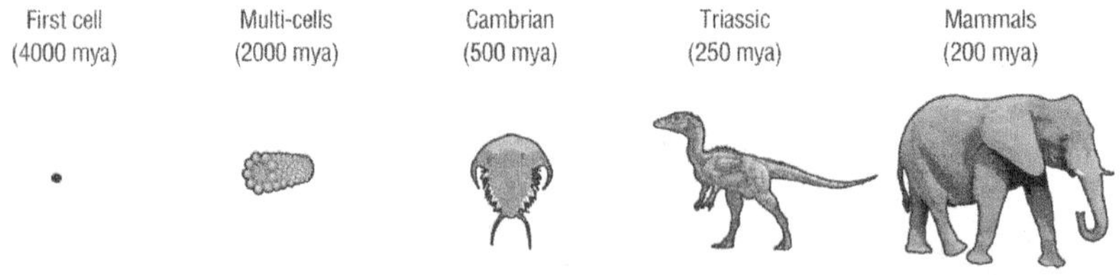

These new "body plans are hard to understand in [Darwinian] terms since they involve radical changes," say Gunter Wagner, the head of evolutionary science at the University of Vienna. "A significant reorganization of existing form, function and systems requires major innovations, [which] biologists need to explain."[42]

But these shifts are explained by Complexity Theory and Bak's "self-organizing equilibrium." Calling these leaps "biological big bangs," Eugene Koonin at the US National Institutes of Health says, "the major transitions in biological evolution [all] show the same pattern of sudden emergence of diverse forms at a new level of complexity."[43]

Importantly for evolution, Bak's discovery explains why life changes as a "whole" (as Cuvier said), why internal "organizing forces" determine the form things will take (as Richard Owen said), and why changes occur in abrupt bursts followed by long periods of stability (as Eldredge and Gould said). Each of these discoveries, previously denigrated and dismissed by Darwinists, are now central to the modern understanding of evolution.

The former president of the American Paleontology Association, Steven Stanley, compares these major changes to the quantum leaps that create new types of atoms, calling them "quantum speciation" events.[44] Stanley says this type of change requires "a new evolutionary timetable."

The Environment doesn't Determine New Species

Darwin said that different environments explain the differences in species. But two new discoveries show that environmental differences have little, if any, effect on the form new species take.

First, evolution follows the same rhythmic cycle regardless of changes in the environment. Many times over the last billion years, the atmosphere on Earth has swung wildly from blistering heat to "snowball Earth." And the appearance of new species don't correlate with these changes. To the contrary, they often occur when there has been no change in the environment. For example, Tom Kemp says there is no indication "of any large or abrupt geographical or climatic events" when mammals emerged.[22]

Second, the recovery from mass extinctions always takes about the same amount of time, regardless of the environment. Geologist James Kirchner and biologist Anne Weil carefully studied many mass extinctions and the later rebirth of ecosystems. They found it always takes the Earth roughly the same amount of time to replenish itself from these events.[45]

Why the sameness? "The interdependence of species is the primary determinant of recovery," they write. "Species are interdependent, so the destruction of one" requires reworking every relationship. To recover, everything must be brought into a new equilibrium. This takes roughly the same amount of time regardless of the environment or amount of extinction.

Weil and Kirchner say this standard recovery time is a previously unrecognized "general property" of natural history.[45] It is also another example that Bak's "self-organizing equilibrium" (re-balancing relationship) is central to most processes in nature.

Stress Can Trigger Duplication of the Entire Genome

So if environmental change doesn't lead to new species, what does?

Like all of biology, the details are complex, but the short answer is stress.

We know stress causes individual genes to duplicate and specialize. Occasionally stress has caused the *entire genome* to duplicate itself. When this happens, the genes reorganized themselves in ways that led to the emergence of entirely new creatures – the origin of new species.

How large-scale gene duplications led to new species is seen in the sudden appearance of flowers 140 million years ago. Flowers were so unlike anything prior they seemed to appear magically, enveloping the Earth in a sudden burst of dazzling new colors. Darwin called their origin an "abominable mystery."

Today that mystery has some answers. Penn State biologist Claude dePamphilis, head of the Floral Genome Project, shows flowers emerged from the duplication of an entire plant genome, which transformed green leaves into colored petals.[46] As the duplicated genes rearranged themselves in new ways, new properties emerged - startling colors, exotic new forms and remarkable new ways to connect with the eco-system around them.

Duplication and Specialization of 'Regulator Genes'

Duplication of an entire genome explained other major transitions in evolution. In the Cambrian Explosion, when life leapt from squishy worms to animals with back bones and shells, the genome abruptly doubled, and then doubled again. At the same time, there was a sudden duplication of the RNA system, life's second genetic code.[47]

This huge increase in genetic material occurred in rapid sequence, and led to the biggest explosion in new types of life in the history of evolution.

In 1999, James Valentine, David Jablonski and Douglas Erwin proposed the radical idea that the Cambrian Explosion emerged primarily from duplicating "regulator" genes (the genes that Barbara McClintock said govern evolution). They wrote:

> "It is likely that much genomic *repatterning* occurred during the Early Cambrian, involving both key control genes and regulators," leading to the appearance of "novel body plans."[48]

Scientists had known for decades that McClintock's mobile "regulator" genes can rearrange themselves, but the immensity of this process is only now being understood. Moving through the genetic landscape, mobile genes are "self-transforming," says French geneticist Cristina Vieir.[49] They "contain...all the instructions needed to cut themselves out of their host DNA and splice themselves into another spot." They even capture entire genes "and move them wholesale to new parts of the genome."[49]

These discoveries moved McClintock's mobile genes from the Darwinist dustbin to the center of evolution. Christian Biémont, Director of Research at the French Laboratory of Biometrics and Biological Evolution, says that mobile genes "were long dismissed as useless, but they are emerging as major players in evolution. They were discovered by Barbara McClintock in the 1950s, but it has taken half a century to begin to understand how they act and the effects they can have."[49]

Because mobile genes can change rapidly, they become "powerful producers of the raw material of evolution," says Biémont. "As well as validating McClintock's concept of [mobile genes] as controlling elements, this reveals [they] could have a role in the reorganization of genome structure."

Geneticists today call mobile genes "the agents of open source evolution,"[51] and "a driving force in evolution."[52] There is wide evidence, says Biémont, that mobile genes "can serve as a source of new... genes that allow populations to adapt and species to evolve."

The process of rearranging genes also explains how new traits and properties can appear with new species. Biologists Ricard Solé and Stuart Kauffman believe that when a genome doubles, the new patterns that appear are so complex they cross a threshold, creating a vast range of remarkable new traits.[53]

We can sum up the history of evolution as the duplication and rearrangement of genes and other molecules to form increasingly complex (and cooperative) relationships to cope with stress. From the Big Bang and the origin of atoms, through the behavior of bacteria and the origin of plants, flowers, animals and mammals, evolution emerges when nature forms more complex relationships that lead to more effective ways to cope with stress and achieve greater stability.

Chapter Nineteen

HOW DARWINISTS RESPONDED TO THESE DISCOVERIES

The more genomics... and other disciplines reveal about life, the more obvious it becomes our present understanding is not up to the job. We now gaze on a biological world of mind-boggling complexity that exposes the shortcomings of familiar, tidy concepts.[1]

Roger Highfield, Editor, New Scientist, 2009

Today, many universities have whole institutes devoted to Complexity research. Leading scientists believe it will define the scientific agenda for the 21st century, with the potential for dramatic discoveries and advances.

How did Darwinists respond to this breakthrough in knowledge? As they had to every major discovery of the last 150 years. They attacked it. "I do not know what observations complex systems is trying to explain," wrote John Maynard Smith. "It is a theory looking for a question to answer."[2]

Richard Dawkins compared Complexity to a "New Age cult... swimming in bogus scientific language, [and] regurgitated, half-understood jargon."[3]

Once again, the Darwinists were at war with the evidence, and the real processes of nature.

Darwinists Deny Reality: Words Mean Their Opposite

Repeating their time worn pattern, in addition to attacking the new discoveries, Darwinists began churning out new fictional explanations.

Because Complexity Theory shows Darwin had evolution backwards, Darwinists began to claim that nature is the opposite of how it appears.

They went from making up "just so stories" to an *Alice in Wonderland* description of nature in which words mean their opposite.

When it became clear that the appearance of new species occurs in rapid leaps, not slowly, the Darwinists redefined slow to mean fast.

In *The Blind Watchmaker*, Richard Dawkins invented "variable speedism," claiming there are two types of "slow" in nature: fast-slow, in which change is rapid, and ultra-slow, where there is little or no change. Dawkins said this meant Darwin's claim that evolution is gradual explains both the abrupt bursts of change in evolution, and eons with no change.

In "Darwin's Enduring Legacy," biologist Kevin Padian made a similar claim.[4] "It is widely and correctly recognized that Darwin promoted gradual change," wrote Padian, "but what did he mean by 'gradual'?" Going back to the origin of "gradual" in the Latin word "gradus," meaning "step," Padian argued that when Darwin said evolution is "gradual," he really meant it occurs in "relatively rapid" steps. Padian put it this way:

> The "conceptual tension between 'slow and steady' and 'step-like' is the basis of one of the most important evolutionary ideas of the twentieth century: punctuated equilibrium, [the idea that] species may not change...for much or most of their history, then alter relatively rapidly.
>
> If this turns out to be the predominant pattern of evolution... *as now seems to be the case*, Darwin's view of the plurality of evolutionary tempos...will be vindicated."

Punctuated change vindicates Darwin's gradual evolution?!?

This, of course, is nonsense. Niles Eldredge wrote often of the bitter battles over punctuated equilibrium which, he said, "stood the standard interpretation of Darwin on its head [and] ignited a furious debate about the true nature of evolution."[5] For decades the Darwinists insisted it was wrong (browbeating Eldredge and Stephen Jay Gould every step of the way).

Nor, as Padian claimed, is there any "conceptual tension" between "gradual" and "relatively rapid." They mean opposite things. Nor did Darwin have a "plurality of views" about the speed of evolution. He had one view: it's slow. Darwin opposed the idea that evolution is "relatively rapid," writing:

> "Species...sometimes falsely appear to have come in abruptly... which if true, would have been fatal to my views. But such cases are certainly exceptional."[6]

Entire sections of *The Origin* were written specifically to refute the idea that evolution changes in steps, a point even Thomas Huxley made in 1860, writing "Mr. Darwin...embarrassed himself with the aphorism 'Natura non facit saltum' [nature does not jump]. We believe nature does make jumps."

Darwin staked his entire theory on slow change: "If it could be demonstrated that any complex organ existed which could not possibly have been formed by numerous successive, slight modifications, my theory would absolutely break down."[7]

Darwinists Deny Reality: Facts Mean Their Opposite

In the 1990s, Darwinists walked further into this strange world where things mean their opposite. In Darwin's evolution, natural selection should ruthlessly eliminate any harmful trait. But this doesn't happen. Harmful traits and fatal diseases persist for thousands, even millions of years.

Natural selection should also eliminate those exaggerated sex traits – giant antlers, elaborate peacock feathers and the like – that are useless as weapons and only weigh an animal down, often threatening survival.

Why didn't natural selection remove these dangerous traits? In the 1930s, Ronald Fisher explained this by inventing "explosive evolution," the idea good traits keep evolving until they became harmful. No evidence for it ever materialized, and the idea faded away.

Over the years, Darwinists offered other stories to explain the persistence of harmful traits, but none held water.

In the 1990s, however, they came up with a truly *Wonderland*-ish idea: the "Handicap Principle."[8] "Handicaps" evolved as a way for a male to demonstrate his overall strength to potential mates. The idea was that if the male can survive with the handicap, he must be exceptionally fit.

In other words, harmful traits evolved as a way to show strength.

This meant that both the *presence* and *absence* of harmful traits were now offered as proof of natural selection. Natural selection's power to *eliminate* harmful traits is proven by the *existence* of harmful traits! Only among Darwinists could this pass for science.

Modern Darwinism is littered with such contradictions. "Survival of the fittest" means a single, most fit creature should evolve in each locale. This is the "Diversity Paradox" we covered earlier: how can many different species emerge from a "selection" process powerfully driven to create a single dominant species?

In what is now standard Darwinian double-talk, George Williams, a founder of New Darwinism, explained this paradox by saying, "evolution takes place not so much because of natural selection but to a large degree *in spite of it*."[9] Evolution operates *in spite of* natural selection?!

Slow means fast, gradual means "relatively rapid," harmful is helpful, evolution occurs not because of natural selection but "in spite of it." Darwinism sounded increasingly like a Rube Goldberg contraption.

Darwinists Deny Reality: Nature is an Illusion

As the evidence against their theory mounted, Darwinists went beyond redefining words, to an effort to redefine basic reality.

Darwin had the evolution process backwards. What do you do if you have reality backwards? In an audacious move, Darwinists declared that reality is the *opposite* of what we see. They said that much of nature is an illusion, and actually the opposite of how it appears. Only when it is viewed through a Darwinist lens can the illusion be penetrated and the truth seen.

The idea that nature is an illusion – and reality is the *opposite* of what we see – is a favorite claim of Richard Dawkins, Darwinism's leading voice for the last 40 years.

Dawkins formula is simple. He begins with an important aspect of nature that contradicts Darwinism. He then explains that the evidence we see is an illusion. Nature is really the opposite of how it appears. He then proceeds to reverse the meaning of evidence and events until voila, they all reemerge to mean their opposite.

Dawkins' most important book, *The Selfish Gene*, does this.

A big problem for Darwinism is explaining all the cooperation in nature. All animals cooperate to some extent (it's hard to reproduce alone). Many live in large social groups where cooperation is a way of life. Most share food, protection and social support. Even vampire bats and the most predatory wolves will share food with hungry members of their own group.

This cooperation should not exist if evolution is driven by a competitive struggle and natural selection. Dawkins' solution was to claim the cooperation is an illusion. In *The Selfish Gene* he writes:

> "Like successful Chicago gangsters, our genes have survived...in a highly competitive world... I shall argue that a predominant quality to be expected in a successful gene is ruthless selfishness... However... there are special circumstances in which a gene can achieve its own selfish goals best by fostering a limited form of altruism."[10]

That, in a nutshell, is the essence of *The Selfish Gene*: cooperation is really a disguised form of competition. What appears to us as a natural world filled with cooperation is an illusion.

Because the natural world is the opposite of what Darwin claimed, Dawkins must fall back on this "reality is an illusion" routine often. For example, animals have many traits which are helpful to the group, but not to themselves as individuals. Social insects, for example, devote their entire lives to taking care of others.

While this is the extreme, most creatures have some traits that benefit the group as much, or more, than themselves. Among birds, monkeys, marmots and other animals, individuals who see a predator let out a cry that allows the group to escape, even though it directs the predator's attention to them.

Even bacteria, like slime-mold, will sacrifice for the good of the group. When slime-mold cells are hungry, Darwin's natural selection says the "most fit" should grab all the food, letting the others die. But they don't. Instead, they unite in the thousands and share with each other.

All of this is the opposite of what should happen under Darwinism, so Dawkins defines it all as another illusion. Astonishingly, he says biologists fail to grasp this cooperation is an illusion because they don't properly understand evolution:

> "A group...whose individual[s] are prepared to sacrifice themselves for the welfare of the group is the theory of 'group selection,' long assumed to be true by biologists not familiar with the details of evolutionary theory."[11]

The problem, however, is not that biologists fail to understand evolution, but that Darwinists are in denial about basic biology.

Another major problem for Darwinists is epigenetics, the modern name for Lamarck's direct inheritance. In 1999, the growing evidence for epigenetics forced Dawkins to deal with what he called this "Lamarckian Scare."

> "I use the word 'scare,'" he wrote, "because, to be painfully honest, I can think of few things that would devastate my world view more than a demonstrated need to return to the theory of evolution... attributed to Lamarck. It is one of the few contingencies for which I might offer to eat my hat."[12]

Dawkins' dilemma is that if Lamarck is correct, Darwinism, by definition, must be wrong.

A crucial difference between the modern version of these two theories involves the eggs and sperm (called "germ-line" cells) which carry genes from one generation to the next. Darwinism says these cells are "insulated" and unaffected by changes in the body. Thus changes in the parent can not pass to their children. Lamarckism says these cells are not insulated, and can incorporate changes in the parent's body, which are then inherited by their children.

Now the resolution of this issue is pretty straightforward. These "germ-line cells" either can or cannot incorporate genetic changes from the body. As we saw in Chapter 17, thirty years of research and thousands of studies show *they can* incorporate such changes. These results aren't hiding under a bushel basket. Many scientists explicitly put the words "changes in germ-line cells" in the titles of the research papers they publish.

Dawkins deals with all this mountain of empirical evidence by calling it an illusion. The "inheritance of acquired characters, per se, is not Lamarckian," he says. "Lamarckism is really Darwinism in disguise."[13]

I won't go into the convoluted process Dawkins uses to conclude, in his words, that "Lamarckism is really Darwinism in disguise," but you get the picture. In Dawkins world, anything which contradicts Darwinism is an illusion, the exact opposite of what it seems: cooperation is competition in disguise, traits that benefit the group are really selfish traits in disguise, Lamarck's direct inheritance of changes is Darwinism in disguise, etc.

Inventing False Facts About Darwinism

Dawkins will even reverse the reality of Darwinisms key ideas. In a 2007 book review of *The Edge of Evolution* he wrote, "the crucial passage [in *The Edge*] is this: 'By far the most critical aspect of Darwin's multifaceted theory is the role of *random* mutation.'" Dawkins then says:

> "What a bizarre thing to say! ...Darwin set no store by randomness. New variants might arise at random, or they might be acquired characteristics... Far more important for Darwin was the nonrandom process whereby some survived but others perished. Natural selection...alone...explains the elegant *illusion* of design that pervades the living kingdoms and explains, in passing, us."[14]

As we know, directionless (or random) evolution was central to Darwin, it was the key claim he felt would eliminated God from evolution. Secondly, modern Darwinists *do* say that randomness is central to their theory:

- Ronald Fisher: Evolution occurs through "random survival" and "random extinction," and "the principle of Natural Selection depends on a succession of favourable *chances*."[15]
- Theodosius Dobzhansky: "Mutation is a *random* change of the gene."[16] "The modern theory of evolution is that...mutations occur by *chance* and are undirected."[17]
- Kevin Verstrepen (Harvard) and Oliver Rando (University of Massachusetts): "According to classical evolutionary theory...variation originates from *random* mutations."[18]

Even Dawkins himself says evolution is random. "Complexity cannot spring up in a single stroke of chance," he writes, "but a whole series of tiny chance steps...can build up almost limitless complexity."[19] "Chance" is just another way to say random.

How can Dawkins so easily make such conflicting claims? One explanation may come from Allen Orr, Kearns Professor of Biology at the University of Rochester, who says Dawkins "has a preordained set of conclusions at which he's determined to arrive. Consequently, Dawkins uses any argument, however feeble, that seems to get him there and the merit of various arguments is judged largely by where they lead."[20]

While spinning this strange world, Dawkins loudly condemns anyone who disagrees with his ideology as unscientific and "anti-evolution." *New York Times* science reporter Nicholas Wade calls Dawkins "the Savonarola of science, condemning the doubters of evolution as 'history-deniers' who are 'worse than ignorant' and 'deluded to the point of perversity.'"[21] (Savonarola was briefly the dictator of Renaissance Florence, using street gangs to brutally enforce his puritanical ideology.)

Wade goes on to say that Dawkins' approach "is not the language of science, or civility." That is true. But these attacks are a good way to distract from the fact that your theory contradicts reality.

Explaining Life's Origin: 'Special Circumstances'

It became so impossible to square Darwinism with the real world, that some Darwinists moved beyond the idea that nature is an illusion, to something more a kin to magic.

Evolution's biggest mystery is the origin of life, which appeared four billion years ago. One of the many perplexing puzzles about this is that it happened rapidly, immediately after the earth cooled and the oceans formed. This meant there was no gradual build up from inert atoms to animated cells. Dozens of complex abilities were required for the first cell to function, including the capacity to capture and transform light, and to energize, repair and reproduce itself. They all had to appear at the same microscopic spot, and at the same instant, for life to begin.

In addition, a cell can exist only if it is encased inside a membrane. However, a membrane is built from complex molecules which can form *only from inside a membrane* that already exists. Any scientific explanation of the origin of life must deal with these facts, not obscure them.

To account for all this in Darwinian terms, John Maynard Smith appealed to something called "special circumstances"– events that can't be explained by natural laws, but happen for inexplicable reasons, a bit like magic. Smith said that to explain the origin of life, "special circumstances – in fact a series of them – must have been needed for these innovations [to appear]."[22]

Darwinism not only requires quasi-magical "special circumstances," it needs a whole "series" of them to explain the origin of life.

EXPLAINING LIFE'S ORIGIN: THE ALIEN EXPLANATION

Francis Crick took another approach to explain life in Darwinian terms, saying it appeared quickly because it was "deliberately transmitted to the earth by intelligent beings on another planet."[23] Aliens explain life on earth.

As Crick told the story, life began in a galaxy far, far away. Over billions of years, random changes and natural selection led to the evolution of an advanced civilization which came to Earth and left some "seeds of life."

Calling the process "directed panspermia," Crick said an alien "spaceship would carry large samples of a number of microorganisms" and deposit them on planets like earth.

Crick used aliens to account for other aspects of evolution which Darwinism couldn't explain. For example, if evolution is random, new life should have evolved several times. But it didn't. Life evolved only once.

Also, if evolution is random, different genetic codes should have appeared. But they didn't. Life has a single gene code. "It's a little surprising," wrote Crick, "that organisms with somewhat different codes do not exist."[24]

To explain these crucial evolutionary events in a way compatible with Darwinism, Crick turned to aliens. Life evolved only once because it was brought by aliens only one time.

Aliens explained all the hard questions about evolution which Darwinism couldn't, like the single genetic code. "The universality of the code follows naturally from [the fact that] all life on Earth represents a clone derived from a single extraterrestrial organism."

Crick wrote an entire book (*Life Itself*) using aliens to account for the many mysteries of evolution that can't be explained by Darwinism.

A generation later, Richard Dawkins also used aliens to bridge the gap between Darwinism and the origin of life. (His version of the "aliens did it" can be seen in a video on YouTube.)

This shows the stark difference between Darwinism and true science. These ideas aren't science, of course, but science fiction. Or perhaps something else.

As we saw, Stephen Jay Gould called modern Darwinism a "cult." Lynn Margulis said history will ultimately judge it to be "a twentieth-century religious sect."[25] Many others agreed. From Thomas Morgan in the 1920s ("Darwinians seem less concerned with the advancement...of evolution than expounding Darwinism as a dogma."), through Carl Woese in the 1990s (evolution was captive to "a quasi-scientific movement, who...effectively disguised the fact that their primary concern with it was ideological, not scientific."), a persistent complaint among scientists is that Darwinism is more like a fanatical religious sect than science.

Explaining Life's Origin: The Scientific Explanation

Self-assembly, with its naturally forming relationships, abrupt transitions and emergent properties, offers a truly scientific explanation for the mysteries that surround the origin of life. Parts of the puzzle have already emerged.

Today we know how atoms form, then naturally combine into molecules, which can unite to form the 20 amino acids – the "building blocks of life." We also know these amino acids naturally assemble themselves into proteins. As biologists recently demonstrated, "proteins self-assemble based on their molecular properties."[26]

We are also beginning to glimpse the process proteins use to assemble themselves into a cell. At the California Institute of Technology, scientists have shown how a few parts use self-assembly to naturally form a hollow "Bucky ball," revealing how the first cell membrane might have appeared. (Cal Tech has a short video of the process on YouTube).

Thus, self-assembly takes us to the door step of life.

The astonishing emergent properties which appear when a particular organization pattern is formed supply another piece of the puzzle. There is nothing inherently "alive" in molecules, so we know life is a product of how molecules are *organized*. Only emergent properties can explain how lifeless molecules could be organized in a pattern that produced the first pulsating cell – a thing which could breathe, learn, move, and reproduce itself.

This is far – very far – from explaining the start of life. How thousands of proteins could form at precisely the same place, and then assemble into the complex pattern that created the *first* cell, we don't have the faintest idea. That is far beyond our current conceptual skills.

But self-assembly and emergent properties puts us on the right track to eventually explain what happened scientifically. They also make the origin of life compatible with the other major transitions in evolution, all of which occurred in an explosive burst of new organization patterns, greater complexity and emergent properties.

(Life may have began through nature's "higher organizing principles," like those described by David Pines and Robert Laughlin, but ones that remain to be discovered and will almost certainly be profoundly different from anything we imagine today.)

THE CHOICE WE FACE

What does all this mean at a higher level?

What is at stake in evolution is more than arcane academic debates. Our concept of evolution influences everything from vital medical research to the concept of human nature we use to raise our children, run society and solve major social problems.

Is nature fundamentally cooperative or competitive? Is evolution a predictable process, or random? If evolution were rerun, would we end up with essentially the same world we see today, or one that looks and acts fundamentally different?

Tomorrow's medical discoveries depend on getting the answers right today. Our understanding of human history and basic ideas about human nature, society and social processes will be dramatically different based on the answers.

If evolution occurs as Darwinists say, then we take one approach in our attempts to cure cancer, prevent disease, raise healthier children and solve social problems. If evolution occurs as Complexity Theory says, then we take *exactly the opposite* approach to these efforts.

As a society we face a stark choice.

Will we try to explain evolution with concepts like the Weismann barrier, "variable speedism," selfish genes, the "Handicap principle" and similar ideas, or with Nobel Prize winning discoveries like those of Barbara McClintock, Jean-Marie Lehn and Robert Laughlin?

Will we stay mired in a "science" which claims evolutionary change is unpredictable, or will we use discoveries like epigenetics and self-assembly which show change is predictable to revolutionize medicine, industry and the social lives of the world's poor?

Will journals like *Nature,* and the popular press like *Time*, continue to promote Richard Dawkins ("the Savonarola of science") as evolution's leading voice? Or will they turn to the scientists leading the breakthroughs about evolution in genetics, biology, self-assembly and other areas of science? The answers matter. They profoundly affect all of us, the people we care about, and the society we will hand our children.

Chapter Twenty

A Powerful New Model of Evolution

ENERGY, EYES AND EVOLUTIONARY CLOCKS

What is happening now in evolutionary theory is as exciting and foundational as during the early days.[1]

David Sloan Wilson, Binghamton University, 2008

Complexity Theory, for the first time, allows us to grasp the true wonder of evolution. We can see how a few simple processes, all present at the beginning of life, are duplicated and transformed into the intricate, beautiful, and sometimes bizarre traits we see today.

The power of the Complexity Theory of Evolution is seen in its ability to explain and unite three crucial aspects of life – energy, eyes, and evolutionary clocks – under a single ability possessed by the first cell, the capacity to transform light.

The Ability to Transform Light

At its most basic, life exists because cells like algae can absorb light and turn it into "food." Surprisingly, cells absorb individual packets of light (called photons) which they transform into sugar and store for future use.[2] This provides the energy that fuels the entire chain of life. The energy you are using to read this book comes from photons.

Absorbing light is not remarkable. Any rock can do that, getting hotter as it soaks up sunshine. What made the first cell different from everything that had existed prior, was its ability to *transform* light, to convert energy into matter. It later converts that matter back into energy to power its activities. This extraordinary capacity may be *the* decisive ability that transformed a group of lifeless molecules into the first pulsating cell, the first life.

TRANSFORMING LIGHT LEADS TO CLOCKS AND REGULATION

The ability to transform light explains the origins of biological clocks.

Absorbing light changes the cell's chemistry. The intensity of light varies over the course of the day (and the month and year), creating slight variations in cell chemistry. Early in evolution, the cell used these variations to create biological clocks – internal equivalents of the cycles of the sun and seasons – as a way to regulate the rhythms of life essential for survival.[3]

Tracking these differences in light later told animals when to feed (the morning sun tells jellyfish to swim up to feed), when to store food for the winter, and when to hibernate. Variations in light from the moon were used to coordinate the creation of new life, regulating reproduction in creatures as diverse as simple sponges, up to human beings.

Dartmouth geneticist Jay Dunlap says that fifty years ago, little was known about how biological clocks evolved. Darwinists claimed the clocks in each animal had evolved separately, driven by random genetic changes and natural selection. They "simply ignored" hard-to-explain evidence, says Dunlap, and claimed each clock is a "fundamentally different mechanism."[3]

This never made sense to geneticists who kept digging for answers. They eventually discovered that the *same* "clock" genes (called cryptochromes) are at work throughout the animal world – in flies, fish, frogs and men.[3] As they looked deep into the past, they also found the same "clock" genes in the earliest animals, like sponges and coral, synchronizing everything from feeding to their mating cycle.

Once a year, the coral along Australia's Great Barrier Reef all spawn at the same moment. "It's just magical," says Australian geneticist Bill Leggat. You "sit in front of an individual coral and watch the pink sperm bundles get slowly pushed out of the corals' mouth and float away."[4] It's like watching "a slow symphony," he says. Millions of coral, all spawning at the same time, creating vast waves of color across the ocean.

Coral, however, lack even primitive eyes or brains. So how is this underwater symphony orchestrated? How do millions of coral synchronize across a thousand miles of ocean? In 2007, an Australian group that included Leggat, uncovered the remarkable secret.[5]

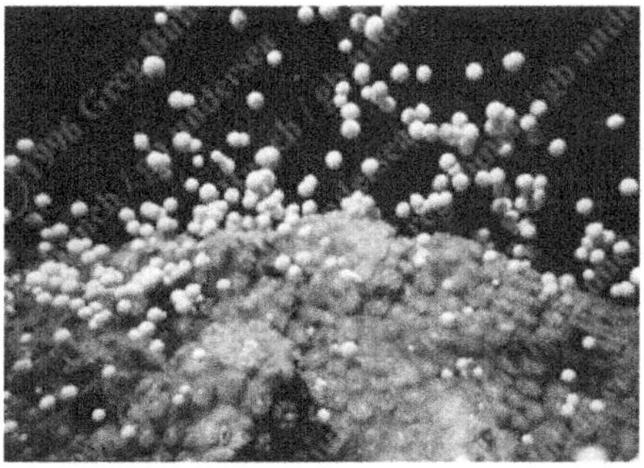

Coral spawning

Ocean view of spawning

Clocks Evolve from Stress, Duplication and Specialization

We know that when a cell is stressed, it makes copies of the genes connected to the stress, and then tinkers with them in search of an adaptive solution. Early in evolution, before the ozone layer fully developed, the intense ultraviolet radiation of the midday sun damaged the cell's DNA (just as it gives us the worst sunburns).

In response, early cells duplicated a light sensitive gene (cryptochrome 1 or *cry1*), and specialized it to react to the ultraviolet light that appears at midday. When those light waves hit, the gene became active, making a protein which absorbed those rays, protecting the cell's DNA. This allowed the cell to continue absorbing other types of light, and making food, without damage.[5] (The process is similar to when intense sun rays hit our skin. At first they burn. But the skin produces pigment which absorbs the intense rays and protects the skin. The added pigment makes the skin look darker, giving us a "suntan.")

Over time, the cell made more copies of this gene, specializing each to be sensitive to a different wavelength of light, including different intensities of moonlight. This turned these genes into "clocks," sensing the time of day, phases of the moon, and even light changes over the course of the year.

These clocks are "the Swiss timing mechanism of biology," says Ove Hoegh-Guldberg, Director of Marine Science at Australia's University of Queensland.[5] This sensitivity allows coral to know what phase the moon is in, and synchronize their mating. Coral repurposed light sensing abilities that existed billions of years earlier to coordinate their reproduction.

These same clock genes are today found in plants, insects, animals and humans, performing a remarkably wide range of tasks. "A single toolbox of clock parts is carried by the entire animal kingdom," says geneticist Michael Young at Rockefeller University.[6] Animals have slight variations in their tool kits, says Young, but "easily overshadowing the differences is a compelling conservation of clock genes...The same 'clock works' is surely marking time throughout the animal kingdom."

It's extraordinary that all the complex rhythms of nature evolved from one mechanism that existed at the beginning of life, and was duplicated, specialized, and repurposed to perform thousands of tasks. Who would have imagined that the daily feeding of jellyfish, monthly ovulation in women, annual spawning of coral, global migration of birds and annual hibernation of bears are all regulated by the same mechanism?

Transforming Light Leads to Eyes and Sight

The same process that produced bio-clocks led to the evolution of eyes.

As one of biology's most complex creations, eyes became the focus of many evolution discussions – from William Paley to Darwin, Dawkins and Stephen Jay Gould. The question was always the same: how did such an intricately engineered piece of biology evolve?

Darwinists said the eye arose from a long string of random changes. This meant that eyes, and the ability to see, evolved independently in each animal group, perhaps as many as 60 different times. Each animal group evolved a unique eye, driven by random changes and natural selection.

As Darwinists acknowledged, given the complexity of the eye this story always sounded improbable, and new evidence now proves it is untrue. "Natural selection has not forged many eyes completely from scratch," writes evolutionary biologist Sean Carroll. "There is a common genetic ingredient to making each eye type."[7]

Early in evolution, some animals stopped turning light into food and began turning it into signals. They duplicated their light sensitive genes and proteins (called opsins), and specialized them to respond to changes in the light reflected off of things around them. At the University of California, for example, they found that primitive animals with no eyes, called sea hydra, can still "see" another animal by using opsins to detect the light reflected off of them.[8]

Repeated duplication and specialization of opsins eventually allowed them to detect the smallest changes in reflected light. The different opsins were later enclosed in a single cell. All eyes later "evolved from this common photo-receptor cell" says Detlev Arendt, head of the European Molecular Biology Laboratory and an authority on eye evolution.[9]

This universal origin of eyes explains why they all share the same underlying processes. The basic physics of light, which all eyes must use, channeled the evolution of all eyes along the same path, converging on the same solutions to sight, regardless of the environment or natural selection.

"Despite the stunning variety of photo-receptor organs," says Neil Shubin, "every animal uses the same kind of light-capturing molecule to do this job."[10] Clams, insects, eagles and humans all use variations of the same light-sensitive opsins in their eyes.

(These light-sensitive molecules were repurposed for the most surprising jobs. University of California biologist Todd Oakley, who pioneered many of these discoveries, found that squids, using the same genes found in eyes, make themselves "invisible" to other animals by "lighting up" their underside to match the sunlight coming from above.[11] Oakley says that "for us, as evolutionists, the message is that photo-reception can do other things besides just facilitate vision. It can do unexpected things...other functions for light sensitivity we may not be thinking of."[12])

Restoring Sight In Blind Fish via New Relationships

Understanding that all eyes evolved by modifying "common genetic elements" led to other breakthroughs.

In 2008, Richard Borowsky, head of New York University's Cave Biology Research Group, startled biologists when he restored sight to blind fish.[13]

When animals are no longer exposed to light, as happens with cave fish and underground moles, they lose their ability to see. But as Borowsky found, the genes for sight don't disappear.

The cave fish lost their sight because without any use, their "sight" genes began to turn off. Fish living in different caves, however, turned off *different* "sight" genes. When Borowsky mated blind fish from different caves, the "on" genes in one group restored the "off" genes in the other group, restoring some sight in their young.

When Borowsky mated blind fish with normal fish, the genes were restored in a single generation, and the offspring emerged with perfect sight.[13] No random gene changes or natural selection was involved in this restoration. But the discovery that unused genes "turn off" does confirm Lamarck's idea of evolution through use or disuse. Without stimulation and stress, any ability is lost.

UNITING EVOLUTION UNDER A FEW SIMPLE PROCESSES

"Similar stories to that of eyes can be told for limbs, hearts, and body segments," says Todd Oakley and his colleague Michael Rose. "Naturally, many biologists now question the assumption organs... evolved multiple times from scratch, each specifically fashioned by powerful mechanisms of natural selection."[14]

Rose and Oakley caution that evolution also involves other mechanisms, leading to what they call "tinkering as well as duplication and divergence at multiple biological levels."

Nonetheless, these discoveries have proven powerful. Over the last decade, scientists have begun to unite a wide range of evolutionary developments under a few simple processes that were present at *the beginning of life*, and evolved through duplication and specialization.

The ability to unite these processes provides science with a road map for uniting the fields of development, genetics and evolution.[15]

For example, all biological clocks use a little "molecular motor" made from two molecules – actin and myosin. These molecules pull each other back and forth, marking time much like the swinging pendulum of a grandfather clock.

When biological clocks respond to changes in light, this molecular motor moves back and forth, triggering a cascade of other chemical signals that lead to eating, sleeping, the fertility cycles of plants and animals, and the vast seasonal migrations of birds, beasts and ocean life.

The same actin-myosin motor shows up in many of biology's key functions, all of which date to the beginning of life:

- **Reproduction**: The back and forth action of actin-myosin divides a cell in two, creating reproduction. In later evolution, this action separates DNA into two parts, allowing for sexual recombination.[16]
- **Development**: During development, the same back and forth movement of actin-myosin divides a blob of tissue into different sections, allowing separate body parts to form.[17]
- **Movement**: This little motor's back and forth action contracts and expands the whole cell, allowing it to move around. It also pulls the hair-like fibers that sprout from cells (called cilia) back and forth, propelling the microscopic creatures you see in a glass of pond water.
- **Muscle control**: The back and forth action of actin-myosin also creates the muscle contractions in arms and legs that allow us to walk, run, pick things up, text or type an email.[18]

This shows the power of the Complexity Theory of Evolution. It can unite the ability to transform light, with the ability to make "food," regulate life's rhythms, reproduce, divide tissue, and make muscles move back and forth. It also reveals that these processes are the same in every type of *animal*, for every type of *adaptive change*, and in every type of *environment.*

THE COMPLEXITY THEORY OF EVOLUTION

Complexity Theory emerged from the discovery of a previously hidden order in nature. The rules which create that order also lead to a previously hidden order in evolution.

1. Relationships. Our new understanding of evolution begins with the fact that there is no new stuff in the universe, only new relationships. Evolution emerges from combining old parts in new ways. "Looking back through billions of years of change," writes anatomist Neil Shubin, "everything innovative or apparently unique in the history of life is really just old stuff that has been recycled, recombined, repurposed, or otherwise modified for new uses. This is the story of every part of us, from our sense organs to our heads, indeed our entire body plan."[19]

2. Self-Assembly. Self-assembly is nature's basic process for creating relationships, using a few simple rules such as "opposites attract" and "likes repel." The rules for making things are also the rules for changing them. Evolution is simply change over time. Thus, self-assembly explains how things are made, how they change and how they evolve. Evolution is the story of how self-assembly made the universe complex.

3. Patterns. Each set of relations creates a unique pattern. These patterns contain the information – the "plan" – for making everything in the universe, from atoms to flowers and fish. Each new pattern becomes both the platform and the instructions for creating the next level of complexity.

The patterns created by these relationships also function as a previously unrecognized "code of evolution." They provide both the platform and the signals (the instructions) that determine the next level of evolution. Genes are only one part of this much larger, and more complex process.

4. Equilibrium and Cooperation. From tiny particles to whole ecosystems, when nature creates relationships it strives for balance (or equilibrium). This effort is at the heart of every natural process, from the dance between two atoms, to chemical reactions, life in an Amazon rain forest, and the vast balancing act of the universe itself.

The key to achieving balance is cooperation. Starting with quarks (which never exist alone and always come in pairs), every new creation in nature requires at least two parts. The relationships they form are inherently collaborative. This means that evolution is overwhelmingly a cooperative process. Martin Nowak, director of Harvard's Evolutionary Dynamics Program, says the need for cooperation is a "basic force" in evolution.[21]

5. Stress. At every level of nature, stress alters how cooperative relations form. These changes create new patterns, which lead to the formation of new things. At a simple level we saw this with atoms. In biology, stress causes molecules (like genes and proteins) to form a slightly different pattern from the original, which is better able to deal with the stress (this is nature's duplication and divergence process). The new patterns created by stress push nature up the ladder of complexity.

6. Emergence. New patterns lead to the emergence of new properties not possessed by (or reducible to) anything that existed prior. Emergence explains how quarks can combine in increasingly complex ways to eventually form dragon-flies and jaguars, with astonishing properties (like learning, memory and feelings) unimaginable in quarks. At higher levels of complexity, entirely new laws of nature can emerge, transforming reality itself and what we think of as the "everyday world."

These discoveries reveal three important things about evolution:

1. Evolution is Restricted. The rules of self-assembly limit which relationships can form, which in turn limits what can exist. These restrictions explain the widespread "convergence" seen in evolution – why the same traits and abilities evolved over and over again, regardless of ancestors or environment. Biologist Paulien Hogeweg calls this the "interlocking of self-organization and evolution."[20] Eyes are just one example. Arms, legs, wings, heads and hearts all evolved much the same way as eyes.

2. Evolution is Predictable. These rules also mean that what nature can create is predictable, which makes evolution predictable. As Cambridge paleontologist Simon Conway Morris has said often, "Far from being accidental, evolution is remarkably predictable"

(It's important to note that while evolution is predictable, it is not "determined." Nature works within ranges (like the "plasticity" we saw earlier). This means that while the same types of things always form, each individual can be unique. For example, over billions of years, an uncountable number of snowflakes have fallen on earth. While they always form the same six pointed crystal, Kenneth Libbrecht, an expert on crystals at Cal Tech, showed that the details of each snowflake are unique. He says it is likely that every snowflake that has ever existed was unique.)

3. Evolution is Universal. The rules of self-assembly are universal, which means that what can exist is universal. This explains why the same types of things appear throughout the universe– why the same atoms, stars and galaxies always evolve. And why it is likely that the same life forms (with the same emergent properties) will always evolve. If there is life on another planet, it almost certainly looks and acts much like life on Earth.

These rules mean that if evolution were rerun, it would produce the same types of atoms, stars and life forms – the same plants, animals and people – as we have today.[21]

The key idea underlying all these discoveries is that evolution did not result from natural selection, or any similar process. It emerged automatically, based solely on the physical properties and basic laws of nature that existed at the beginning of the universe. As old Thomas Huxley put it, "The whole world, living and non-living, is the result of definite laws [and] forces possessed by the molecules of which the universe was [first] composed."

And as Huxley clearly understood, these laws made everything that later evolved predictable. As we will see next chapter, the laws of self-assembly unite most, if not all, of natural history under a single process in which change is predictable.

Chapter Twenty-One

A Powerful New Model of Evolution

EXPLAINING, INTEGRATING AND PREDICTING

We are witnessing a major paradigm shift in the life sciences.[1]

James Shapiro, University of Chicago, 2007

Remember, we can tell a good scientific theory by its ability to explain, integrate and predict. Complexity Theory and self-assembly do that. What Plate Tectonics did for geology, Complexity Theory does for evolution.

It explains evolution's deepest mysteries.

It integrates natural history into a single coherent picture, uniting the evolution of stars, planets, atoms, molecules and minerals with the evolution of plants, animals and ecosystems.

It also makes evolution predictable, with testable claims and empirical answers. This makes evolution truly scientific.

Explaining Evolution

We've already seen some of the mysteries that self-assembly explains: how adaptive changes emerge, how new species arise, why nature goes through that strange "recapitulation" process, and how animals can have the same genes yet be different. It also answered many other questions:

- From the simple: why grass stops growing at a certain height, but begins to grow again if we cut it,
- To the complex: how a cell, which can be anything from a toenail to a

neuron, knows exactly what it should be,

- To the everyday: how the body replaces millions of cells a day, and gets each one *exactly right.*

Discovering that the "plan of life" is in a set of relationships also solved the riddle of how animals evolve to larger or smaller sizes with *everything* scaling in *perfect proportion.* As the development process unfolds, each part responds to everything around it, explaining how thousands of complex relationships all adjust together.

These relationships help explain the mystery of aging. Mediterranean sea grass lives for 100,000 years. Cedar, sequoia, and baobab trees live for thousands of years.[2] Whales, tortoises, sea urchins and other animals can live hundreds of years. Why do our bodies age after a few decades?

The answer is that cancer and aging share a common biology. The genes that stop cancer cells from multiplying also suppress the production of normal cells.[3] These genes allow us to live longer, but leave the body with increasingly older cells which result in wrinkled skin and weak muscles.

This reveals the importance of properly understanding evolution. With this knowledge, we can now look at adjusting the body's natural defenses to slow our aging, without increasing the risk of cancer.

Self-assembly resolved other old mysteries. When Janis Antonovics asks students in his Evolutionary Mechanisms class to explain why we don't have helpful traits like wings, he says "I'm met with quizzical stares."[4] Arms are common. So are wings. And both are pretty handy. But despite billions of years, and millions of species, not a single animal ever evolved *both* arms and wings. Why not?

Self-assembly resolved the mystery. The constraints on how skeletons can form mean that only some body shapes are possible. Animals, like us, can have only two front limbs – it's either arms or wings, take your pick.

EXPLAINING EVOLUTION: WHY DOES SEX EXIST?

Another mystery self-assembly partly resolved is sex. Sex gets a lot of attention from plants, animals and people. But it's a risky way to exchange genes, especially mysterious given that nature has far simpler ways to do the same thing. So why did sex evolve?

Key tenets of Darwin's theory, like competition and survival of the fittest, argued *against* sharing helpful traits with others, and he found sex baffling. "We do not even in the least know...why new beings should be produced by the union of the two sexual elements."[5]

In need of an explanation, Darwin invented "sexual selection," the idea that animals select mates with the best traits. However, as Victorians knew, breeding the "most fit" animals can be counterproductive, sometimes producing very weak young. Today we know why.

There's a dark side to "good" genes, says geneticist Adam Chippendale at Canada's Queens University.[6] Genes that are helpful to sons can harm daughters. Genes good for daughters can hurt sons.[7] These harmful effects are common, thus any benefit from sexual selection would be completely lost within a few generations.

Darwin's theory had so many problems that his inability to explain sex drew little notice for many years. But with the rise of Darwinism in the 1950s, explaining this powerful aspect of evolution became imperative.

The first attempt came from William Hamilton. Conjuring up an image out of a 1950s horror movie, he described the early world as overrun by flesh eating, blood sucking parasites. Hamilton said sex evolved to create defenses against these microscopic monsters. Sex reshuffled genes and occasionally, by chance, this shuffling would produce a new parasite defense.

Of course the idea that genes are randomly shuffled during sex was preposterous. It would lead to biological chaos.

The Darwinists had few other explanations, and sex remained a mystery.

SEX AS SELF-REPAIR

With the advent of self-assembly, however, sex began to give up some of its secrets.

Self-assembly has a built in repair process that works at all levels of biology, including genes. The most common process for fixing damaged DNA is "template repair." One side of the DNA automatically pulls in the molecules needed to repair the damage on the other side.

Over time, the template repair processes became more elaborate, replacing increasingly larger sections of damaged DNA. This led University of Arizona biologist Harris Bernstein to a radical idea. A side effect of template repair might be the recombination of genes in sex.

Called the "repair hypothesis," Bernstein said that as life became more complex, the template repair processes expanded until eventually it could match one entire side of the genome to another.[8]

DNA Damage Repair

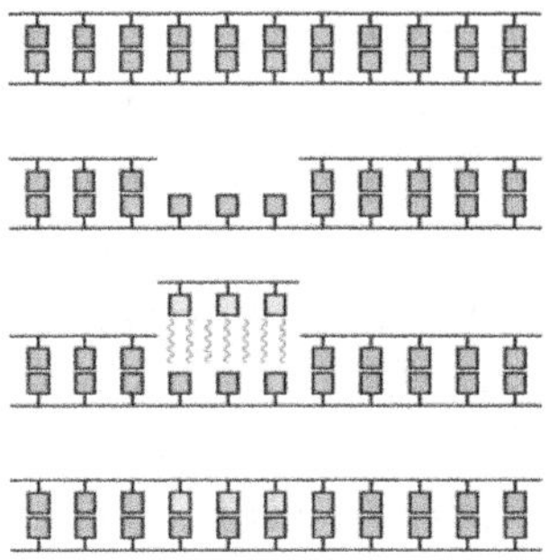

One side of DNA naturally attracts the correct molecules to repair itself.

Combining DNA in Sex

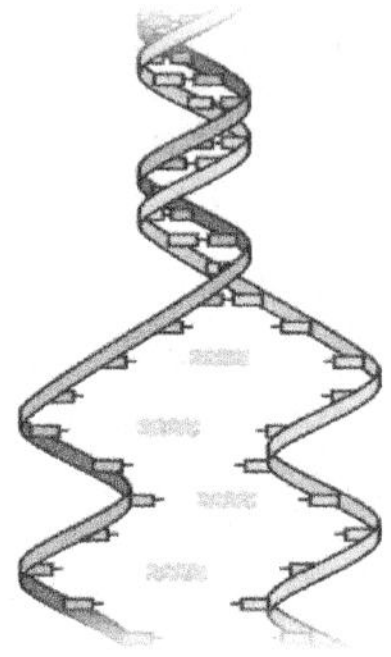

Combining DNA strands is a variation of DNA repair process: Matching parts naturally attract each other.

Thanks to Lynn Margulis it was now well known that the first complex cells emerged when one cell absorbed another, and put the second set of genes to work on specialized jobs. That is essentially what happens in sex. The egg absorbs the entire sperm cell, and then combines the two sets of genes through a version of the DNA self-repair process.

Sex as Cooperation

The idea that sex is an extension of the cell's basic self-repair process helped explain other mysteries, such as the widespread ability of creatures to *change their sex.*

Fish are particularly adept at these changes. Among the first to study them was Robert Warner, head of Ecology, Evolution and Marine Biology at the University of California, Santa Barbara. "Gender can be surprisingly labile, [changeable]" says Warner. Fish "may function first as one sex, then another over the course of their lives." Surprisingly, he says, "these changes can be socially controlled."[9]

A group of fish called the Cleaning Wrasse are typical. If a group loses its male, within hours the largest female begins to act like a male. Within ten days she – now he – can produce sperm. How can something as basic as sex reverse itself, and do it in ten days no less?

Complexity Theory held some answers. We know that stress changes the body, and social stress is as powerful as any other type. We also know that every cell contains the material to create both males and females. It appears that the lack of a male creates enough social stress to trigger changes in one of the females. Once begun, these changes cascade through her body at a blistering speed.

This type of reversal is surprisingly common in nature, making a mess of Darwin's "sexual selection" idea. "The whole context for Darwin's theory of sexual selection is dissolving," says Stanford biologist Joan Roughgarden, one of the world's leading experts on animal sexuality.[10] "Darwin is incorrect in the particulars, but more importantly, is inadequate even as an approach."[11]

"An unbelievable amount of sexual diversity is very common," say biologist Paul Vasey, who also studies animal sexual behavior. "I see it every day. Traditional evolutionary theories [i.e. Darwinism] are inadequate and impoverished to account for what is going on."[11]

A great deal is still to be explained about sex, but discoveries like this put research on the right track. In particular, they remove the idea that sex is primarily a competitive activity. Yes, some creatures compete for mates. But as ants, bees, birds, fish and many other species show, this is only one aspect of sex, and may have emerged late in evolution.

Much sexual behavior involves cooperative bonds that extend far beyond the short term urges of mating, showing that its origins and purpose are far more complex than those simple explanations.

Integrating Evolution: The Same Process Throughout Nature

One of Complexity's most remarkable achievements is integrating the entire natural world under one evolution process, based on the same laws. This is a comprehensive explanation for evolution at every level of nature.

The breadth of the integration is astonishing, covering evolution in every area of science, from physics and chemistry to cosmology, geology, paleontology, biology, ecology, genetics, neurology, psychology, sociology, anthropology and economics.[14]

For the first time since the Darwinists took control, evolution is no longer at war with the evidence, or with the other sciences. No more imaginary processes. No more "just so stories" that change at the drop of a hat. "Part of the current maturing of biology," said Richard Strohman, "is the surrender of simple 'storybook' explanations for how life works, and the acceptance that life is beginning to appear more like a complex adaptive system."[15]

After 60 years of bitter debate (within science), biological evolution is once again integrated with other sciences, and consistent with the basic process of nature seen in physics, chemistry and the other life sciences. Explanations no longer confuse, but enlighten. When new discoveries are made, they no longer conflict with evolution, they confirm it. This is the science of Newton and Einstein, full of the "ah ha" moments that leave scientists exclaiming "so that's how it works!"

Integrating Development and Evolution

An important effect of this integration is reuniting evolution with the development process. Because development can't be explained in Darwinian terms, the Darwinists said it is unrelated to evolution, and disconnected the two fields for 50 years.

Stephen Jay Gould's work in the 1980s rekindled the long suppressed idea that evolution and development are two sides of the same process.[16]

Discoveries in the 1990s confirmed this connection, revealing that the same processes are involved in both. This led to the emergence of evolutionary developmental biology, or "Evo-Devo," which says that evolution and development share the same underlying process.[17] Both change by creating *new relationships* between genes, not by making new genes.

This is "a profound departure from evolutionary biology's long obsession with genes," says biologist Allen Orr.[18] With Evo-Devo, he says, "evolution works not so much by changing genes, but by changing...the *patterns* of gene expression."

Changing these patterns also alters development, intimately linking the two processes. This is not some bit of abstract textbook biology. It changes our understanding of how birth defects develop, how to deal with infant diseases, and what can disrupt brain development and learning abilities.

Integrating Structure and Adaptation

Complexity Theory also allows evolution to reclaim its scientific roots, lost under Darwinism, by integrating "adaptation" and "structure." This restores three great ideas about evolution:

1. John Baptist Lamarck's idea of acquired traits, today called epigenetics, which explains how adaptive traits emerge.
2. Richard Owen's idea that the laws of structure explain the origin of new species.
3. Barbara McClintock's idea that adaptive evolution and new species are made possible by rearranging mobile "regulator" genes.

Self-assembly particularly validates Owen's discoveries. He described an "architectural logic" to life, which today we recognize as the laws of self-assembly. He spoke of a "polarizing force," and an "organizing force," which determine what can evolve. In his words, the "organizing principle would... mold specific forms."[19]

Owen said that the "essential character of a part" leads it to form a "predetermined pattern," and this pattern then determines "all [future] modifications"– as clear a statement of how self-assembly determines evolution as any modern scientist would make.[20]

INTEGRATING EVOLUTION WITH THE BASIC LAWS OF NATURE

Richard Owen's work points to another important discovery. Many developments in evolution are not determined by adaptation, but by the rules of physics and chemistry. "The old notion of natural selection as an omnipotent force in biological evolution has given way," says Stephen Miller at the University of Minnesota Biotechnology Institute.[21] "Adaptive processes are constrained by physical, chemical, and biological requirements."

These "requirements" determine what can evolve, explaining why nature converges again and again on the same genes, networks, and biological systems.[22]

Physical requirements have been determining the course of evolution from the beginning of life. The Earth was originally enveloped in carbon dioxide, which the first cells used to help absorb light and make food. In the process they broke the carbon dioxide apart, releasing oxygen into the atmosphere (making life on land possible) and carbon into the soil (which formed the backbone of all later life).

These two changes completely altered the chemistry of the oceans and atmosphere, and set the course of all future evolution. And both were determined solely by basic physics.

The first complex life also came to exist through similar physical processes. Cells naturally cast off chemicals that allow other cells to sense their presence, and unite when needed.[23]

The chemical reaction of these unions leads automatically to the formation of a "net" (an "extra cellular matrix") that surrounds the cells and holds them together.[24]

This surrounding "matrix" creates other chemical reactions that automatically lead different regions to form inside the group. The different chemical properties of each region cause these identical cells to change and begin to specialize in one task.[25]

These new areas become the specialized body parts of later life. For example, some cells specialize in reproduction, creating spores. Over time, these spores evolve into the seeds used by plants to reproduce.

Similar physical processes (not genes) led to the evolution of the first individual body parts during the Ediacara era, and to other complex traits that evolved later, like the skeleton.

Skeletons are one of nature's most important developments, allowing the spongy worms that crawled the ocean floor to begin to swim and eventually crawl onto land. Recent discoveries have revolutionized our understanding of skeletons, showing they evolved through the same self-assembly process which makes everything else in nature.

Stuart Newman, who pioneered many of these discoveries, shows that during development, the skeleton forms based on a set of physical signals which emerge as the embryo unfolds.[26] We saw how this process creates arms. Newman had earlier shown that similar physical signals determine where bone cells emerge *through the entire body*. Chemical signals from the surrounding tissue then determine each bone's size and shape, creating the skeleton. When first introduced, Newman's concept was so novel it was initially met with skepticism, but it has now become mainstream science.[27]

These physical processes explain why many species, despite their radically different forms, repeatedly evolved similar traits (called "convergent evolution," over time they all converged on the same form).

The pelvis, for example, is a key evolutionary innovation that allows animals to bend and twist. How a skeleton is able to bend is determined by the laws of physics, not natural selection, explaining why *all creatures*, from flies to fish to man, evolved *the same basic pelvic form*.[28] Once evolution abandons Darwinism and embraces the laws of physics, explaining evolution becomes easy and makes common sense.

Explaining How the Turtle Got Its Shell

These breakthroughs about the role of physical properties in evolution explained one of nature's more peculiar creations – the turtle shell.

Unlike any other animal, turtles push their ribs outside the body and fuse them together, encasing their body in solid bone. Darwinists claimed the shell evolved as protection against predators. But just watching a turtle shows that shells make them slow and vulnerable.

Biologist Scott Gilbert unraveled part of the mystery when he found that the signals which push ribs out (creating the rib cage in other animals) don't stop in the normal place in turtles, but continue pushing.[29] Once outside the body, natural bone chemistry triggers a process that automatically fuses the ribs together, forming the shell.

The evolution of this process unfolds based purely on bone chemistry, with no "natural selection" involved. Its evolution also happened rapidly, explaining why the turtle shell left no "intermediate" species.[29]

Why the turtle began to push its ribs outside the body is still unknown, but it may not be as strange as once thought. University of Chicago anatomist Neil Shubin says the first "hard thing" to evolve (the first bone so to speak), was a tooth. It enabled an otherwise soggy worm to take a bite out of something larger than itself.[30] Shubin says the first skulls to evolve look remarkably like a set of teeth pushed to the outside of the head and fused together, encasing the brain. Maybe the turtle was trying to do the same thing, just in a different place.

Predicting Evolution: Convergence

Self-assembly goes beyond explanations, to make powerful predictions about evolution. We saw that it predicts how adaptive changes occur, and how new species emerge.

It also predicts evolution will show "convergence," the idea that the same traits evolve in different species. The universal constraints on self-assembly will lead the same trait to repeatedly appear.

Convergence shows up everywhere in nature. A group at Oxford University, for example, discovered that plants evolve similar leaves because they share the same genes and same underlying "leaf-making" processes.[31]

Stanford's Kevin Boyce and Harvard's Andrew Knoll found that other shared elements place "strong constraints" on all aspects of plant evolution, putting many plants on what they call "parallel trajectories" in evolution.[32]

Animals evolve similar traits for the same reason: they share the same underlying biology. Hidden inside every bird wing, for example, are ancient dinosaur bones. This shared bone system, coupled with the basic laws of aerodynamics, means that all birds will evolve wings that look and function much the same. These constraints forced birds to evolve the same wing design regardless of their environment or other issues.

Predicting Evolution's Next Step

The ability to predict evolutionary change has profound implications for everyday life. The University of Rochester, in "Scientists Begin Predicting Evolution's Next Step," said discoveries by biologist Barry Hall had led to a model of evolution "so accurate it can be used to predict how a strain of bacteria will become resistant to antibiotics - giving researchers a possible tool to create drugs to which bacteria cannot adapt."[33]

"Fighting bacteria with antibiotics has always been done the same way," said Hall, one of America's leading geneticists. "We make a drug and after a while, the bugs adapt to it," he says. "But if we can predict how they're going to get around our treatments, we can work out a way to make that route impossible for them. We can cut them off at the pass."

As we will see, these discoveries are transforming many other areas of life, from medicine, to the development of previously unimaginable materials, to new insights into protecting the environment and helping nature heal itself.

PART 5

HOW THE NEW EVOLUTION CHANGES EVERYTHING

What does all of this mean for us?

The revolution in physics which emerged in the 1920s, raced through the sciences in the 1950s, and later exploded into the electronics, computer and communication revolutions that we live in today.

The discovery of Complexity in the 1980s, has swept through the other sciences over the last two decade, and today places humanity on the threshold of a powerful and transforming era. This new understanding of nature allows us to harness nature's *own processes* to change, well, *everything*.

Chapter Twenty-Two

HEALTH, HEALING AND WELLNESS

Most biological characteristics arise from complex interactions. A key challenge for biology in the twenty-first-century is to understand that web of interactions. This will revolutionize our view of disease.[1]

Albert-László Barabási, Northeastern University, 2004

The most immediate effect of these discoveries are in medicine. They reveal how to harness the body's natural "self-assembly" process to repair damaged hearts and other body parts, an advance already moving from laboratories into hospitals. They also hold the potential for dramatic breakthroughs, such as understanding how viruses evolve and cancers spread.

Beyond this, these discoveries alter our basic concept of disease – what it is and how to prevent it. And they change our concept of how to fight disease by activating the body's own defense processes.

They even alter our basic concept of health – how we define it and maintain it. This opens the door to a *different type* of health care, with major consequences for improving the quality of life.

The Body's Natural Self-repair Process

At the simplest level, self-assembly has a built-in self-repair process. Damaged parts naturally attract the right molecules to repair themselves. A broken bone shows how doctors can activate this process to repair the body.

All bones have a mild magnetic field, with a positive and negative end, just like a magnet. A broken bone alters this magnetic field in a way that pulls calcium into the broken spot. In essence, it creates an "I need calcium" signal at the broken spot. As calcium molecules arrive, they are pulled into the open space, healing the bone. When the gap is filled and the parts reconnected, the bone's magnetic field is restored to its normal state, and it stops sending "I need calcium" signals.

A similar process repairs other body parts.[2] The damaged area creates a series of "I need" signals. Conveyed by chemical messengers, these "need" signals echo throughout the body, causing other systems to produce what is needed. Once in the bloodstream, a string of chemical processes pull the "needed" material to the damaged tissue. Once there, natural attractions pull them in to repair the damage. This is how a torn muscle or cut finger heals itself. Scientists hope to harness this process to heal damaged ligaments, muscles and other tissue without surgery.

In the near future, it will be possible to harness the same processes to rebuild damaged hearts, diseased livers and perhaps even injured brains. In the not too distant future, hardware, like dialysis machines, will be replaced by naturally repaired kidney tissue.

Some of these developments are already a reality.

The idea body parts can assemble themselves is what allowed scientists to create the throat part for Claudia Castillo mentioned in the Introduction, in 2008.[3] She quickly regained a normal life, able to raise her two children, walk up stairs, and even go dancing. "This is the first time a tissue-engineered whole organ has been transplanted into a patient," said Dr. Martin Birchall, a member of the transplant team.[4] "In 20 years' time it will be the commonest operation – it will transform the way we think about surgery."

Birchall was right, only it's happening faster than anyone imagined. In 2012, doctors created a new windpipe for 30-year-old Christopher Lyles, who had a type of throat cancer normally considered inoperable. Paolo Macchiarini, director of Sweden's Advanced Center for Translational Regenerative Medicine said, "what we did is surgically remove the malignant tumor. Then we replaced the trachea with a tissue-engineered scaffold."[5]

Doctors then placed cells in the "scaffold," which grew naturally into a new trachea. Because Lyles' own cells were used, there was no need for drugs to prevent his body from rejecting the new organ, a common problem in transplants. "He went home in very good shape," said Macchiarini.

The Body's Natural Regeneration Process

Understanding that the body is built through self-assembly makes it possible to create an entirely missing part. In 2008, Angela Irizarry was born with only one heart chamber. In the US, 3,000 babies a year are born this way, 68,000 worldwide. Without surgery 70% die before their first birthday.

In 2012, doctors at Yale University grew a second heart chamber inside Angela's chest. "We're inducing regeneration," says Yale pediatric surgeon Christopher Breuer, who led the procedure.[6] "We're making a blood vessel where there wasn't one."

Previously weak and listless, after the procedure Angela became like any other energetic four-year-old. Angela's case is a milestone, says MIT regenerative-medicine pioneer Robert Langer. "It gives you hope that when you combine cells with a scaffold and [put] them in the body, they will do the right thing."[6]

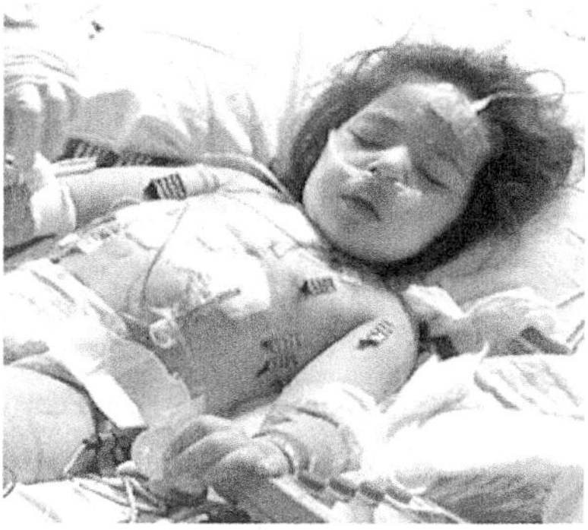

ANGELA IRIZAARY
PRE OPERATION

ANGELA IRIZAARY
POST OPERATION

These "discoveries relate to what determines a cell's fate – is it going to be a heart cell, a liver cell, a brain cell," says Richard Klausner, former head of the National Cancer Institute.[7] "This whole field took a turn" in 2007, he says. Scientists discovered they did not need embryo cells, but could take *any cell*, reverse it back to its original form, and then turn it into any new type of cell they chose. One group turned them into nerve cells, another into heart muscles, and a third into pancreas cells which produced hormones like insulin (representing a possible cure for Type 1 diabetes). "I find it so wild," says Klausner, we went "from nobody being able to do this to everybody being able to do it."[7]

The process Klausner describes takes place simply by changing the signals a cell receives. In the right context, the cell responds to the surrounding signals and begins creating a new organ. Once started, the naturally unfolding relationships of development do the rest.

Anthony Atala, director of the Wake Forest Institute for Regenerative Medicine, who pioneered many of these advances, said:

> "One day, and it's not science fiction, a doctor will take a skin cell and grow you a new kidney. It's not that far off. I think, as we master these tricks, we're going to get better and better at this, not only in how to make organs but in how to make them more efficiently."[8]

Michael Friedlander, Director of Neuroscience at Baylor University believes we may eventually be able to reverse brain damage, by getting cells to recreate their initial role in wiring up the brain's connections.[9]

Steps in this direction have already emerged. Chitin is one of those molecules that nature puts to many uses. One type of chitin, called chitosan, forms the membrane wall of a cell. When a cell gets injured, the inner membrane is exposed and it begins sending out an "I need" signal that attracts chitosan, which then is pulled in, patching the broken cell wall.

At Purdue University's Center for Paralysis Research, biologist Young Nam Cho wondered if this process could repair nerve cells. Working with damaged spinal cords in animals, she found that a single injection of chitosan in the damaged area restored sensation to animals in a mere 30 minutes.[10] Cho is now researching whether this process can help paraplegic animals walk. If successful, it may eventually help hundreds of thousands of paralyzed people repair damaged nerves and regain control of their bodies.

FIGHTING VIRUSES AND BACTERIA

This new concept of evolution is also altering the fight against disease.

Viral diseases like smallpox have plagued humanity for eons. Centuries ago, waves of viral disease carried by rats swept through Europe, killing one-third of the population. Societies took decades, even a century, to recover.

The other great killer of humankind, bacteria – cholera, tuberculosis and a dozen others – killed thousands every year.

The early 20th century immunization revolution dramatically reduced deaths from viruses. In 1928, Scottish biologist Alexander Fleming discovered that penicillin kills bacteria, and by World War II "antibacterial" drugs were killing bacteria of every type, saving millions of lives.

By 1950, doctors and scientists were predicting that the twin scourges of humanity, viruses and infectious bacteria, would soon be wiped from the face of the earth.

But that didn't happened.

Although antibiotics were astonishingly effective when first introduced, over time they became increasingly less effective. In 2012, a lethal new strain of the tuberculosis bacteria emerged that is resistant to *every* antibiotic we have. It's only a matter of time before other untreatable strains emerge.

And while we've had some success in combating viruses (for example, polio and small pox), other viruses have proven impossible to kill off.

The great influenza pandemic of 1918 killed 50 million people, 10 times more than all the combat deaths of World War I.[11] New viruses appear each year. If a particularly lethal strain emerged today, creating a pandemic like 1918, deaths worldwide could easily reach 300 million – equal to the population of the United States – and we would be helpless to prevent it. We can't even stop the virus that causes the common cold.

BACTERIA AND VIRUS RESISTANCE

With such a promising start, why are we losing the battle against these diseases?

The problem began in the 1950s, when researchers started using the wrong model of evolutionary change: random mutation. University of Virginia microbiologist Robert Kadner explains. "Most antibiotic resistance in ...bacteria are not the result of a mutation."[12] Instead, a bacteria recombines its genes until it finds a form that matches the antibiotic. The new combination attaches itself to the antibiotic and renders it inactive.

These recombinations are not random, but targeted to specific antibiotics. One strain of pneumonia bacteria generated an astonishing 700 new gene combinations in 40 years – that's a new one *every three weeks.*[13]

This "shows the remarkable ability of this pathogen to evolve over very short time scales by frequent [gene] recombinations," says Brian Spratt at Imperial College, London.[14] "The consequences of such rapid evolution for the emergence and spread of antibiotic resistance...are clear."

Researchers in the 1950s held the same erroneous idea about virus genes, believing they change randomly. But that proved untrue.

Viruses alter their genes in specific ways designed to avoid specific vaccines. Michael Sargent, at Britain's National Institute for Medical Research, says these targeted changes are caused by a type of RNA "that is extraordinarily mutable and able to alter [itself] to evade any immunity."[15]

The New Evolution explains how bacteria develop their resistance:

- Stress leads bacteria to alter their genes (as John Cairns described).
- Bacteria rearrange mobile genes to create a combination that inacti-

vates specific antibiotics.
- Resistance spreads rapidly because bacteria actively exchange the new gene combinations (as Carl Woese described).

Viruses become immune to vaccines using the same three-step process – stress alters genes, the altered genes are recombined to inactivate vaccines, and these recombinations are widely exchanged.

Overcoming Bacteria and Viral Resistance

In simple terms, bacteria and viruses defend themselves the same way our immune system defends us – rearranging genes to create defenses.

Understanding that microbes develop resistance by rearranging existing genes radically alters how researchers approach disease. "What would be ideal is a table for infectious agents," says Leonid Margolis at the National Institute of Health.[16] It would have "the microbe, the immune factors it affects, the receptors it uses and the signalling systems it incorporates." This would give scientists a road map for disabling them.

A similar idea is being tested to combat "juvenile" diabetes, which is caused by a virus. In the United States, juvenile diabetes infects 80 people every day, 30,000 a year. Worldwide new cases top 500,000 a year. Although juvenile (or Type 1) diabetes is manageable, treatment is expensive and requires constant monitoring. Many children can't afford it or get inadequate supervision, with tragic consequences, including early kidney failure, blindness, heart disease and death.

Juvenile diabetes is caused by a virus that inserts its DNA into the body's insulin-producing cells. The immune system then identifies those cells as a virus and kills them. The body makes new insulin cells every day, but because they contain virus genes, the immune system kills them.

With our new understanding of genetic change, it may be possible to eliminate the virus genes in these cells. Once they are gone, the immune system would stop killing the new insulin producing cells the body makes, and eliminate juvenile diabetes as a disease. Belgian biologist Harry Heimberg has discovered how to stimulate the body's repair process to produce insulin cells at ten times the normal rates, offering the potential for a rapid recovery once the virus genes are eliminated.[17]

Epigenetics: Transforming our Concept of Disease

The new science of epigenetics is poised to transform other aspects of medicine. Tulane University geneticist Melanie Ehrlich, whose lab first linked cancer to epigenetic changes, says, "epigenetics is one of the fastest-moving areas of science, period."[18]

Epigenetics further alters our concept of disease, and how to approach treatment. For example, research in Germany, France and Finland shows that new antibodies created by a mother's immune system can be directly inherited by her children, providing them with disease protection from birth, a discovery with breathtaking potential.[19]

Hilmar Lemke and Hans Lange at Germany's Biochemical Institute, say this research "reveals a Lamarckian dimension of the immune system."[20]

Path breaking work on Lamarck's type of inheritance and the immune system by biologists Ted Steele, Eva Jablonka, Mary Jane West-Eberhard and others, which was once attacked, now forms the foundation for this exciting new approach to disease.

EPIGENETICS: TURNING OFF DISEASE

Epigenetic changes can also alter the "on-off" switches which activate genes. Andrew Feinberg at Johns Hopkins, says that a strong argument can be made that these epigenetic switches are the cause of some cancers. Feinberg has identified three gene switches which, if reversed, halt the growth of some brain tumors.[21]

In a remarkable discovery, scientists have found that drugs which change these switches can be targeted to specific cancers, viruses and diseases. McGill University geneticist Moshe Szyf shows that it may soon be possible to simply "turn-off" some cancers by flipping the switches on their genes.[22]

Geneticists have also found that what they call "messenger genes" can travel through the body and alter other genes. At the Johns Hopkins Institute for Basic Biomedical Sciences, for example, Nancy Craig is investigating the *Hermes* gene.[23] Named after the messenger of the Greek gods, *Hermes* is unusually versatile. It can pick up a gene fragment, carry it to a new place and insert it into another gene, altering what the gene does. *Hermes* was recently used to insert a "fluorescent" gene into a small fish whose young were then born with bright, "glow in the dark" colors.

Craig says that "understanding how *Hermes* and [similar] genes work holds clues to fighting bacterial infections, improving gene therapies and tackling disease-carrying insects."[24]

Mobile genes provide a safer way to treat cancer than chemotherapy or radiation. This is another development that has also moved out of the laboratory and into hospitals. When fully mastered, messenger genes will become microscopic medical tools, delivering lifesaving solutions to the body's most difficult to reach places. The fictional medicine of the movie *Fantastic Voyage* is becoming today's reality.

Another promising approach is to use epigenetics to enhance the body's natural cancer-suppressing mechanisms.

The body is constantly hit with viruses, but the immune system fights them off. Adrian Hayday and Jessica Strid at King's College London, and a team at Yale led by Mike Girardi, have identified the molecule that activates this natural anti-cancer response long before some tumors form.[25] If this mechanism can be enhanced, it may be able to detect other, more stealthy cancer agents that invade the body.

The results from epigenetics are so promising that in recent years the National Institutes of Health have allocated billions to expand research in these areas.

EPIGENETICS: PREVENTING DISEASE

These discoveries also hold great potential for *preventing* disease.

Biologists Stephen Baylin at Johns Hopkins and Peter Jones at UCLA are using epigenetic markers to predict which liver cancer patients will respond to particular anticancer drugs. In 2009, German pharmaceutical company Epigenomics, introduced an epigenetic screening test for colon cancer and is at work on a similar test for breast cancer. In 2013, pharmaceutical company Predictive Biosciences introduced an epigenetic-based test for prostate cancer.[26]

Epigenetic changes can also halt the spread of disease carried by insects like mosquitoes and medflies by disrupting their developmental process.[27]

Another approach to disease prevention involves changes in diet, the same change that transformed Randy Jirtle's mice. Researchers at the University of California School of Medicine found that changes in diet epigenetically altered 501 genes related to prostate cancer. If inherited, changes like these could reduce prostate cancer for generations to come.[28]

EPIGENETICS: EXTENDING LIFESPAN

Epigenetic changes may also extend our lives. In 2011, Stanford geneticist Anne Brunet showed that tiny worms (called nematodes), each with identical genes and living conditions, can have different lifespans based solely on epigenetic changes. Some lived dramatically longer than normal, and *passed this benefit on to their descendants.* The "inheritance of lifespan is...associated with epigenetic changes," says Brunet. "Manipulation of specific [genes] in parents can induce...longevity in descendants."[29]

EPIGENETICS: UNDERSTANDING CHILDHOOD DISEASES

Epigenetics may also shed light on asthma, ADD and similar conditions, which were largely absent two generations ago but today plague many families. Since 1980, for example, cases of childhood asthma have accelerated dramatically, reaching epidemic levels. Around the same time, Reye's syndrome in children was linked to aspirin use, leading doctors to switch to acetaminophen for children's fevers.

In 1998, University of Wisconsin immunologist Arthur Varner proposed that the shift to acetaminophen drove the increase in asthma. Dozens of studies have now confirmed the link.[30] In 2008, an ISAAC study (International Study of Asthma and Allergies in Childhood), covering 200,000 children in 31 countries, found that using acetaminophen for fevers in the first year of life, later led to a 50% increase in asthma symptoms.[31] A 2009 study showed that children who took acetaminophen doubled their risk of developing breathing problems in the year that followed.[32]

High temperatures can alter a gene's "on-off" switches. The question being asked is whether using acetaminophen for a fever changes the "on-off" switches related to asthma.

In a similar development, Wolf Reik at England's Babraham Institute, found that merely handling an embryo changes the "on-off" switches on some genes, altering the baby's development.[33] This has profound implications for mothers like Stephanie Mullins, who used *in vitro* fertilization to get pregnant. Her son, Ciaran, was born with Beckwith-Wiedemann Syndrome, a rare disorder linked to epigenetic changes. Research shows that babies conceived *in vitro* have a 300% greater chance of developing Beckwith-Wiedemann Syndrome.

In a chilling discovery, Reik found that these changes are sometimes inherited, meaning the disorders can appear for generations into the future.

EPIGENETICS: THE SOCIAL IMPLICATIONS

One of the most surprising findings of the Epigenetic Revolution is social. The first indication of what this means came from Britain's famous Avon Study, which tracked the health of thousands of children, matching it to their parents' behavior prior to their conception.[34] The connections were stunning. Fathers who began smoking before puberty (not uncommon in the early 20th-century), had sons with consistently higher rates of obesity, diabetes and other diseases. (Surprisingly, their daughters showed no ill effects.)

Smoking by a mother before pregnancy impairs the ability of both her daughters and sons to later have children of their own.[35]

At the Institute of Child Health in London, Lars Olov Bygren and Marcus Pembrey, using 200 years of meticulous records, found that a man's health was affected by the amount of food his *grandfather* ate. A series of droughts led to wide swings in the amount of food available in the prior two centuries. Bygren and Pembrey found that men live significantly *longer* if their grandfathers had only a *small amount* of food to eat in the years prior to the grandfather's puberty. Conversely, men live shorter lives if the grandfather had *large amounts* of food during those years.[36]

Why it skips a generation is unknown, but the effect is clear. If, during his pre-puberty years, a man eats *small amounts* of food, his grandsons *live longer*. If he eats *more food* in those years, his grandsons have *shorter lives*.

Bygren and Pembrey also found a link between the amount of food parents ate in their pre-puberty years, and the levels of heart disease and diabetes their sons experienced as adults.

Epigenetics: The Mental and Emotional Effects

Our emotional experiences also alter our genes, and the changes can be inherited, affecting the mental abilities and emotional states of our children.

Abuse and neglect during childhood alter the genes related to thinking and emotions. One of the leading researchers in this area, neuroscientist David Sweatt, says "childhood maltreatment and early trauma leave lasting imprints on the mechanisms of cognition and emotion."

Remarkably, these emotion-driven changes are inherited by the next generation. "Early maltreatment produces persisting changes," says Sweatt, "potentially underlying lifelong, and *trans-generation*, changes."[37]

Research by Duke University neurogeneticist Idan Shalev, shows that children who are physically abused, or who even witness violence, sustain genetic damage similar to that caused by old age. The more violence, the faster the genes "age." Shalev estimates that a child exposed to violence loses up to 10 years of life compared to one who is not. The "aging" of genes skyrockets for children who experience multiple types of violence.[38]

Other emotional experiences that change genes include:

- **Depression:** Separation from the mother often triggers "depression-like conditions" which alter genes.[39] (Dietmar Spengler, Max Planck Institute.)
- **Suicide**: People severely abused as children and who later commit suicide show unique epigenetic changes.[40] (Michael Meaney, McGill University.)
- **Brain Development**: Early stress in rats alters the genes involved in brain development.[41] (David Sweatt, University of Alabama.) Some epigenetic changes appear able to trigger mental retardation. (Huda Zoghbi, Baylor University.)
- **Autism**: People with autism show an epigenetic change in the gene that makes oxytocin, a hormone that influences social bonding.[42] (Jessica Connelly, University of Virginia.)

Epigenetic changes can also have positive effects. Tufts University biochemist Larry Feig has shown that mice raised in stimulating environments have smarter young with better memories. In other words, a mother's childhood experience alters her offspring's ability to learn and remember.[43]

Remarkably, this change appears in the next generation even if they themselves *never have a stimulating childhood*. "What is so unique is that... the beneficial effect reached into the next generation," says neuroscientist Dean Hartley, who worked with Feig. "The offspring had improved memory even without an enriched environment."[44]

Other positive benefits of epigenetic changes include:

- **Coping Skills:** Rats raised by attentive mothers deal with stress much better than those raised by negligent mothers, and this coping ability is inherited by the next generation.[45] (Michael Meaney, McGill University. Initially rejected by many journals, this is now one of science's most highly cited articles)
- **Self-control**: Mothers who are highly attentive have daughters with less promiscuous sexual behavior.[46] (Frances Champagne, Columbia University. Champagne says, "doting mothers have prudish daughters, whereas the daughters of inattentive rats cavort around like mini Mae Wests.")

Understanding both the harmful and helpful effects of epigenetic changes gives us powerful tools to change the emotional damage done to children. Most epigenetic changes can be reversed. If we can unravel the process, we may be able to reverse the unfortunate effects of childhood abuse, and prevent it from cascading into future generations.

EPIGENETICS: THE CULTURAL EFFECTS

Scientists have discovered that even culture can change genes.

University of British Columbia cognitive anthropologist Joe Henrich says that "a rising tide of evidence from epigenetics" shows that culture can create "lasting heritable" changes, in both individuals and entire groups. "Many in the behavioral sciences have yet to take seriously the implications of epigenetic inheritance and culture-gene co-evolution," says Heinrich.[47]

Anthropologists now believe that some genes "co-evolved" with human culture as it progressed. Changes in culture lead to new behavior, which then altered people's genes. John Hawks, a leading expert on the co-evolution of culture and genes shows, for example, that human gene change "accelerated greatly" 40,000 years ago, in tandem with the cultural explosion at the time, which included first symbols and a major leap in the complexity of social life.[48]

EPIGENETICS: THE LONG-TERM IMPLICATIONS

"These discoveries are shaking the modern biological and social certainties about genetics and identity," says science journalist Ethan Watters. "We commonly accept the notion that through our DNA, we are destined to have particular body shapes, personalities, and diseases. 'Gene as fate' has become conventional wisdom."[49]

Epigenetics proves this is untrue. Our genes don't control us, we control them, and have been for eons. The evidence is clear that our *behavior* alters our genes, and the genes of future generations. Our *mental, emotional* and *cultural* experiences (ideas, beliefs, language) also alter our genes. If these changes are passed to future generations, as seems to be the case, this overturns every old idea about human evolution and history. Genes didn't create us, we created our genes, or at least how they act. "Epigenetics is proving we have some responsibility for...our genome," says Randy Jirtle. "Everything we do... can affect our gene expression and that of future generations. Epigenetics introduces the concept of free will into our idea of genetics."[49]

Understanding that our experiences – physical, mental, emotional – are inherited by future generations *reverses* the idea that our behavior is hard wired by genes. It's hard to imagine a more powerful concept for transforming society: the choices we make change not just the society we hand our children, but the genes they receive.

Chapter Twenty-Three

SELF, SOCIETY AND THE FUTURE

Four Empowering Ideas

Complexity Theory, and the new concept of evolution it contains, affects every aspect of our future.

In *Complex Systems: Science for the 21st Century,* a group of leading scientists outline the remarkable implications. "Can we harness the exquisite complexity of Nature," they ask? The answer is a resounding "Yes," With this new understanding we can uncover "nature's design rules and try to mimic green plants which capture solar energy," discover how nature repairs damaged tissue and how it creates complex structures and new materials. "Nature tells us it can be done," they say, "and provides us with models."[1]

These discoveries also explain the life cycles of oceans, landscapes and ecosystems, and offer insights for keeping them healthy. They reveal how to develop more protein-rich foods, how to cheaply tap renewable energy sources, and how to create futuristic new materials that are both stronger *and* more eco-friendly than anything previously imagined.

"These concepts may seem fanciful," they say, "but with the revolution now taking place the possibilities seem endless. Nature has done it. Why can't we?"[1] The answer is we can, with benefits that are broad, tangible and immediate.

Complexity gives us something else, perhaps even more important. These scientists say the deep understanding of nature it reveals provides society with new wisdom, telling us "what can be achieved, what problems can be solved and where we can set our goals in exploiting this understanding to benefit society."[1]

Using nature to understand what can and cannot be done is what the ancients called the "wisdom of nature." Complexity allows us to recapture some of that wisdom for guiding ourselves, our societies, and the world we will hand to our children.

OPENING THE DOOR TO A NEW FUTURE

Complexity Theory has already transformed scientific research and our efforts to understand nature. Two of the leaders in this revolution, physicists David Pines and Robert Laughlin, say that with this transformation, the central task of physics "is no longer to write down the ultimate equations but rather to catalog and understand emergent behavior in its many guises. We call this physics of the next century the study of complex adaptive matter."[2]

More broadly, say Pines and Laughlin, "we are now witnessing a transition from the science of the past, to the science of the future, the study of complex adaptive matter...with its hope of providing a jumping off point for new discoveries, new concepts, and new wisdom."

These breakthroughs are also revolutionizing business, industry, and agriculture. To focus on just the Princeton University Center for Complex Materials, researchers there are using "the self-assembly of atomic, molecular and supramolecular systems" to create new materials, like "electricity-conducting plastics which could dramatically lower the cost of manufacturing solar panels."[3]

Others at the Princeton Center are working on how "the best features of different materials can be combined to form a new material." One group at the Center found that the materials they created allowed electrons to "act like miniature superheroes, relentlessly dodging imperfect surfaces, sometimes moving straight through barriers," yielding many new abilities.[4]

One researcher even created "a miniature magic carpet made of plastic that has taken flight."[5]

These are not just laboratory developments. The Princeton Center has what they call an active "industrial collaboration program," which reflects their commitment, "to explore the science and technology of materials, and... forge collaborations with industry to develop a host of new products and technologies," including "self-assembled nanostructures and quantum materials."[6]

At hundreds of universities worldwide, developments like these will accelerate economic growth (allowing millions to exit poverty), while using truly eco-friendly products and processes.

The recent "Inspiring Matter" Conference in London offered a glimpse into the future, with new materials that ranged from self-cleaning glass to self-adapting clothes. "Unconstrained by...perceived limitations, potential new applications and commercial opportunities could be opened up by artistic and scientifically creative minds working together."[7]

Textile designer Bradley Quinn, for example, "inspired by self-assembly," showed fabric that responds to light, heat and cold, and automatically adjusts to meet the wearer's needs. Mike Davies, known for his work on the Paris Pompidou Centre and London's Millennium Dome, offered a vision of future buildings with "biomimetic skins that respond to the environment, can harvest energy, and understand and respond to the needs and moods of their inhabitants."[7]

A broad range of new discoveries about the environment (for example, how oceans and ecosystems naturally replenish themselves) will give us more effective ways to protect the Earth, without compromising the economic growth needed to lift billions out of poverty, and fund universal education and expanded healthcare.

MIND AND MAN: A NEW CONCEPT OF HUMAN NATURE

The New Evolution also gives us a more scientifically accurate picture of human nature, allowing us to resolve some social problems.

A long-standing question is whether humans are unique. Darwinists argue that the human mind reduces to chemistry and then the activity of the atoms. Because consciousness is not a property of atoms, they say human awareness and Free Will cannot exist – they are just an "illusion of biology."

But, as we have seen, cognitive scientists say the human mind is unique. Complexity science can explain how humans evolved from an animal, and yet can possess unique abilities not found in animals. It shows that emergent properties appear when nature becomes more complex. This explains how animals can learn and remember, although the atoms that make up animals can do neither. It also explains how human consciousness can emerge from biology, yet be different from anything found in biology (or animals).

This realization leads to a far-reaching re-conception of human evolution. Evolutionary theorist Elliott Sober says the human mind leads to a radically different evolution process than animal biology. Animals don't ponder the future, says Sober, but people do.[8]

A distinguishing feature of human evolution is how *little* it involves biology. Steve Jones, the former head of Genetics and Evolution at University

College London says "humans are remarkable in the number of things they do and in how little they are specialized [biologically] to do them. Our evolution is unique in that it is mainly in the mind."[9]

These discoveries transform our concept of human nature.

Over the last decade, simplistic biological ideas of human behavior have been replaced with an understanding that people think about the consequences of choices *before they act*. This concept of human behavior, called "Rational Choice Theory," now dominates the social sciences. In 1900, humans were seen as instinct-driven and irrational, especially in groups. By 2000, scientists understood humans are thoughtful and rational, *especially in groups*. Groups provide a wealth of knowledge that makes people less prone to error and *more* rational. Groups make people smarter.

Despite the picture often seen in the media, we humans are also unusually cooperative. The essence of human society, from the beginning, has been cooperative exchange and collective action. "Reductionist sociobiologists argue human biological limitations constrain society," says Richard Lewontin, "but, in fact, social organization overcomes individual limitations:"

> "No human being can fly by flapping her or his arms and legs. Yet we do fly because of the existence of airplanes, airlines, pilots, fuel production, radios – all products of social organization. Most important, it is not society that flies, but individuals who have acquired a property as a consequence of socialization"[10]

For humans, social progress arises strictly from these collective efforts, involving the contributions of many people with widely different abilities.

DEBUNKING GENETIC DETERMINISM, RACISM AND EUGENICS

The New Evolution also debunks ideas of genetic determinism, racism and eugenics that continue to permeate social thinking. It shows that humans all share the same genes, and that we control those genes, able to change how they act by altering our own behavior.

It also discredits racist eugenic claims that remain strong in some academic circles, especially in England. If selective breeding were simple, every Kentucky Derby winner would sire the next Derby winner. But, as any horse breeder can tell you, nature doesn't work that way. The vast interdependence of all parts in biology means breeding to improve one characteristic inevitably produces problems elsewhere.

This means the Social Darwinists – Galton, Fisher, Crick, Dawkins –

have fundamentally misunderstood biology and evolution. Increasing the elites, and reducing the supposedly "less fit," would have the opposite effect they intend. As genetics has taught us, breeding the "strongest" often leads to weaker children, an effect confirmed not just in animals, but throughout history by many military conquerors, royal families, and titans of industry.

Removing the False Choice Between Science and Religion

These discoveries also offer an opportunity to resolve the social battle that plagues evolution. As we've seen, those who believe in God overwhelmingly believe evolution is true. The former Archbishop of Canterbury, George Carey, summed up this view, saying "I'm an evangelical Christian, but I have no difficulties in believing that evolution is the best scientific account we have for the diversity of life on our planet."[11]

The problem is not with evolution, but with Darwinism. There are no big battles over astronomy's claim that the universe has been evolving for 14 billion years. No one disputes that the sea floor is evolving, pushing continents around and creating mountain ranges, earthquakes and tsunamis. Nor is there a battle over the evidence that shows oceans, lakes, forests and savannahs evolve over time. Even the earliest humans knew this, seeing it in the streams and woodland around them.

The hidden issue in the evolution dispute is the question of what it means to be "human." Religion and Darwinism hold diametrically opposite views, forcing some into a false choice between belief in God or evolution. Complexity Theory's emergent properties and modern neuroscience decisively resolve this dispute in favor of human uniqueness. This does not prove the existence of God, or any religious tenet. But it does remove the false Darwinist claims about Man and human nature.

Complexity Theory may allow those who now oppose evolution to reconnect with Christianity's long-held beliefs about evolution

- From the early Christian belief that Genesis is a metaphor for events which occurred through natural processes, and that human interpretations of scripture cannot over ride what is found in nature.
- Through Medieval "Natural Theology," the belief that both Dogma and Nature are God's creations and thus must be compatible,
- Up to the Victorian belief that evolution leads to an awe-inspiring concept: God as a Divine engineer "who had devised natural laws so delicately balanced they would run forever."

Today, some see Complexity Theory as reinforcing this belief. Lee De-Raud put it this way: "Any deity worthy of a graven image can cobble up a working universe, complete with fake fossils in under a week – hey, if you're not omnipotent, there's no real point in being a god. But to start with a big ball of elementary particles and end up with the duckbilled platypus without constant twiddling requires a degree of subtlety and the ability to think things through – exactly the qualities I'm looking for when I'm shopping for a Supreme Being." Cheeky, but it makes a point.

PARTICLES TO PLATYPUS

Some people think the new discoveries bridge evolution and Christian beliefs. Stanford University evolutionary biologist Joan Roughgarden, for example, says that "what evolutionary biologists are finding through their research and thinking actually promotes a Christian view of nature" – widespread cooperation and mutual care-taking.[12]

Some of evolution's most important discoveries have come from people of strong religious faith. The "Big Bang" idea, which describes the origin and evolution of the universe, was conceived by Georges Lemaître, a Belgian cosmologist and Catholic priest who taught physics at the Catholic University of Louvain.

Lemaître and Albert Einstein were friends. For many years Einstein considered him one of only four or five people who truly understood Relativity Theory. In 1927, using Einstein's equations, the 33-year-old Lemaître grasped that the universe had a beginning – a birth – 14 billion years ago.

The Universe emerged from what he called a "primeval atom"– a bit of matter so small it could fit through the eye of a needle, yet so unimaginably dense it contained all the matter needed to form the trillions of stars that now exist.[13] Lemaître said that in a great explosion, this "primeval atom" began expanding at millions of miles a minute, creating the cosmic world we see today.

GEORGES LEMAÎTRE

Einstein and most other physicists opposed the idea for decades, believing the universe was static. But Lemaître persisted against the opposition. The term "big bang" originated in 1947, with a physicist who sarcastically told a BBC interviewer it was absurd to think that the universe had begun billions of years ago in a giant explosion, a "big bang."[14]

In June 1966, ill and a few days from dying, Lemaître was told that the "background radiation" astronomers had found buzzing in the cosmos the year before, was indeed a remnant of the Big Bang, confirming his theory. But Lemaître had never doubted it was there. He was sure of his science. He also believed the evolution of the universe was a magnificent confirmation of his faith.

FOUR EMPOWERING IDEAS

In the years ahead, Complexity Theory and the New Evolution will transform every area of life, and lead to discoveries we can't even imagine today.

Darwinism is about how nature controls us. Complexity is about how we can control nature. Darwinism is all about how we can change *nothing*. The New Evolution is all about how we can change *everything*. These discoveries empower us to change the future in ways unimaginable even two decades ago. Now, we can even precisely re-engineer our own genes.

More importantly, they will change how we think, transforming how we envision life and think about ourselves, society, and the future of humankind. As Robert Laughlin and David Pines said at the beginning of this section, this is the science of the future, pointing to "new discoveries, new concepts, and new wisdom."

The new discoveries and concepts are already here. It is the new wisdom which we must find.

The New Evolution reveals four truths about nature which transform how we think about ourselves, society and the future of humankind. These truths also hold new wisdom for dealing with the extraordinary power these discoveries place in our hands.

1. Cooperation is King

The first is about cooperation. The great diversity and splendor of nature emerge only through cooperative relations. They lead to increased complexity, new adaptive changes, and the emergence of remarkable new properties. Martin Nowak, director of Harvard's Evolutionary Dynamics Program, calls cooperation a "basic force," perhaps the basic force, in evolution.

In 2010, at the University of California, Santa Barbara, scientists decoded the genome of one of evolution's earliest animals, a marine sponge dating back 650 million years. "This is a milestone," said Ken Kosik, co-director of the University's Neuroscience Research Institute. Kosik explained that:

> "The evolutionary origin of animals was marked by the ability of individual cells to assume specialized properties and work together for the greater good of the entire organism. The sponge represents a window on this... momentous event."[15]

Nature is sending an important message here. Over and over, from the origin of atoms, to the formation of life and emergence of vast ecosystems, we see that cooperation is the key to greater complexity and new creations. In the natural world, cooperation is king.

This is also true for human society. While competition has an important role, making us stronger and more innovative, competition can accomplish these things *only from within* a community that is fundamentally cooperative.

Individual achievements are important. But they have value only from inside a community. (If you have a billion dollars or create a great invention but are the only person on earth, what value do they have?) It is social cooperation that makes these achievements valuable, combining them with many others to create human progress – greater knowledge, economic growth and more individual choices and freedom.

2. Constraint Creates Freedom

Second, cooperation comes at a price: constraint. Cooperation requires committing to a relationship, and that comes with restrictions.

We think of constraints as the rules about what *cannot* be done. But this is only half the picture. One of nature's great truths is that constraint leads to increased complexity and new abilities giving us greater freedom.

Constraints allow relationships to build one on another, bringing the natural world into existence. Constraints are needed to form everything from the first atom, to wildflowers and the human brain. A constraint like gravity brings time and space into existence, allowing nature (and us) to have a past, a present and a future.

Constraints make it possible for a group to accomplish what cannot be done alone. They allow individual cells to unite and form plants and animals, with specialized skills and adaptive abilities far beyond what an individual cell could create.

Constraints also lead to the emergence of abilities which *only groups* can possess, like collective action, distributed intelligence, and specialization. In sailing a four-masted schooner, no one on board knows how to do all the jobs, not even the captain. Most have only a small piece of the knowledge needed to run the ship. Yet it sails perfectly.

This is called "distributed knowledge," or "distributed intelligence." It means *everyone* on-board is dependent on *everyone else*. A man alone can sail only a few miles. Together they can sail to every part of the globe. But only if they agree to be bound in a set of relationships, where each does his or her job.

Constraint also leads to specialization, allowing individuals to get better and more creative at their jobs. Alone, a man is a jack of all trades and master of none. But in a group he can refine his skills and tools to become a master craftsman.

Few people know that constraint is also what changes the biological response of animals into human rational choices. Animals don't delay gratification, but humans do.[16] "Children develop specific psychological strategies to maintain self-control in pursuit of future goals," says Columbia University psychologist Walter Michel.[16] Cooperation in groups depends on that self-control. The larger the group the more self-control is required. The path from primitive village to ancient civilizations and modern society was paved with increasing self-control and constraint.

We can think of this as the "Constraint Paradox." Constraint creates freedom. *More* constraint creates *more* freedom. It is one of the great truths of nature.

3. Relationships are Destiny

The Constraint Paradox points to the third empowering idea: relationships. At *every stage* of evolution (from the Big Bang to the modern world), and *every type* of evolution (from simple to complex – atoms to ecosystems), relationships determine what we are, and what we can do.

Darwin said geography is destiny. Nature says relationships are destiny. The idea is as powerful as it is simple. The relationships we form – physical, mental, emotional – determine our future. At first this may seem ordinary, even pedestrian. What is extraordinary is that Complexity now tells us this is *the basis of all existence*. It is how everything comes to exist.

Humankind, with our unique consciousness, adds an extraordinary new dimension to relationships. With consciousness we can move beyond the relationships inherent in nature, creating things that come from our imagination, like aqueducts, airplanes and electron microscopes.

We can also form relationships with each other based solely on ideas, like "social rules," "individual rights," "justice" and many others. This allows us to create a uniquely human existence not possible with biology alone, a life based on ideas and the mental relationships we create with each other. These concepts and "relationships of the mind" determine human destiny.

4. We Can Control Our Destiny and Future

Self-assembly also returns order and direction to evolution.

In a world that often seems chaotic, where people desperately seek understanding and a sense of direction, self-assembly provides that. Evolution is predictable, which means that (within some limits) we can *know the future of evolution*. With our new understanding of evolution, we can *change that future*.

Beyond change, we can *create a new type of future*, based on ideas. Working within the rules of nature, we can imagine relationships with no counterpart in nature, and use them to craft new ways to transform ourselves, our societies and our children's futures. This is perhaps the most powerful message of the Evolution Revolution. Evolution does not control us, we control it. We now control our own evolutionary destiny.

THE LISTS

SIX SURPRISING DISCOVERIES ABOUT EVOLUTION

1. Relationships

The universe began with a few simple particles. All that came later evolved by *rearranging* these particles in new ways. A few simple rules from physics determine which relationships can form between these particles, and thus determines *everything that will later evolve,* from atoms to animals.

2. How Evolution Works

Nature's parts have properties which attract each other (like magnets) causing them to pull together - to "self-assemble" or "self-organize"- into naturally forming relationships. (See chapter 14.)
In a major breakthrough, scientists found that each new set of relationships serves as both the platform and the instructions for building the next level of relations, pushing nature up the ladder of complexity.
This discovery unites the entire natural world - from stars, planets and galaxies, to plants, animals and eco-systems - under a single evolution process, all evolving based on the same set of laws.
This is why leading biologists, like Richard Lewontin, now call for "*a biology that is relational,*" and an evolution based on "the laws of interaction" between relationships. ("Organisms are a consequence of the interactions between bits and pieces of matter, so it is laws of interaction that we must study.")

3. Matter: Physical and Non-physical

Nature is entirely material. But matter is not always physical in the traditional sense. Nature includes many elements that do not have three dimensional form, such as forces and "fields," or properties like conductivity and color. (Color does not exist in nature's basic particles, but emerged only when nature became more complex.) Even the basic laws of nature have no physical dimensions, yet they are real, and influence how physical things behave.

Physicists have long known that *non-physical aspects* of nature both exist and *influence the physical parts* of nature. Scientists have now discovered this is also true in chemistry and biology, a finding that profoundly transforms our ideas about evolution and life. (See chapter 14.)

4. Emergent New Properties

The whole is more than the sum of its parts. At each higher level of complexity, new patterns lead to the appearance of new properties which cannot be reduced to any property at lower levels. Called "emergent properties," many of them have no physical dimensions, yet they influence the physical parts of nature.

Nobel prize winning physicists, like Robert Laughlin, say that all the laws of nature did not exist at the beginning of the universe. Some *new laws of nature* emerge only at higher levels of complexity, transforming the evolution process as they appear. (See chapter 14.)

5. Predictability, Direction and Evolution

Everything in nature is governed by laws with predictable effects. This means that how atoms, stars and galaxies will evolve is predictable. Even the evolution of the Universe is predictable – it has direction, it is headed somewhere. These are undisputed scientific facts. Science now knows that evolution is predictable at *every level of nature*, including biology. The types of life which can exist, and how they will evolve, is predictable. It also means that the evolution of life (like the Universe) has direction, it is headed somewhere. (What this direction "means" is not a question that can be scientifically tested, and thus is not a question science can answer. It is a question for philosophy.)

6. Predictability is Not Being the Same

The predictability of nature does not mean everything will be the same. As we saw with snowflakes, while the *types* of things nature makes are always the same, each *individual* item can be unique.

The Darwinist Wars Against the Evidence

Darwin had the evolution process backwards. Thus all the evidence, and every new discovery contradicted his theory. The response by both Darwin and modern Darwinists was always the same:

1. **Attack** the evidence as wrong. (Their theory was superior to any fact.)
2. **Invent**: When attack fails, they explained the facts with imaginary processes for which there was no evidence (the list is long).
3. **Exclude**: When their imaginary processes were disproven, Darwinists declare the evidence was irrelevant to evolution.

In recent years, as denying the facts has become increasingly difficult, Darwinists (like Richard Dawkins) claim the evidence is an illusion. Nature is really the *opposite* of how it appears. In a world where reality is the opposite of what we see, then Darwinism is true.

Darwin's wars against the evidence included most of biology:

1. **Laws of Adaptation** and Lamarck's idea of "use and disuse."
2. **Laws of Structure** and their constraints on new species.
3. **Laws of Development** and their constraints on evolution.
4. **The Fossil Record** and revelation that evolution is abrupt and fast.

Every major scientific discovery in the 20th century also contradicted Darwinism, leading modern Darwinists to attack them, including:

5. **Genes** and the idea that Mendel's laws play a role in evolution.
6. **Molecular Biology and Genetics** of the 1950s.
7. **Developmental Biology** and the idea developmental constraints limit evolution.
8. **Punctuated Equilibrium**. The idea new species appear in short, rapid bursts, followed by long periods of little or no change.
9. **Horizontal Gene Transfer** and **Symbiogenesis**. The evidence each species is not genetically isolated, nor uniquely formed by natural selection.
10. **Decoded Genome**. The evidences showing
    a) most genes existed at the beginning of life, and evolution occurred when they duplicated themselves and specialized
    b) evolution occurs when "regulator" genes create new development patterns, not when new protein-making genes appear.

The Restoration List: Discoveries Previously Dismissed by Darwinists, But Now Shown To Be True

Below are a few of the scientists and ideas dismissed by the Darwinists that are now known to be true. (Outside of Darwinist circles, many of these ideas were never considered wrong.)

- **Jean Baptiste Lamarck** and his Laws of Adaptation based on use and disuse of the body. Today, use and disuse are broadly defined as stress (or the lack of it), which triggers changes in biology that are directly inherited through epigenetic processes.
- **Richard Owen's** Laws of Structure and the physical constraints they place on change and the evolution of new species.
- **Barbara McClintock** and her three transforming ideas:
    1. Mobile genes manage development, moving from place to place, telling the embryo what to do and when to do it.
    2. During development, the job of a "regulator" gene is not fixed, but changes based on the context.
    3. Stress alters when and where "regulator" genes are active, changing the traits they make.
- **Susumu Ohno** and the idea that adaptive evolution occurs when genes duplicate themselves and then specialize on some aspect of their prior job.
- **Richard Goldschmidt** and his discovery that the entire genome can duplicate itself and undergo a rapid, total reorganization, leading to new species: "The change from species to species is not a change involving more and more additional atomistic changes, but a complete change of the primary pattern… into a new [pattern]."
- **Stephen Jay Gould**, **Niles Eldredge** and Punctuated Equilibrium, the idea that new species emerge in abrupt, rapid leaps, followed by long periods with little or no change.
- **The Laws of Development** and their constraints on evolution, today called Evo-Devo (evolutionary developmental biology).

Physics and Self-assembly

An important part of this restoration is reconnecting evolution with the other physical sciences. Dobzhansky once said that "nothing in biology makes sense except in the light of evolution." As usual, the Darwinist had things backwards. Nothing in evolution makes sense except in the light of biology. The laws of physics, chemistry and biology determine much about evolution, as is reflected in key discoveries from the past, for example:

- Jean-Baptiste Lamarck's 1809 discovery that stress creates evolutionary change in biology. This is an analog to Fred Hoyle's discovery that the stress of gravity leads to evolutionary change in atoms and stars.
- Richard Owen's 1849 structural laws, which included a "polarizing force" and an "organizing force" that determined what can evolve. In his words, the "essential character of a part" leads it to form a "predetermined pattern," which then determines "all [future] modifications."
- Gregor Mendel's genetic laws from 1900.
- Barbara McClintock's mobile genes and evolution model from 1950.

The role of physics and chemistry in evolution is also reflected in many recent discoveries. At every level, their laws make sense of what is known about evolution. As Stephen Miller at the University of Minnesota Biotechnology Institute said, "the old notion of natural selection as an omnipotent force in biological evolution has given way... Adaptive processes are constrained by physical, chemical, and biological requirements." To list just a few examples that walk us up the evolutionary ladder of life:

- Richard Lewontin's idea of evolution as "the laws of interaction" and "constrained relationships."
- Max Taylor's discovery (not covered in this book) that the physical properties of nucleic acids determine how DNA combines, and thus which genes can exist and which proteins they can make.
- Stuart Newman's discovery that physical properties largely determined how the skeleton evolved, and that physical signals determine where bones will form during development.
- The discovery by Clifford Tabin and others that physical processes explain how arms, legs and fingers develop.
- Chris Cherniak's discovery that physical processes explain how complex organs, like the human brain, evolved and develop.

The Removal List: Darwinists' Claims with No Basis in Fact

In his attempt to make the facts fit his theory, Darwin invented many imaginary processes that were later shown to be untrue, and were removed from the scientific discussion about evolution, including:

- **Blending**: The idea that traits change like modeling clay.
- **Gemmules**: The idea that elements in the blood carry traits and are subject to natural selection.
- **Pangenesis:** Darwin's theory of inheritance based on "gemmules."

Modern Darwinists continued to invented imaginary processes that have also proven untrue, and should likewise be removed from the scientific discussion of evolution, including:

- **The Weismann Barrier**: The idea that changes in the body (somatic changes) don't effect the cells used in inheritance (germ-line changes), thus Lamarck's direct inheritance of changes is impossible.
- **The Central Dogma:** The Weismann Barrier updated for modern genetics, fabricated to refute Lamarck's idea of direct inheritance. Today, a vast body of evidence shows that somatic changes *do* alter germ-line genes, and that the direct inheritance of changes in biology is widespread in both plants and animals.
- **Random Gene Change**: The idea that new traits are a product of random gene mutations. In Dobzhansky's words, "the modern theory of evolution is that... mutations occur by *chance* and are undirected."
- **Directionless Evolution**: The idea that random gene changes make the emergence of new traits random and evolution directionless.
- **Kin Selection**: The idea that creatures cooperate in proportion to the genes they share. (All animals share *the same genes*.)
- **The Selfish Gene**. The idea that individual genes engage in a competitive struggle for survival in which the "losers" are eliminated. In evolution, genes are never eliminated, only rearranged.
- **Nature is an Illusion**: Because Darwin had evolution backwards, Darwinists claim that nature is an illusion, and reality is the *opposite* of what scientific experiments show.
- **Natural Selection**: The evidence for Darwin's natural selection has simply never materialized. If it is evolution's driving force, millions of examples should exist. But in Darwin's lifetime none were found. The few examples offered later (Peppered Moths, Darwin's Finches) proved untrue. Today, at best, a few instances exist. It may turn out, that on closer inspection, these also fail the test of science.

DEFINING AND PROVING DARWINIAN EVOLUTION

Darwinism says evolution occurs in a two-step process:

1. A random gene change.
2. Natural selection.

A random gene mutation leads to a slight variation in a trait. Environmental differences lead creatures with the mutated gene (and small trait variation) to survive, while those with the original gene (and trait) are eliminated. The random appearance of new genes and adaptive changes, and elimination of the older gene and trait requires evolution to be an extremely slow, gradual process.

This is often distilled to "variation and selection," and much of the confusion over evolution emerges from the unscientific use of these two terms.

In most scientific research, the claim that trait variations are a product of random gene mutations is assumed, but never demonstrated. The problem is, if biological differences don't come from random mutations, then one must *look beyond genes* to find the source of evolutionary change.

There is a similar problem with natural selection, which is also often just assumed, but never demonstrated. It isn't shown that after a random mutation, another gene (and its related trait) were eliminated, and that this occurred hundreds of times in succession.

Elimination is the central mechanism of Darwinian evolution. If there is no elimination, then both the mutated and non-mutated genes would persist, along with the traits they create. (Rather than natural selection, Darwin would have more accurately called his process "natural elimination.")

To scientifically establish Darwinian evolution a person would have to demonstrate four things:

1) That a random genetic mutation occurred,

2) This mutation led to a specific (tiny) trait change,

3) An older gene, and its related trait, were eliminated, and

4) That a long series of such mutations and eliminations, over hundreds of generations, led to the gradual change of the trait.

In addition, to be truly scientific, this demonstration would need to be replicated by other scientists.

Lets see if Darwinism can be proven using the "classic" example of the beak changes in Darwin's finches (from Chapter 17). In 1999, Ernst Mayr summed up the Darwinian explanation this way:

> "The development of the heavy [finch] bill was a slow process, probably involving dozens of small mutational steps, each one surviving only if proving its usefulness in the actual test of selection. It must be admitted, however, that it is a considerable strain on one's credulity to assume that finely balanced systems such as certain sense organs (the eye of vertebrates, or the bird's feather) could be improved by random mutations.... However, the objectors to random mutations have so far been unable to advance any alternative explanation that was supported by substantial evidence"[1]

Mayr's account of these changes is demonstrably false.

First, beaks don't change from a series of random gene mutations. Quite the opposite. Beak changes always involve *the same genes*. These genes have been around for millions of years, and are used to create the jaws of many animals, not just finches. In fact, nature uses these same genes to build a wide variety of bones in the body. For example, while the gene *BMP4* is normally associated with forming the skull, it also determines the shape of a beak. "Different [beak]shapes arise depending on where and when...*BMP4* is turned on during development."[2]

In addition, beak shapes are determined by several elements (called "prominences"), including the nasal prominence (the nose), maxillary prominence (the upper jaw) and mandibular prominence (the lower jaw). Geneticist Cheng-Ming Chuong, at the University of Southern California, has shown that "during development, these prominences are *proportionally coordinated* to compose a unique beak."[3] The "proportional coordination" of these elements is accomplished through a complex network of chemical signals, such as *BMP4* which determines beak width, and *CaM*, which determines beak length.

This means that changing the shape of a beak requires *simultaneously altering* the actions of dozens of genes in the signal network. The idea this occurs through a series of random mutations, all occurring simultaneously, in just the right genes, and just the right sequence, over hundreds of generations, defies both mathematical probability and basic biology.

Second, there is no indication of the gradual change over hundreds of years, nor the eliminate process, central to Darwin's natural selection. On the contrary, Peter and Rosemary Grant's work shows the changes occurred rapidly, sometimes in just a few generations. "From 1972 to 2001, *Geospiza fortis* (medium ground finch) and *Geospiza scandens* (cactus finch) changed several times in body size and two beak traits... The evolutionary responses...to the two strongest...episodes occurred in the *following generations*."[4] In other words, the changes appear quite rapidly.

More importantly, there is no evidence that with each change a gene mutation appeared AND an older gene was eliminated from the finch population. There is no evidence for the central mechanism of Darwinian evolution, the elimination of a "less fit" gene and its trait.

In short, random variation and selection (or more accurately variation and elimination) can't explain what causes these beak changes to appear.

They can't explain what causes an *entire complex signaling network*, with thousands of molecules, to change *simultaneously*, and *rapidly*. Nor why, since the genes that make beaks make many bones, changes occur in some bones, but not others.

While the answer to these question can't be found in the "variation and selection" process, they are explained by Lamarck's evolution. In the modern version of his process, stress leads to *multiple, simultaneous* changes in gene *expression*, which are specifically targeted to the *areas under stress*. This explains, when a trait is altered, how changes in the other body parts are "proportionally coordinated" with each other, and how these complex changes can be directly inherited by the next generation through epigenetics.

As Lamarck predicted, if the stress is removed, the changes are reversed. This explains how the finches' beaks were able to swing back and forth in size within just a few years.

Most adaptive changes attributed to Darwin's "variation and selection" face the same problems. On close inspection, one cannot point to specific biological evidence (vs statistical calculations) which shows:

1) that a random genetic mutation occurred,
2) that this led to a specific trait change,
3) that an older gene and it related trait were eliminated, and

4) that a long series of such mutations and eliminations, over many generations, led to the gradual change of a trait.

On the contrary, a wide body of evidence clearly demonstrates that most adaptive changes, like those seen in the finch beaks, result from:

1) stress creates a change in biology or the body,

2) which alters the switches that control gene activity, changing the time, place and amount of that activity (called gene "expression"),

3) and these change are then directly inherited by the next generation through one of the body's many epigenetic mechanisms.

The Darwin Industry
The Continued Attempt to Rewrite History and Science

Many scientists today object to what they call the "Darwin Industry," the exaggerated claims, or outright untruths, advanced by those deeply invested in the Darwinist social agenda.

As we saw, much of what is written today about the history of evolution, and Darwin's role in it, is untrue.

A similar problem surrounds the "science" attributed to Darwin. An example of this are the claims in "Darwin's enduring legacy," the article mentioned in Chapter 19, where Kevin Padian says that Punctuated Equilibrium vindicates Darwin's ultra-slow evolution.[1]

In that same article, Padian credits Darwin with ten great "legacy" discoveries in science. This includes "deep time," the discovery the Earth is millions of years old. Padian writes that in establishing "deep time," Darwin "calculated the period necessary to build up and wear down many of the major rock formations of England... It was no longer possible to accept that Earth was 6,000 years old, as some Biblical scholars estimated."

Leave aside the fact that Darwinists can never manage to discuss evolution without dragging in creationism, the claim that Darwin played a major (or even minor) role in establishing "deep time" seems to me entirely unsupported by the facts.

The idea of "deep time," was established long before Darwin, based on a hundred years of geological research, capped off by Charles Lyell's *Principles of Geology* in 1830. Darwin's small contribution did little to establish this idea. In *Earth Time: Exploring the Deep Past from Victorian England to the Grand Canyon*, Douglas Palmer tells us that:[2]

- At Edinburgh University, Darwin took a geology course from Robert Jameson. "Jameson's performance as a teacher of geology severely 'turned off' from the subject a young English student called Charles Darwin... Darwin subsequently remarked that Jameson's lectures were 'incredibly dull' and that as a result he determined never 'to read a book on geology or in any way study the science.'"(p22)
- "Darwin seriously overestimated the possible duration of part of the geological record because he was trying to make calculations based on data from small, post-glacial rivers." (p 345)
- The famed physicist, Lord Kelvin, using the laws of physics, estimated

> the earth was at least 200 million years old. Kelvin thought Darwin's work on the earth's age, and other topics, was wholly unscientific. Palmer says, "scientifically, Kelvin was a strict empiricist and despised the 'woolly thinking' and vague speculation without hard data of the kind that Darwin was using." (p 402-403). (Kelvin, using geology and the laws of physics, estimated the Earth was more than 200 million years old.)

Padian, toward the end of his article asks, "has any single individual made so many lasting contributions to a broad area of science as Darwin did to biology?" Given that 1) historians show Darwin added little to Victorian biology, and 2) modern Darwinists excluded much of biology from any role in their version of evolution because the two were at such odds (see Chapter 11), Padian's question seems to me entirely misleading, implying something for which I find little, if any, evidence.

While Darwin was noted for his niche work on barnacles and pigeons, few Victorian biologists thought he had made any major contributions to their field. When *The Origin* was published, Huxley could count only three biologists who supported Darwin: himself, Dalton Hooker and John Lubbock.[3] All three were members of the *X-Club*, and saw *The Origin* more as a battering ram against religion than a contribution to biology.

Nor did biologists' opinion of Darwin improve over time. As we saw, at the 1894 meeting of the British Association for the Advancement of Science, Darwinism was still facing stiff opposition from biologists. Nothing in the 100 years since the1894 BAAS meeting has changed the assessment that Darwin's contributions to biology were, at best, marginal.

For many biologists, Darwin's ideas (like "blending") revealed a surprisingly simplistic understanding of biology's complex processes.

There are numerous examples of Darwin's poor understanding of biology. We saw several in his work with Richard Owen. Another was the scathing assessment of the eminent biologist Samuel Haughton, ("all that is new in [Darwin's] paper is false, and all that is true is old."

In another, particularly sad example, Darwin believed he had solved a major mystery about development: why the embryos of many animals look similar, while the adults look so different. "Hardly any point gave me so much satisfaction," Darwin wrote, "as the explanation of the wide difference... between the embryo and the adult animal."[4] Late in his life, however, a disappointed Darwin wrote:

> "No notice of this point was taken... and I recollect expressing my surprise on this... Within late years several reviewers have given the whole credit [for solving this mystery] to Fritz Muller and [Ernst] Haeckel, who undoubtedly have worked it out much more fully, and in some respects more correctly than I did... it is clear that I failed to impress my readers."[4]

Darwin was right, he had failed to impress his readers. Even in old age, he still failed to grasp the difference between the "woolly thinking and vague speculation" he offered, and the "much more fully and correctly" worked out cause and effect explanations used by other biologists.

Like so many in the Darwin Industry, Padian ends his article by dragging creationism into the picture: "It is for this vision, one that liberates humans from the conceit of special creation, that Darwin was honoured by interment in Westminster Abbey."

As we know, there was little belief in special-creation among Victorians. And as to Darwin's burial in Westminster Abby, this reflects the fast footwork and power of Thomas Huxley at the time, not to Darwin's standing in the scientific community which, as we have seen, was at best modest.

Attacking New Discoveries

An important part of the Darwin Industry is to attack anyone whose discoveries conflict with Darwinism. The attacks generally follow the same three steps. The Darwinists:

1. Pick out a small error or omission and then use it to discredit an entire body of work.
2. Equate criticizing Darwinism with criticizing belief in evolution.
3. Equate criticizing Darwinism with supporting creationism.

A typical example were the attacks on James Shapiro's highly regarded book, *Evolution: A view from the 21st Century*. Shapiro, one of America's leading geneticists, dismisses Darwinism as largely irrelevant to evolution. Many reviewers echoed the assessment of Oxford University's Denis Noble, who wrote "this book is a magnificent analysis of the key questions... Jim Shapiro has new insights on all the central issues of evolutionary theory."

Darwinists, like Lawrence Moran, however, bypassed the "magnificent analysis" and instead launched a three step attack:[5]

First, use a small omission to discredit an entire body of evidence. Moran wrote that "none of the 1162 references in [Shapiro's] book is to Stephen Jay Gould's... spandrels paper...[which is] quite troubling." Shapiro's book has over 1000 references. Moran finds one not included, and uses that to call entire work into question. In fact, Gould's "spandrels paper" was not central to Shapiro's analysis.

Next, equate criticizing Darwinism with being against evolution. Moran writes, "these days, there seems to be a class of evolution critics who are determined to overthrow the Central Dogma... The Central Dogma has never been overthrown or seriously challenged."

Never been seriously challenged?! As we saw in Chapter 17, this is simply untrue. "Central Dogma *Reversed*" was the 1970 headline in the journal *Nature*. Today, hundreds of studies show the Central Dogma is wrong. In every type of plant and animal, body changes alter the genes or gene regulation in germ-line cells, exactly the process the Central Dogma said was impossible.

Third, equate criticizing Darwinism with supporting creationism. "Shapiro... is widely admired in the "intelligent design" community," says Moran. "And there's a good reason for this. This book is highly critical of old-fashioned evolutionary theory (neo-Darwinism) using many... silly arguments."

What are the "silly arguments" Shapiro uses to discredit Darwinism? Merely all the major discoveries of modern biology. Moran provides the list himself: "Shapiro trots out all the usual suspects: epigenetics, regulatory sequences, transposons, (endo)symbiosis, chromatin, evo-devo, horizontal DNA transfers, abrupt changes in the fossil record (saltation), polyploidy, and the demise of the Central Dogma... All of these new discoveries apparently conflict with neo-Darwinism."

Well, yes, they do. More than that, they disprove Darwinism.

That has always been the problem with Darwinism, it contradicts the evidence. That is why Shapiro and many others believe science needs a new model of evolution, one that matches the discoveries of modern biology.

Making up the Evidence

Even more remarkable than the personal attacks and lack of evidence, is that the Darwin industry simply makes up facts. Even though most people are unfamiliar with the details of biology or evolution, they assume no scientists would just make up evidence to support their claims. But look at this and judge for yourself.

James Shapiro wrote a series of columns which said the evidence from modern biology largely eliminates natural selection and gradual change as factors in evolution. Arch-Darwinist Jerry Coyne attacked Shapiro writing, "Given the fossil evidence of transitional forms ... such a statement is simply embarrassing, and is identical to ones you'll see in the creationist literature."[6]

The problem is that, if Darwinism is true, millions of fossil "transitional forms" should exist. But they don't. Very few exist. And as we saw, even these few are being challenged by increasingly accurate genetic analysis.

In response to Shapiro's assertion that genetic change is not random Coyne says "we know of no evidence for mutations occurring non-randomly or adaptively." But as we have seen, a great deal of evidence flatly contradicts this.

Unable to refute Shapiro's evidence, Coyne, astonishingly, says the highly regarded Shapiro "simply doesn't understand" evolution. "I often find molecular biologists fail to grasp natural selection," wrote Coyne.[6]

The idea that biologists don't understand evolution is the same argument made by Ernst Mayr in the 1960s, and by Darwin in the 1860s. But as we have seen, the problem is not that biologists don't understand evolution, but that Darwinists don't understand biology, and often deny the established facts of biology.

Other Darwinists joined in the attack on Shapiro. Keith Roragen wrote "where genome changes come from in the first place is irrelevant... Darwinism isn't concerned with how genomes change."[6] This is remarkable, ignoring the fact that random genetic change has been at the center of Darwinism for over half a century, as Ronald Fisher, Ernst Mayr, Theodosius Dobzhansky, and Jerry Coyne's comments (two paragraphs above) show.

Random genetic change is called both central to modern Darwinism and "irrelevant." This gives the appearance that Darwinists just make up their claims as they go along.

Shapiro's response to all this was to call for some facts.

> "The curious responses to my position exposed a fundamental difference in understanding of biological functions be-

> tween molecular geneticists and these particular proponents of [Darwinism]....There is simply no way... natural selection can account for how [genetic] structures and networks originated... and changed over time. That is why [their] assertions are so perplexing. Keith, and Jerry have stated that I am mistaken (and dangerously misleading others). If they wish to maintain that stance, then I think they need to explain why in detail. Fiat assertions and personal attacks are no substitute for rigorous argument."[6]

That is true, but unsubstantiated claims and personal attacks have been at the center of the Darwinists program from the beginning.

MAKING UP THE EVIDENCE IN TEXTBOOKS

The Darwin industry extends to textbooks, which continue to perpetuate Darwinian ideas long known to be untrue.

For example, one recent textbook, summarizing the key principles of evolution, said that "higher taxa [what we commonly call "new species"] arise by the prolonged, sequential accumulation of small differences, rather than by the sudden mutational origin of drastically new 'types.'"[7]

Overwhelming evidence, however, shows this is untrue. If Darwinism is true, there should be thousands of examples of small changes (micro-evolution) leading to the origin of "higher taxa" (macroevolution). But there is little or no evidence of this. Douglas Irwin, a senior scientists at the Smithsonian Institution, explains the view of most biologists: "Historically, the neo-Darwinists... stressed the predominance of micro-mutations in evolution. But that doesn't explain new species. Macro-evolution is more than repeated rounds of micro-evolution."

The same textbook goes on to promote other disproven Darwinists ideas:

> "The fossil record includes many gaps [which] may be explained by the incompleteness of the fossil record. But the fossil record also included examples of gradations from apparently ancestral organisms to quite different descendants. These data support the hypotheses that the evolution of large differences proceeds incrementally."

As we saw, going back to Darwin, the evidence all shows new species don't appear incrementally, but abruptly. Niles Eldredge and Ian Tattersall, forcefully made this point, saying "the fossil record flatly fails to substantiate finely graded change." New species appear "with a bang, often with no evidence of prior evolution," they wrote. "All the evidence shows that the record is real: the gaps we see reflect real events in life's history. With gradualism we are faced more with a great leap of faith... than any hard evidence."

University of California paleontologist James Valentine, a leading authority on the fossil evidence, says much the same thing. For example, the changes during the Cambrian Explosion were "abrupt and extensive," far too great to result from "flaws in the fossil record."

Even Darwinists like Ernst Mayr acknowledge that "intermediates between species are not observed," instead "all species are separated from each other by bridgeless gaps."

In true Darwinist form, this textbook quickly begins equating Darwinism with evolution and then pitting Darwinism/evolution against "the creationist movement" and the idea of "special-creation." The message this textbook sends as I read it, is that you either support Darwinist ideas or you are a creationist. I believe this is a false choice designed to do one thing: insulate Darwinism from the type of factual critique to which all scientific theories are normally subjected.

While this textbook eventually discusses other ideas about evolution, it calls them speculative "hypotheses." For example, in covering Punctuated Equilibrium, the textbook says "It remains to be seen whether punctuation and stasis is the most common pattern in the fossil record or not." (p 86)

This ignores the fact that even Darwinists now call Punctuated Equilibrium "one the most important evolutionary ideas of the twentieth century," and acknowledge it appears to be "the predominant pattern of evolution."

Is it any wonder that ordinary people are thoroughly confused as to what is scientifically true about evolution when they read a textbook like this, and then read leading scientists like Irwin, Eldredge, Tattersall and Valentine, whose discoveries directly contradict the textbook?

Unfortunately, it appears that many in the Darwin Industry is not interested in resolving these issues, keeping the science of evolution hopelessly confused.

Modern biology, however, is now lining up against the Darwin Industry. One reviewer of Shapiro's *Evolution: A view from the 21st Century*, said this:

> "The book's contention that natural selection's importance for evolution has been hugely overstated represents a point of view that has a *growing set of adherents*. (A few months ago, I was amazed to hear it expressed, in the strongest terms, from another highly eminent microbiologist.) My impression is that evolutionary biology is increasingly separating into two camps... On the one hand are the [Darwinian] evolutionary biologists who... believe that selection has a "creative" and crucial role in evolution, and on the other, there is a growing body of scientists (largely those who have come into evolution from molecular biology, developmental biology or developmental genetics, and microbiology) who reject it. In contrast to Victorian scientists, who regarded Darwinian natural selection as "incapable" of creating... biological complexity, modern sceptics tend to regard it as of "trivial" importance: the "right" variant for the right place and time arises and, presto, the population changes!"[8]

Among scientists, the curtain seems to be closing on Darwinian magical genes, which "presto" appear at just the right place and time to explain some aspect of evolution. Apart from Darwinists, science has never much liked "presto" as a cause and effect explanation for what happens in nature.

ACKNOWLEDGMENTS

My appreciation to best-selling author Elizabeth Kaye for tireless hours editing and encouraging. Thanks to screenwriter Henry Rosenbaum for reading the earliest drafts and offering many helpful suggestions, including going to the middle of the manuscript and saying "Start here!" And my thanks to Mark Levine for many years of encouraging support.

Thanks to Will Hopkins and Mary K. Baumann for their design skills, many good ideas, and never failing good humor and encouragement. Also, thanks to Michael Kline for many illustrations.

Thanks to Chase, Bryn, Paige and Brooke for enduring my absences during many long days. My deep thanks to "Mare," whose remarkable copy editing eye and unconditional encouragement were essential to bringing this book across the finish line. Most of all, I am grateful to Judy, whose love, enduring support and always warm kitchen hearth, made the journey possible.

CREDITS

Richard Owen portrait, courtesy Indiana University Archives. Charles Darwin with his sister Catherine portrait, courtesy Darwin Heirloom Trust and English Heritage Library. Microraptor illustration courtesy Xing Xu and Institute of Vertebrate Paleontology and Paleoanthropology, China. *Anchiornis* illustration by Michael DiGiorgio courtesy Yale University. Tree of Life illustration adapted from "Family Tree of Life" in *The Evidence of Evolution* by Nicholas Hotton III, Smithsonian. Portions of "Charles Darwin and the Children of the Evolution" by Dennis Sewell courtesy *The Time Literary Supplement.* Spawning coral image courtesy National Oceanographic and Atmospheric Administration, credit Emma Hickerson. Coral spawn floating in the ocean image courtesy Australian Department of Environment and Conservation. Angela Irizarry images courtesy Yale University

The "Ladder of complexity," "Pangea," "Hierarchy of life," "One cell absorbs another," "Emergent properties," "Limits on what can exist," "Brain development," "DNA self-repair," "Overlapping genes," "Rebalancing" "Major leaps in evolution," "DNA recombination as self-repair" and other illustrations are (c) copyright 2013 by Alan Bennett, all rights reserved, not to be reproduced without written permission.

REFERENCES

ABBREVIATIONS

CUP: Cambridge University Press, Cambridge, UK and New York

HUP: Harvard University Press, Cambridge, MA

JCS: Journal of Cambridge Studies

JHUP: The Johns Hopkins University Press, Baltimore, MA.

MITP: Massachusetts Institute of Technology Press, Cambridge, MA

NYRB: The New York Review of Books

NYT: The New York Times

PNAS: Proceedings of the National Academy of Sciences of the United States

PUP: Princeton University Press, Princeton, New Jersey.

UCP: University of Chicago Press, Chicago, IL

YUP: Yale University Press, New Haven, CN

INTRODUCTION

1. Rose, D (2008) Claudia Castillo Gets Windpipe Tailor-Made From Her Own Stem Cells. The Times. 19 Nov 2008.
2. Adams, S (2011) First transplant of a fully synthetic organ carried out. The Telegraph, 07 Jul 2011.
3. Smith, R (2010) British boy becomes first in the world to have stem cell transplant. The Telegraph, 19 Mar 2010.
4. Cho, Y, R Shi, and RB Borgens, (2010) Chitosan produces potent neuro-protection and physiological recovery following traumatic spinal cord injury. J. Exp. Biol, 213:1513-1520.
5. Daily Mail (2010) Stem cells used to rebuild 10-year-old boy's windpipe in groundbreaking surgery. 20 Mar 2010.
6. Lawrence Berkeley National Laboratory staff (1999) Complex Systems: Science for the 21st Century. Lawrence Berkeley National Laboratory.
7. BBC Radio 4 (2008) The Six O'clock News, 18 April 2008.
8. Chang, K (2008) Edward N. Lorenz, a Meteorologist and a Father of Chaos Theory, Dies at 90. NYT, 17 April 2008.
9. Safina, C (2010) Darwinism Must Die So That Evolution May Live. The New York Times, 30 Apr 2010.
10. Woese, RC and N Goldenfeld (2009) How the Microbial World Saved Evolution from the Scylla of Molecular Biology and the Charybdis of the Modern Synthesis. op. cite.
11. Moeller, R (2002) New Theory of Cell Evolution Rejects Single-Ancestor Doctrine. Scientific American, 19 June 2002.
12. Yoon, CK (2009) Genes Offer New Clues in Old Debate on Species' Origins. The New York Times, 10 Feb 2009.
13. Laughlin, RB (2005) A Different Universe: Reinventing Physics from the Bottom Down. Basic Books, NY. p 168.
14. Lewontin, RC (1998) Survival of the Nicest? NYRB, October 1998.
15. Goldscheider, E (2009) Evolution Revolution. Wisconsin Alumni Magazine, Fall 2009.
16. Hall, TD (2007) A Brief History of Evolutionary Science. http://acad.depauw.edu/~thall/hp1.HTM
17. Rosenfield, I and E Ziff (2006) Evolving Evolution. NYRB, 11May 2006. Emphasis added.
18. Shapiro, JA (2007) Bacteria Are Small But Not Stupid: Cognition, Natural Genetic Engineering and SocioBacteriology. Studies in the History and Philosophy of Biology and Biomedical Science, 38:807-819.
19. Laughlin, R, D Pines (2000) The Theory of Everything. PNAS, 97:28-31.
20. Lawrence Berkeley National Laboratory staff (1999) Complex Systems: Science for the 21st Century. op cite.

CHAPTER 1 A NEW CONCEPT OF TRUTH

1. Nabta, is 6,000 years old, 1,000 years older than Stonehenge, and is located near the Egyptian-Sudanese border.
2. Trewavas, A (2006) A Brief History of Systems Biology. The Plant Cell, 18:2420-2430.
3. Laplace, PS (1814) [2007] A Philosophical Essay on Probabilities, F. Truscott, F. Emory (eds.). Cosimo Classics. p 3-4.

4. Baron D'Holbach (Paul-Henri Thiry) (1770) The System of Nature.
5. Chomsky, N (1959) A Review of B. F. Skinner's Verbal Behavior. Language, 35:26-58.
6. Holland, J, K Holyoak, R Nisbett, P Thagard (1989) Induction: Processes of Inference, Learning and Discovery. MITP. p 7.

Chapter 2 How Science Redefines Truth

1. Overbye, D (2008) Dark, Perhaps Forever. NYT, 3 June 2008.
2. Kuhn, T (1962) The Structure of Scientific Revolutions. UCP.
3. Planck, M (1949) Scientific Autobiography. Philosophical Library.
4. In Our Time (2008) Plate Tectonics. BBC Radio 4. 24 Jan 2008.
5. Kious, W.J., R.I. Tilling (nd) Historical Perspective. This Dynamic Earth: the Story of Plate Tectonics. U.S. Geological Survey.
6. Egger, A.E. (2003) ."Plate Tectonics I: The Evidence for a Geologic Revolution. Visionlearning. Vol. EAS-1 (1).
7. Egger, AE (no date) The Evidence for a Geologic Revolution. Vision Learning Library. Emphasis added.
8. Maugh, T.H. (2009) Hormone therapy nearly doubles risk of death from lung cancer. Los Angeles Times. 20 Sept 2009. "The findings come from the Women's Health Initiative, a large study originally begun in 1991 to demonstrate, in part, that administering estrogen and progestin could relieve debilitating symptoms of menopause and reduce the risk of heart attack and stroke.... The findings suggest that the hormones do not cause lung cancer, but that they accelerate the growth of existing tumors, making them more aggressive and more likely to metastasize."
   Chlebowski, RT, et al (2009) Oestrogen plus progestin and lung cancer in postmenopausal women (Women's Health Initiative trial): a post-hoc analysis of a randomised controlled trial. The Lancet. 374:1243-51. ""Although treatment with oestrogen plus progestin in postmenopausal women did not increase incidence of lung cancer, it increased the number of deaths from lung cancer."
9. Marjoribanks, J, C Farquhar, H Roberts, A Lethaby (2012) Long term hormone therapy for perimenopausal and postmenopausal women. Intervention Review, Published Online: 11 Jul 2012. "Main results: Twenty-three studies involving 42,830 women were included...
   In relatively healthy postmenopausal women (that is generally fit, without overt disease) combined continuous HT [hormone treatment] significantly increased the risk of a coronary event."
10. Ganti, AK (2009) Another nail in the coffin for hormone-replacement therapy? The Lancet, 374:1217-1218.
11. http://www.cancer.gov/cancertopics/factsheet/Risk/menopausal-hormones
12. Dennett, D (1995) Darwin's Dangerous Idea: Evolution and the Meanings of Life. Simon & Schuster, NY.
13. Goldscheider, E (2009) Evolution Revolution. Wisconsin Alumni Magazine, Fall 2009.

Chapter 3 Darwin Didn't Discover Evolution

1. Secord, J (2009) The Secret History of Victorian Evolution. JCS, 4:23-36.
2. Endersby, J (2007) Evolution Myths. The Times Literary Supplement, 14 Mar 2007.
3. Endersby, J (2007) Evolution Myths. op cite.
4. Evans, J (nd) Anaximander. Encyclopedia Britannica http://www.britannica.com/EBchecked/topic/23149/Anaximander Evans is the Director of the Program for Science, Technology, and Society, University of Puget Sound, Tacoma, Wash.
5. Augustine (1982) The Literal Meaning of Genesis, JH Taylor (ed). The Newman Press. p 89-90.
6. Carroll, WE (2000) Creation, Evolution, and Thomas Aquinas. Catholic Education Resource Center.
7. Aquinas, T, Physica Book 2, Lecture 14.
8. Richards, RJ (2002) The Romantic Conception of Life: Science and Philosophy in the Age of Goethe. UCP p 212.
9. Leibniz, GW (1753) Lettre Prétendue in AO Lovejoy, Great Chain of Being: A Study of the History of an Idea (1936) 144-5.
10. Maupertuis, PL (1748) Histoire de l'academie des sciences et belle lettres de Berlin. p 267-294.
11. Burnett, J (1773-1792) The Origin and Progress of Man and Language. (6 volumes).
12. Kingdon, J (1996) Self-Made Man: Human Evolution From Eden to Extinction. Wiley, NY. p 220.
13. Wade, N (2009) Darwin, Ahead of His Time, Is Still Influential. NYT. 9 Feb 2009.
14. Secord, J (2009) The Secret History of Victorian Evolution. JCS, 4:23-36.
15. Grant, RE (1826) Observations on The Nature and Importance of Geology. Edinburgh New Phil. J, 14:270-84.
16. Lawrence, W (1819 [2007]) Lectures on Physiology, Zoology and The Natural History of Man. Kessinger Pub. p 348.
17. Chambers, R (1844 [1994]) Vestiges of the Natural History of Creation, J. Secord (ed.). UCP.
18. Secord, J (2009) The Secret History of Victorian Evolution. JCS, 4:23-36.
19. Secord, J (2001) Victorian Sensation: The Extraordinary Publication, Reception, and Secret Authorship of Vestiges of the Natural History of Creation. UCP. p 132 and 129.
20. Secord, J (2009) The Secret History of Victorian Evolution. JCS, 4:23-36.
21. Secord, J (2001) Victorian Sensation. op cite. p 168.
22. Packard, AS (1901) Lamarck, the Founder of Evolution: His Life and Work. Longmans, Green,

NY. p 303.
23. Ospovat, D (1995) The Development of Darwin's Theory: Natural History, Natural Theology, and Natural Selection, 1838-1859. CUP. p 10.
24. Bryson, B (2003) A Short History of Nearly Everything. Doubleday, NY. p81.
25. University of California, Berkeley (nd) Richard Owen (1804-1892) http://www.ucmp.berkeley.edu/history/owen.html
26. Rupke, N A (2009). Richard Owen. UCP. p 1.
27. Owen, R (1849) On the Nature of Limbs. John Van Voorst, London. p 86.
28. Owen, R (1849) On the Nature of Limbs. John Van Voorst, London.
29. Coates, M (2008) In Retrospect: Diagnosing Deep Similarity In Nature. Nature, 451:631.
30. Owen, R (1849) On the Nature of Limbs. John Van Voorst, London. p 86.
31. Owen, R (1849) On the Nature of Limbs. John Van Voorst, London. p 83.
"We have been accustomed to regard the vertebrate animals as being characterized by the limitation of their limbs to two pairs, and it is true that no more diverging appendages are developed for station, locomotion and manipulation. But the rudiments of many more pairs are present in many species. And though they may never be developed as such in this planet, it is quite conceivable that certain of them may be so developed, if the Vertebrate type should be that on which any of the inhabitants of other planets of our system are organized. The conceivable modifications of the vertebrate archetype are very far from being exhausted by any of the forms that now inhabit the earth, or that are known to have existed here at any period."
32. Darwin, C (1857) Letter to JD Hooker. Darwin Correspondence Project. http://www.darwinproject.ac.uk/entry-2117
33. White, TD, et al (2009) Ardipithecus ramidus and the Paleobiology of Early Hominids, Science, 326:64, 75-86. Human Evolution Research Center and Department of Integrative Biology, University of California, Berkeley.
34. Maugh, TH (2009) Fossils Radically Alter Ideas About the Look of Man's Earliest Ancestors. Los Angeles Times, 2 Oct 2009.
35. Lovejoy, CO (2009) "Reexamining Human Origins in Light of Ardipithecus ramidus," Science, Vol. 326. p. 74.
36. Rupke, N (1994) Richard Owen: Victorian Naturalist, YUP, New Haven, CT. pp. 172 ff.
E.g. Jamniczky, H (2008) Phenotypic integration patterns support an account of homology as a manifestation of evolvability. Evolutionary Biology 35:312-316.
Brigandt, I (2007) Typology now: homology and developmental constraints explain evolvability. Biology and Philosophy 22:709–725.
37. Amundson, R (2007) The Changing Role of the Embryo in Evolutionary Thought: Roots of Evo-Devo. CUP. p. 76-106.

CHAPTER 4 DARWIN'S IDEAS WERE OLD

1. Darwin, C (1887) The Life and Letters of Charles Darwin: Including an Autobiographical Chapter, Volume 1, John Murray, London. p 32. "I became convinced...that my father would leave me property... though I never imagined that I should be so rich a man as I am."
2. Malthus, TR (1798) An Essay on the Principle of Population, as It Affects the Future Improvement of Society: With Remarks on the Speculations of Mr. Godwin, M. Condorcet, and Other Writers. Johnson, London.
3. Darwin, CR (1859) On the Origin of Species by means of Natural Selection. John Murray, London. p 63.
4. Bell, T (1858) J of Proceedings of the Linnaean Society, Linnaean Society of London, Longman, Brown, Green, London. p. viii–ix. Reproduced by BiblioBazaar, 2009.
5. Wallace, A (1858) On the Zoological Geography of the Malay Archipelago. "Communicated by Charles Darwin, Esq., F.R.S &L.S." J. of Proceedings of the Linnaean Society. Longman, Brown, Green, London. p 180. (reproduced by BiblioBazaar, 2009).
6. DeArce, M (2011) The parallel lives of Joseph Allen Galbraith (1818-90) and Samuel Haughton (1821-97): religion, friendship, scholarship and politics in Victorian Ireland. Proceedings of the Royal Irish Academy 112C:1-27. Darwin recounted this in his autobiography and put Houghton's response in the past tense: "all that was new in the paper was false, and all the was true was old."
7. Desmond, A, J Moore, J Browne (2007) Charles Darwin. OUP. p 67.
8. Mitchell, S. (2008) Daily Life in Victorian England, Second Edition, The Greenwood Press. p 237-8.
9. Browne, J (2010) Charles Darwin: The Power of Place. Knopf, NY. Kindle Location 4814.
10. From of a printing of 1250 copies, 150 were given away, 500 were sold in retail stores and 600 were picked up by a subscription lending library.
11. Desmond, A., J. Moore (1994) Darwin: The Life of a Tormented Evolutionist, W. W. Norton & Company, NY. p 500.
12. Granovetter, M and R Swedberg (eds) (2001) The Sociology of Economic Life. Perseus Books, Cambridge MA, p 293.
13. Secord, J (2009) The Secret History of Victorian Evolution. JCS, 4:23-36.
14. Darwin, C (1887) The Life and Letters of Charles Darwin: Including an Autobiographical Chapter, Volume 2, John Murray, London. p 186.
15. Rupke, N (1985) Richard Owen's Hunterian Lectures on Comparative Anatomy and Physiology. Med Hist, 29:240.
16. Rupke, N (1985) Richard Owen's Hunterian Lectures, op cite.
17. Editors (2009) Evolution: The next 200 years. New Scientist 28 Jan 2009 Emphasis added.

18. Lewontin, RC (1998) Survival of the Nicest? NYRB, Oct 1998.
19. Martin, L and C Coon (2010) Infection Protection and Natural Selection. Science 330:602-603.
20. Gratten, J, et al (2008) A Localized Negative Genetic Correlation Constrains Microevolution of Coat Color in Wild Sheep. Science 319:318 -320.
21. Judson, O (2008) An Original Confession. NYT, 8 July 2008.
22. Ospovat, D (1995) The Development of Darwin's Theory. op cite. p 26. Also see pg 241, endnote 66.
23. Amundson, R (1998) Typology Reconsidered. op cite.
24. Ospovat, D (1995) The Development of Darwin's Theory: Natural History, Natural Theology, and Natural Selection, 1838-1859. CUP. p 6.
Ospovat, D (1978) Perfect Adaptation and Teleological Explanation. Studies in History of Biology 2:33–56.
25. Darwin, C (1887) The Life and Letters of Charles Darwin: Including an Autobiographical Chapter, Volume 1, John Murray, London. p 76.
26. Kingsley, C (1886) The Water-babies.
27. Endersby, J (2007) Evolution Myths. The Times Literary Supplement, 14 Mar 2007.
28. Secord, J (2009) The Secret History of Victorian Evolution. JCS, 4:23-36

Chapter 5 The War Against Religion

1. Desmond, A and J Moore (1994) Darwin: The Life of a Tormented Evolutionist. Norton, NY. p xxi.
2. Gay, P (1996) The Enlightenment: The Science of Freedom. Norton, NY. p 144.
3. Gay, P (1996) The Enlightenment: The Science of Freedom. Norton, NY. p 163.
4. Darwin, speaking of Grant during his time in Edinburgh, wrote "I knew him well," and often accompanied him on long walks. Darwin, C (1887) The Life and Letters of Charles Darwin: Including an Autobiographical Chapter, Volume 1, John Murray, London. p 34.
5. In Our Time (2009) Darwin, The Genius of Evolution. BBC Radio 4, 5 Jan 2009.
6. Browne, J (2009) http://www.ed.ac.uk/news/darwin/edinburgh
7. Eldredge, N (2009) What Darwin learned in medical school. The Lancet 373:454-455.
8. Darwin Online: http://darwin-online.org.uk/content/frameset?viewtype=text&itemID=F1497&pageseq=84
9. Darwin Online: http://darwin-online.org.uk/content/frameset?viewtype=text&itemID=F1497&pageseq=84
10. Browne, J (1995) Charles Darwin: A Biography, Vol. 1-Voyaging. Knopf, NY. p 499.
11. Darwin Correspondence Project http://www.darwinproject.ac.uk/death-of-anne-darwin
12. Keynes, R. (2002) Darwin, His Daughter, and Human Evolution, Riverhead, NY.
13. Ospovat, D (1995) The Development of Darwin's Theory. op cite. p 231.
14. BBC Radio 4 (2009) Darwin, The Genius of Evolution. In Our Time, 5 Jan 2009.
15. Paley, W (1802) [1986] Natural Theology; or, Evidences of the Existence and Attributes of the Deity, Collected from the Appearances of Nature, 12 ed. Lincoln-Rembrandt Pub. Paley called the laws of nature the "secondary causes" of evolution, the primary cause being God, who he said created the laws in the first place.
16. Paley, W (1809) Natural Theology. Or Evidences of the Existence and Attributes of the Deity, Collected from the Appearances of Nature. 12th edition. J. Faulder, London. p 419-420. Paley refers to the "second causes" of creation, i.e. the laws of nature such are found in chemistry, and says these may guide the formation of the universe and what exists in nature: "Now where the order of second causes is mechanical, what is here said of mechanism strictly applies to it... The philosopher beholds with astonishment the production of things around him. Unconscious particles of matter take their stations, and severally range themselves in an order, so as to become collectively plants or animals, i. e. organized bodies, with parts bearing strict and evident relation to one another, and to the utility of the whole."
17. Darwin, C (1887) The Life and Letters of Charles Darwin: Including an Autobiographical Chapter, Volume 2, John Murray, London. p 202-203. "The acute champion of Teleology, Paley, saw no difficulty in admitting that the "production of things" may be the result of trains of mechanical dispositions...that is to say, he proleptically accepted the modern doctrine of Evolution; and his successors might do well to follow their leader, or at any rate to attend to his weighty reasonings, before rushing into an antagonism which has no reasonable foundation." NB: The evidence is that Paley's "successors," in the main, also accepted the idea of evolution through natural processes.
18. Darwin, C (1876) Recollections of the development of my mind and character [autobiography]. Cambridge University Library, p 65.
19. Coyne, J (2010) Better All the Time. NYT, 5 Nov 2010. Emphasis added.
20. Brooke, JH (nd) St Edmunds College, Cambridge University, website: http://www.st-edmunds.cam.ac.uk/cis/brooke/ Brooke is a historian of science at Cambridge University.
21. White, P (2003) Thomas Huxley: making the 'Man of Science', CUP.
22. White, P (2003) Thomas Huxley: making the 'Man of Science', CUP. p 69.
23. Wollaston, AFR (1921) Life of Alfred Newton 1829–1907. p 102.
24. MacBride, EW (1934) Huxley, London: Duckworth. p65.
25. Huxley, TH (1854) Review of Vestiges of the Natural History of Creation, tenth edition. British and Foreign Medico-Chirurgical Review 13:425–439.
26. Huxley, TH (1855) On certain zoological arguments commonly adduced in favour of the hypothesis of the progressive development of animal life in time. Proceedings of the Royal Institution 2 (1854–58) p82–85.

27. Brooke, JH (nd) St Edmunds College, Cambridge University, website: http://www.st-edmunds.cam.ac.uk/cis/brooke/
28. Secord, J (2009) The Secret History of Victorian Evolution. JCS 4:23-36.
29. Secord, J (2009) The Secret History of Victorian Evolution. JCS 4:23-36.
30. Bowler, ,P (1988) The Non Darwinian Revolution. TJHUP. pg 68.
31. Huxley, TH (1911) Life and Letters of Charles Darwin, Francis Darwin (ed). Appleton & Co, NY. pp 26–27.
32. Huxley, TH (1862) On our knowledge of the causes of the phenomena of organic nature, London Huxley, L (1900) The Life and Letters of Thomas Henry Huxley. Macmillan, London. p189.
33. Huxley, L (1900) The Life and Letters of Thomas Henry Huxley. Macmillan, London.
34. Huxley, TH (1869) Review of The Natural History of Creation by Dr. Ernst Haeckel. In The Academy, A monthly record of Literature, Learning, Science and Art, 9 Oct 1869, p 13. (Published in The Academy, Volume 1. John Murray, London).
35. Darwin, C (1887) The Life and Letters of Charles Darwin: Including an Autobiographical Chapter, Volume 2, John Murray, London. p 202-203. Chapter 5 "On the Reception of the Origin of Species," where Huxley repeats the 1867 statement, including, a few lines later this: "Having got rid of the belief in chance and the disbelief in design..." Huxley is saying he wants to eliminate the idea that evolution occurs by "chance," and restore belief in the idea evolution has "design" or patterns. He just doesn't want to attribute any of the "design" to God.
36. Huxley, L. (2004) The Life and letters of Thomas Henry Huxley, Volume 1, Kessinger Pub. p 173. Also see Desmond, A, J Moore (1991) Darwin. Penguin Group, NY. p 477–491."The leader in this crusade was Thomas Huxley, a passionate opponent of religion and advocate of radical change."
37. Huxley, TH (1873) Critiques and Addresses, MacMillan, London. p 271.
38. Huxley, L (2004) The Life and letters of Thomas Henry Huxley. op cite. p 173. Huxley spelled the last word "peoplesh." I removed the "sh" for clarity, although note it here.
39. Browne, J (2002) Charles Darwin: The Power of Place. Knopf, NY. p 248.
40. Browne, J (2002) Charles Darwin: The Power of Place. Knopf, NY. p 247.
41. Browne, J (2002) Charles Darwin: The Power of Place. Knopf, NY. p 249.
42. Browne, J (2002) Charles Darwin: The Power of Place. Knopf, NY. p 248.
43. Vernon, J (1970) "The X Club: Fraternity of Victorian Scientists. British J for the Hist of Science, 5: 63–72. Also MacLeod, R (1970) "The X-Club a Social Network of Science in Late-Victorian England. Notes and Records of the Royal Society of London, 24: 305–322.
44. Browne, J (2010) Charles Darwin: The Power of Place. Knopf, NY.
45. Browne, J (2010) Charles Darwin: The Power of Place. Knopf, NY.
46. Browne, J (2002) Charles Darwin: The Power of Place. Knopf, NY. p 249.
47. Desmond, A. (1994) Huxley: Vol. 1 The Devil's Disciple, Michael Joseph, London. p 284, 289–290.
48. Browne, J (2002) Charles Darwin: The Power of Place, Knopf, NY. p 248.
49. Browne, J (2002) Charles Darwin: The Power of Place, Knopf, NY. p 248.

CHAPTER 6 THE WAR AGAINST THE WORKING CLASS

1. Darwin, CR (1871) The Descent of Man, and Selection in Relation to Sex, Volume 1. John Murray, London. p 187.
2. Darwin, CR (1871) The Descent of Man. op cite. Chapter 5.
3. Malthus, T (1798) An Essay on Population. London.
4. Malthus, T (1798) An Essay on Population. London.
5. Darwin, CR (1871) The Descent of Man. op cite. Chapter 5.
6. Burnett, J., 1773-1792, The Origin and Progress of Man and Language (6 volumes).
7. Packard, AS (1901) Lamarck, the Founder of Evolution: His Life and Work with Translations of His Writings on Organic Evolution, Longmans, Green, and Co., NY. p 365-370.
8. Darwin, CR (1871 [1981])The Descent of Man, and Selection in Relation to Sex. PUP. p 35.
9. Darwin, CR (1871 [1981])The Descent of Man, and Selection in Relation to Sex. PUP. p 105.
10. Huxley, TH (1863 [1959]) Evidence as to Man's Place in Nature, University of Michigan Press Ann Arbor.
11. Moore, J (2009) Darwin, The Genius of Evolution. In Our Time, 5 Jan 2009.
12. University Of Chicago Medical Center (2005) University Of Chicago Researchers Discovered That Humans Are A 'Privileged' Evolutionary Lineage. Science Daily, 15 Jan 2005.
13. Gazzaniga, M (2008) Human: The Science Behind What Makes Us Unique, Ecco Press. Gazzaniga is Director of the SAGE Center for the Study of Mind, University of California, Santa Barbara.
14. Carey, S (2009) The Origin of Concepts. OUP. "The human conceptual repertoire is a unique phenomenon on this earth."p 538. Carey is chair of the Psychology Dept., Harvard University.
15. Darwin, C (1887) The Life and Letters of Charles Darwin: Volume I, John Murray, London, pg. 316.
16. Darwin, CR (1871) The Descent of Man, and Selection in Relation to Sex, Volume 1. John Murray, London. p 187.
17. Galton, F (1905) Essays in Eugenics. Emphasis added.
18. Huxley TH (1890) The Natural Inequality of Man. Nineteenth Century Jan 1890, in Collected Essays vol 1, p290–335.
19. Bowler, PJ (2009) Darwin's Originality. Science, 323: 223-226.
20. Polkinghorne, J (2011) Science and Religion. YUP, New Haven. p 7.
21. Temple, F (1884 [1903]) The Relations Between Religion and Science: Lecture IV of Eight Lectures Preached Before the University of Oxford. McMillan, NY.

22. Lucas, JR (1979) Wilberforce and Huxley: A Legendary Encounter. The Historical Journal 22:313–330.
23. Darwin, C (1897) The Life and Letters of Charles Darwin: Including an Autobiographical Chapter, Volume 2, John Murray, London, p. 117.
24. Brooke, JH (2001) Darwinism & Religion: A Revisionist View of the Wilberforce-Huxley Debate. Lecture delivered at Emmanuel College, Cambridge 26 Feb 2001.
25. Huxley, L (2004) The Life and letters of Thomas Henry Huxley, Volume 1, Kessinger Pub.
26. Hooker, JD (1860) Letter 2852: Hooker, J. D. to Darwin, C. R., 2 July 1860. Darwin Correspondence Project.
27. The Darwin Correspondence Project (entry 2852, footnote 9. http://www.darwinproject.ac.uk/entry-2852) says this: "Huxley gave his own account of the proceedings [the Oxford debate] in a letter to Frederic Daniel Dyster about two months later: 'If then, said I, the question is put to me would I rather have a miserable ape for a grandfather or a man highly endowed by nature and possessed of great means and influence and yet who employs those faculties for the mere purpose of introducing ridicule into a grave scientific discussion—I unhesitatingly affirm my preference for the ape' (Jensen 1988, p. 168). See also Appendix VI. The events of the session have been documented in LL 2: 320–3; F. Darwin ed. 1892, pp. 236–42; L. Huxley ed. 1900, 1: 179–89; L. Huxley ed.1918, 1: 521–7; and Jensen 1988."
    It is unclear if Huxley actually made this comment during the debate (it is not mentioned in any of the contemporary material on the debate, including Huxley's own accounts) or if he is embellishing (or inventing) a comment in an effort to bolster what was seen by others (e.g. Joseph Hooker) as a very poor performance by Huxley. Either way, Huxley makes no mention that this comment is in response to anything Wilberforce said, which he surely would have done if it was in response to the comment attributed to Wilberforce.
28. e.g. the Liverpool Mercury, 5 July 1860 makes a short mention, with no sense of a great drama.
29. Huxley, L (1900) The Life and Letters of Thomas Henry Huxley. Macmillan, London. Vol 1, p 202.
30. Gould, SJ (1992) Punctuated Equilibrium in Fact and Theory, in The Dynamics of Evolution, A Somit and S Peterson (eds). Cornell University Press, NY. p 56-57.
31. Laughlin, RB (2005) A Different Universe: Reinventing Physics from the Bottom Down. Basic Books, NY. p 168.

CHAPTER 7 THE WAR AGAINST THE EVIDENCE

1. Conway Morris, S (2003) Lamarck and Natural Selection. In Our Time, BBC Radio 4.
2. Endersby, J (2007) Evolution Myths. The Times Literary Supplement, 14 Mar 2007.
3. Hull, DL (1983) Darwin and His Critics: The Reception of Darwin's Theory of Evolution by the Scientific Community. UCP. p 3.
4. Hull, DL (1983) Darwin and His Critics: The Reception of Darwin's Theory of Evolution by the Scientific Community. UCP. p 3 and p 7.
5. Darwin, C (1897) The Life and Letters of Charles Darwin: Including an Autobiographical Chapter, Volume 2. John Murray, London, p. 72.
6. Darwin, C (1887) The Life and Letters of Charles Darwin: Including an Autobiographical Chapter, Volume 2. John Murray, London, p. 147. Charles Darwin to T H Huxley, December 2, 1860.
7. Darwin, C (1897) The Life and Letters of Charles Darwin: Including an Autobiographical Chapter, Volume 2. John Murray, London, p 146. Charles Darwin to Asa Gray, 26 Nov1860.
8. Darwin, C (1887) The Life and Letters of Charles Darwin: Including an Autobiographical Chapter, Volume 1. John Murray, London. p 76.
9. Darwin, C (1887) The Life and Letters of Charles Darwin: Including an Autobiographical Chapter, Volume 1. John Murray, London. p 82-83. Darwin wrote of himself,
    "I have no great quickness of apprehension... I am therefore a poor critic: a paper or book, when first read, generally excites my admiration, and it is only after considerable reflection that I perceive the weak points. My power to follow a long and purely abstract train of thought is very limited; and therefore I could never have succeeded with metaphysics [the basic nature of things] or mathematics.
    "My memory is extensive, yet hazy: it suffices to make me cautious by vaguely telling me that I have observed or read something opposed to the conclusion which I am drawing, or on the other hand in favor of it... So poor in one sense is my memory, that I have never been able to remember for more than a few days a single date or line of poetry.
    "Some of my critics have said, "Oh, he is a good observer, but has no power of reasoning!" I do not think this can be true, for the 'Origin of Species' is one long argument from the beginning to the end... On the favourable side of the balance, I think that I am superior to the common run of men in noticing things which easily escape attention, and in observing them carefully." [Contrary to Darwin's comment about, *The Origin*, most scientists at the time saw it as confirmation of Darwin's assessment of himself that "my power to follow a long and purely abstract train of thought is very limited."
10. DeArce, M (2011) The parallel lives of Joseph Allen Galbraith (1818-90) and Samuel Haughton (1821-97): religion, friendship, scholarship and politics in Victorian Ireland. Proceedings of the Royal Irish Academy 112C:1-27.
    Haughton's opposition to natural selection, like that of Richard Owen, was built in part on ideas we recognize today as the physical constraints of self-assembly. For example, in Chapter 14, DeArce says Darwin believed the hexagonal shape of a bee honeycomb emerged from natural selection. Haughton however argued "hexagonal prisms emerged not out of an improved instinct for efficient use of materials [i.e. natural selection]... but by *physical*

necessity; if many bees were working together, each being the centre of an imaginary sphere. Inevitably the spheres... will pack in hexagons centred on another sphere, this being *simply a physical constraint, not a selectable instinct.*" (Emphasis added.) In other words purely physical constraints, not natural selection, led to the shape of a honeycomb.
11. Owen, R (1860) Darwin on the Origin of Species. Edinburgh Review 111:487–532.
12. Cosans, C (2009) Owen's Ape & Darwin's Bulldog: Beyond Darwinism and Creationism. Indiana University Press, Bloomington. p xix and p 97-103. "Owen agreed evolution occurs, but argued that it occurs by more complicated means than those outlined in the *Origin*. He argued that anatomical studies uncover all sorts of life processes in addition to genealogical relationships. From Owen's perspective natural selection does not explain major anatomical differences."
13. Cosans, C (2009) Owen's Ape & Darwin's Bulldog: Beyond Darwinism and Creationism. Indiana University Press, Bloomington. p.xiix, xx, and Chapters 4 and 8 "Huxley argued that "ape brains have at least some form of the posterior lobe, the posterior horn of the lateral ventricle and the hippocampus...Hence he asserted that a strictly scientific perspective indicates no great differences between apes and humans."
"In 1851 Owen gave a specific analysis of human and ape brain development [differences]... In a paper before the Linnaean Society in 1957 Owen made even more distinct growth patterns [between apes and humans]. He argued that humans constitute a distinct subclass of mammals by virtue of their brains."
14. Huxley, TH (1861) On the Zoological Relations of Man with the Lower Animals. Natural History Review, 1: 67-84. "If we place A, the European brain, B, the Bosjesman [an African bushman] brain, and C, the orang [orangutan] brain, in a series, the differences between A and B, so far as they have been ascertained, are of the same nature as those between B and C."
Cosans, C (2009) Owen's Ape & Darwin's Bulldog: Beyond Darwinism and Creationism. Indiana University Press, Bloomington. "[Huxley] also argued that the brains of [African] blacks are intermediary between the brains of whites and apes."
15. Owen, R (1862) "On the Zoological Significance of the Cerebral and Pedal Characters of Man," Report of the British Association for the Advancement of Science, pt2, p 116-118.
Rupke, NA (2010) Richard Owen. UCP. (Kindle Locations 4415-4421). "Owen proposed that the human species not only be placed in a genus, a family, and an order of its own but be assigned to a separate subclass, for which he coined the name Archencephala, the "ruler brains." The human brain—Owen stated—differs markedly from that of all other mammals."
16. Rupke, N (2010). Richard Owen. UCP (Kindle Locations 4700-4707). "In 1862 Owen presented his paper "On the Zoological Significance of the Cerebral and Pedal Characters of Man"... reiterating his previous observation that the transition gorilla [to]Negro is accompanied by a jump in the size of the brain [and that a] Negro's brain is as large as the average brain of a Caucasian. And Owen agreed with the antiracist conclusion of the Heidelberg anthropologist Friedrich Tiedemann "that there had been no province of intellectual activity in which individuals of the pure Negro race had not distinguished themselves.""
17. Browne, J (2002) Charles Darwin: The Power of Place. Jonathan Cape, London. p 329-355.
18. Browne, J (2010) Charles Darwin: The Power of Place. Knopf, NY Kindle Location 7852.
19. Browne, J (2002) Charles Darwin: The Power of Place. Jonathan Cape, London. p 329-355.
20. Browne, J (2010) Charles Darwin: The Power of Place. Knopf, NY. Kindle Location 7306-7311.
21. Browne, J (2010) Charles Darwin: The Power of Place. Knopf, NY. Kindle Location 7871-7872.
22. Darwin, C (1887) p. 331 Letter 2893: Darwin, C. R. to Huxley, T. H., 8 Aug 1860.
23. Hull, DL (1983) Darwin and the Nature of Science, in Evolution from Molecules to Men, DS Bendall (ed) CUP. p 63.
24. Darwin, C (1857) Letter to Asa Gray. Darwin Correspondence Project.
25. University of California biologist John Gerhart and Mark Kirschner, head of Harvard's Systems Biology Department say that modern Darwinists, created a similar idea in what is referred to as the Modern Synthesis. "The Modern Synthesis made the concept of adaptation paramount in evolutionary theory. The organism was like modeling clay, and remolding of the clay meant that each of the billions of little grains was free to move a little bit in any direction to generate a new form." Kirschner, MW and JC Gerhart (2005) The Plausibility of Life: Resolving Darwin's Dilemma. YUP. p 31.
26. Lewontin, RC (1998) Higher Superstition: An Exchange. NYRB. 3 Dec 1998.
27. Darwin, C (1868) [(2001] Variation of Plants and Animals Under Domestication, Vol. 2. Adamant, Boston. p 367-368.
28. Darwin, C (1868) [(2001] Variation of Plants and Animals. op cite. p 367-368.
29. Moore, JR (1981) The Post-Darwinian Controversies. op cite. p 141.
30. Darwin, C (1868) [(2001] Variation of Plants and Animals. op cite. p 367-368.
31. Darwin, C (1868) [(2001] Variation of Plants and Animals. op cite. p 367-368.
32. Bulmer, M (2003) Francis Galton: Pioneer of Heredity and Biometry, TJHUP.
33. Darwin, C (1871) Pangenesis. Nature, 3:502-503.
34. Darwin, C (1871) Pangenesis. Nature, 3:502-503. "When I first heard of Mr. Galton's experiments, I did not sufficiently reflect on the subject, and saw not the difficulty of believing in the presence of gemmules in the blood."
35. Second, JA (1986) Controversy in Victorian Geology: The Cambrian-Silurian Dispute, PUP. p 301.
36. Palmer, D (2005) Earth Time: Exploring the Deep Past from Victorian England to the Grand Canyon. Wiley, NY.
37. Rudwick, MJS (1985) The Meaning of Fossils, 2nd edition, UCP. p 200-201.
38. Lewontin, RC (2008) The Triumph of Stephen Jay Gould. NYRB, 55:2.
39. Darwin, CR (1859) On the Origin of Species. Kindle Edition. p 165 and p 149.

40. Darwin, CR (1859) On the Origin of Species. Kindle Edition. p 181-182.
41. Darwin, CR (1859) On the Origin of Species. Kindle Edition. p 168.
42. Moore, JR (1981) The Post-Darwinian Controversies. op cite. p 176.
43. Gould, SJ (1977) The Return of the Hopeful Monster. Natural History June-July, 1977.
44. Darwin, C (1859) The Origin of Species: A Facsimile of the First Edition, Harvard University Press, 1964. p 307.
45. Eldredge, N and I Tattersall (1982) The Myths of Human Evolution, Columbia University Press. p 163.
46. Darwin, C (1861) Introduction, The Origin Of Species, 3rd edition, and prior citations.
47. Darwin, C (2008) Charles Darwin: Evolutionary Writings, JA Secord,(ed). OUP. *The Times* review of *The Descent of Man* appeared on 8 April 1871. I removed one ellipse for clarity. The full text is: "This proposition will not, we think, in the minds either of the mass of men or of the learned, lose its first character of enormous and painful improbability by any amount of that preparatory exercise in cognate speculations on which Mr. Darwin relies... We wish we could think that these speculations were as innocuous as they are unpractical and unscientific, but it is too probable that if unchecked they might exert a very mischievous influence... A man incurs a grave responsibility who, with the authority of a well-deserved reputation, advances at such a time the disintegrating speculations of this book. He ought to be capable of supporting them by the most conclusive evidence of facts. To put them forward on such incomplete evidence, such cursory investigation, such hypothetical arguments as we have exposed, is more than unscientific - it is reckless."
48. Conway Morris, S (2003) Lamarck and Natural Selection. In Our Time, BBC Radio 4, 2003.
49. Bowler, PJ (1992) The Eclipse of Darwinism: Anti-Darwinian Evolution Theories in the Decades around 1900. JHUP. p 3-4.
50. Burkhardt, RW (1980) Lamarckism in Britain and the United States. In The Evolutionary Synthesis, E Mayr and W Provine (eds). CUP. p 346.

CHAPTER 8 REDEFINING EVOLUTION TO MATCH DARWIN

1. Spiro, JP (2008) Defending the Master Race. Univ. of Vermont Press.
2. Sewell, D (2009) Charles Darwin and the Children of the Evolution. The Sunday Times (UK), 8 Nov 2009.
3. Keller, M (2006) The Scandal at the Zoo. NYT. 6 Aug 2006.
4. Galton, F (1908) Memories of My Life. Methuen, London. Chapter 21 "Race Improvement."
5. Galton, F (1873) Galton letter to The Times (UK), 6 June1873. The Collected Work of Francis Galton.
6. Blacker, CP (1945) Eugenics In Retrospect and Prospect. In Occasional Papers on Eugenics, The Galton Lecture. The Eugenics Society and Cassell and Company, London.
7. Gillham, NW (2001) A Life of Sir Francis Galton: From African Exploration to the Birth of Eugenics. OUP, NY. p 348. Nicholas Gillham is emeritus professor of biology at Duke University.
8. Walker, HM (1968) Karl Pearson. International Encyclopedia of the Social Sciences. NY. p 496-503.
9. Porter, TM (2004) Karl Pearson: The Scientific Life in a Statistical Age. PUP. p 74.
10. Pearson, K (1892) [2006] The Grammar of Science. Kessinger Publishing. p 7.
11. Porter, TM (2004) Karl Pearson: The Scientific Life in a Statistical Age, PUP. p 244.
12. Galton, F (1908) Race Improvement, chapter 21 in Memories of My Life, Methuen, London.
13. Pearson, K (1901) National Life From the Stand-Point of Science. Black, London.
14. Pearson, K (1901) National Life From the Stand-Point of Science. Black, London.
15. Galton, F (1908) Race Improvement, chapter 21 in Memories of My Life, Methuen, London.
16. Pearson, K (1901) National Life From the Stand-Point of Science. Black, London.
17. Grant, M (1922) The Passing of the Great Race or The Racial Basis of European History. Charles Scribner's Sons, NY. p 16.
18. Winchester, AM (2008) Genetics. Encyclopædia Britannica. Winchester is Emeritus Professor of Biology, University of Northern Colorado and the author of Genetics and other books.
19. Bateson, W.F, (1894 [1992]), Materials for the Study of Variation: Treated with Especial Regard to Discontinuity in the Origin of Species, TJHUP.
20. Kimura, M, N Takahata and JF Crow (1994) Population Genetics, Molecular Evolution, and the Neutral Theory: Selected Papers. UCP. p 616.
"Soon a bitter conflict arose between the biometric school championed by Karl Pearson and WFR Weldon and the Mendelian school led by William Bateson. The biometric school denied the generality of Mendelian inheritance and claimed, following Darwin, that evolution occurred continuously through natural selection acting on small variations. On the other hand, the Mendelians believe that all heritable variations are discontinuous and therefor evolution occurs in decisive steps."
21. Allen, GE (2008) Thomas Hunt Morgan. Encyclopædia Britannica.
22. Kimura, M, N Takahata and JF Crow (1994) Population Genetics, Molecular Evolution, and the Neutral Theory: Selected Papers. UCP. p 616.
The idea that "evolution occurs by leaps rather than by gradual natural selection soon became very popular among biologists and found many adherents at that time."
23. Antonovics, J (1987) The Evolutionary Dys-Synthesis: Which Bottles for which Wine? American Naturalist 129:321-331. Morgan said in 1909: "The neo-Darwinians seem less concerned with the advancement of the study of evolution than with expounding Darwinism as a dogma... To imagine a use for an organ is for them equivalent to explaining its origin by natural selection

without further inquiry."
24. Box, JF (1978) Preface in R. A. Fisher: The Life of a Scientist, Wiley, NY.
25. Fisher, RA (1930) [2000] The Genetical Theory of Natural Selection, J H Bennett (ed). OUP.
26. Fisher, RA (1930) [2000] The Genetical Theory of Natural Selection, J H Bennett (ed). OUP.
27. Haldane, JBS (1932 [1990]) The Causes of Evolution. PUP
Wright, Sewall (1932). "The roles of mutation, inbreeding, crossbreeding and selection in evolution". Proc. 6th Int. Cong. Genet 1: 356–366.
28. Mayr, E (1997) The Establishment of Evolutionary Biology as a Discrete Biological Discipline. Bioessays19:263-266.
29. Huxley, J (1926) Essays in Popular Science. Chatto and Windus, London. p ix.
30. Huxley, JS (1941) The Uniqueness of Man. Chatto and Windus, London.
31. Hubback, D (1989) Julian Huxley and Eugenics. In Evolutionary studies: a centenary celebration of the life of Julian Huxley, M. Keynes and G.A. Harrison (eds). Macmillan, London.
32. Lifton, RJ (1986) The Nazi Doctors–Medical Killing and the Psychology of Genocide. Basic Books.
33. Schopenhauer, A (1851) Parerga and Paralipomena, Volume II, Section 92.
34. Wistrich, RS (1982) Who's Who in Nazi Germany, Routledge, London.
35. Black, E (2004) Hitler's Debt to America. The Guardian, 6 Feb 2004.
36. Grant, M (1922) The Passing of the Great Race or The Racial Basis of European History. Charles Scribner's Sons, NY. Emphasis added.
37. Arluke, A and C Sanders (1996) Regarding Animals. Temple University Press. p 132-134. "Boundary work – the drawing and blurring of lines of demarcation between humans and animals – was essential to the Nazi paradox... this boundary work was effected in three ways – by morally elevating animals, identifying with them, and animalizing humans." p 133.
38. The New York Times (1935) U.S. Eugenicists Hails Nazi Racial Policy. 29 Aug 1935, p 5.
39. Time (1935) Praise for Nazis. 9 Sept 1935, pp 20– 21.
40. Bird, R and G Allen (1981) Archival Sources in the History of Eugenics. Journal of the History of Biology14:2. p. 351.
41. Mehler, B. (1988) History of the American Eugenics Society. University of Illinois Press, Urbana IL. From chapter 5: "Eliminating the Inferior: American and Nazi Sterilization Programs, Science for the People."
42. Huxley, J, et al (1939) Social Biology and Population Improvement. Nature, 16 Sept 1939. This article is commonly referred to as "The Eugenics Manifesto," even by the journal Nature in which it first appeared.
43. Reich, W (2009) We Are All Guilty. NYT, 13 May 2009. Emphasis added. Reich is professor of international affairs, ethics and human behavior at George Washington University and a senior scholar at the Woodrow Wilson Center.
44. Sewell, D (2009) Charles Darwin and the Children of the Evolution. The Sunday Times, 8 Nov 2009.
45. Lifton, RJ (1986) The Nazi Doctors–Medical Killing and the Psychology of Genocide. Basic Books. p 504.
46. Sakura, O (1998) Similarities and Varieties: A Brief Sketch on the Reception of Darwinism and Sociobiology in Japan. Biology and Philosophy 13:341-357. Sakura is at the Institute of Human Behavioral Sciences, Yokohama National University, Japan.
47. Glad, J (2003) Hermann J. Muller's 1936 Letter to Stalin. The Mankind Quarterly 43:305-319.
48. Walker, S (2008) Stalin's space monkeys. The Independent 15 April 2008.
49. Dunning, B (2010) Stalin's Human-Ape Hybrids. Skeptoid: Critical Analysis of Pop Phenomema 17 Aug 2010.
50. Desmond, A and J Moore (1994) Darwin: The Life of a Tormented Evolutionist. Norton, NY. p xxi.
51. Bowler, PJ (2009) Darwin's Originality. Science, 323:223-226.

Chapter 9 How The Darwinist Triumphed

1. Goldscheider, E (2009) Evolution Revolution. Wisconsin Alumni Magazine, Fall 2009.
2. Goldschmidt, R (1940) The Material Basis of Evolution. YUP. p 205-206.
3. Dobzhansky, T (1940) Catastrophism Versus Evolutionism. Science 92:356-348.
4. Creighton, HB and B McClintock (1931) A Correlation of Cytological and Genetical Crossing-Over in Zea Mays. PNAS 17:492-497.
5. Dimmock, NJ and SB Primrose (1994) Introduction to Modern Virology, fourth ed. Blackwell Publishing. p 80.
6. McClintock, B (1951) Chromosome Organization and Genic Expression. Cold Spring Harbor Symposia on Quantitative Biology 16:13-47.
7. McClintock, B (1987) The Discovery and Characterization of Transposable Elements: The Collected Papers of Barbara McClintock. Garland, NY.
8. McClintock, B (1951) Chromosome Organization and Genc Expression. op cite.
9. Clark RW (1968) The Huxleys. McGraw-Hill, NY.
10. Gould, SJ (1980) The Panda's Thumb. Norton, NY. p 76-84.
11. Cain, J (1994) Ernst Mayr as Community Architect: Launching the Society for the Study of Evolution and the Journal Evolution. Biology and Philosophy 9:387-427.
12. Huxley, J (1942) Evolution: The Modern Synthesis. Allen & Unwin, London.
13. Mann, A (2010) Genetics, Paleontology, and Evolution: An International Conference held at the Princeton Inn. Princeton University http://www.princeton.edu/forbescollege/about/history/internation-conference-on/ "By bringing together in one book scholars from a wide variety of

biological disciplines who had had little or no previous contact and who had in fact harbored some distrust of each other, the articles documented the ways by which varying biological data served to support the Synthetic Theory."
14. Cain, J (1994) Ernst Mayr as Community Architect: Launching the Society for the Study of Evolution and the Journal Evolution. Biology and Philosophy 9: 387-427.
15. Mayr, E (nd) History of The Society for the Study of Evolution. The Society for the Study of Evolution.
16. Muller, HJ (1949) Redintegration [sic] of the Symposium on Genetics, Paleontology, and Evolution. In Genetics, Paleontology and Evolution, G Jepson, GG Simpson and E Mayr (eds). PUP. p 421–445.
17. Dietrich, MR (1998) Paradox and Persuasion: Negotiating the Place of Molecular Evolution within Evolutionary Biology. J of the History of Biology 31: 85-111. "Although there was not a single theory with which all members of the synthesis community fully agreed, there was certainly an evolutionary constriction as proposed mechanisms of evolution such as saltation, orthogenesis, and the inheritance of acquired characteristics were ruled out."
18. Margulis, L (1995) Gaia is a Tough Bitch. The Edge.org http://www.edge.org/documents/ThirdCulture/n-Ch.7.html
19. Beatty, J (1994) The Proximate/Ultimate Distinction, in The Multiple Careers of Ernst Mayr. Biology and Philosophy, 9:333–356.
20. Lewontin, RC (2005) The Wars Over Evolution. NYRB 20 Oct 2005
21. Olson, EC (1960) Morphology, Paleontology, and Evolution. In Evolution After Darwin. Volume 1: The Evolution of Life, its Origin, History and Future, S Tax (ed). UCP. p 523-545.
22. Desmond, A (1995) Forward. In D Ospovat (1995) The Development of Darwin's Theory: Natural History, Natural Theology, and Natural Selection, 1838-1859. CUP. p ix. "Owen, arguably England's greatest comparative anatomist... was still a hate figure. With the Darwinian 'victors' writing history, Owen had never been given a good press.
23. Amundson, R (1998) Typology Reconsidered. op cite.
24. McClintock, B (1973) Letter from McClintock to geneticist Oliver Nelson cited by NLM on the McClintock Papers introductory page. http://profiles.nlm.nih.gov/ps/retrieve/Narrative/LL/p-nid/52
25. Gould, SJ (1977) The Return of Hopeful Monsters. Natural History Magazine, 86:24-30.
26. Wilson, DS (2012) Pugilistic Science. This View of Life. 10 Mar 2012. http://www.thisviewoflife.com/index.php/magazine/articles/pugilistic-science
27. Mazur, S (2009) The Altenberg 16: An Expose of the Evolution Industry. North Atlantic Books, Berkeley, CA p 40.
28. Wright, R (1999) The Accidental Creationist: Why Stephen Jay Gould is bad for evolution. The New Yorker, 13 Dec 1999.
29. Nagel, T (1997) The Last Word, OUP. p 130-131.
30. Rensberger, B (2009) Science Journalism: Too Close For Comfort. Nature, 459:1055-1056.
31. Lewenstein, BV (1992) The Meaning of 'Public Understanding of Science' in the United States After World War II. Public Understanding of Science, Vol. 1. p. 45–68.
32. Huxley, J (ed) (1958) A Book That Shook the World, Pittsburgh University Press, Pittsburgh, PA.
33. Huxley, J (1946) UNESCO: Its Purpose and Its Philosophy, The United Nations Educational, Scientific and Cultural Organization, NY. p 13. Emphasis added.

CHAPTER 10 DELAYED DISCOVERIES, RACISM AND EUGENICS

1. Dobzhansky, T (1968) Introduction. In F. Osborn (1968) The Future of Human Heredity: An Introduction to Eugenics in Modern Society. Weybright and Talley, NY. p v.
2. Steele, EJ and AJ Cunningham (1978) High proportion of Ig-producing cells making autoantibody in normal mice. Nature 274:483-484.
3. Steele, EJ, RA Lindley and RV Blanden (1998) Lamarck's Signature: How retrogenes are changing the natural selection paradigm. Perseus Books, NY.
4. Steele, EJ, (1998) Lateline. The Australian Broadcasting Corporation. 8 Oct 1998. Transcript http://www.abc.net.au/lateline/stories/s13250.htm
5. Giordano R, AR Magnano, G Zaccagnini, C Pittoggi, N Moscufo, R Lorenzini and C Spadafora (2000) Reverse Transcriptase Activity in Mature Spermatozoa of Mouse. Journal of Cell Biology, 148:1107–1113.
Hoenigsberg, H (2002) The future of selection: individuality, the twin legacies of Lamarck and Darwin. Genet. Mol. Res 1:39-50.
6. Steele, EJ, RA Lindley and RV Blanden (1998) Lamarck's Signature: op cite. Back cover.
7. Gurdon, JB (1962) The developmental capacity of nuclei taken from intestinal epithelium cells of feeding tadpoles. J. Embryol. Exp. Morph. 10:622–640.
8. Abbott, A (2012) Cell rewind wins medicine Nobel. Nature 8 Oct 2012.
9. Woese, CR and N Goldenfeld (2009) How the Microbial World Saved Evolution from the Scylla of Molecular Biology and the Charybdis of the Modern Synthesis. op. cite.
10. Gould, SJ (1977)The Return of Hopeful Monsters. Natural History 86:24-30.
11. Chouard, T (2010) Evolution: Revenge of the hopeful monster. Nature 463:864-867.
12. Hall, BK (2001) Commentary. American Zoologist 41:1049-1051.
13. Antonovics J, JL Abbate, CH Baker, D Daley, ME Hood, et al. (2007) Evolution by any other name: Antibiotic resistance and avoidance of the e-word. PLos Biol 5(2): e30. doi:10.1371/journal.pbio.0050030.
14. Linder, DO (no date) State v. John Scopes: The Monkey Trial. http://law2.umkc.edu/faculty/projects/ftrials/scopes/evolut.htm.

15. Hunter, GW (1914) A Civic Biology Presented in Problems. American Book Company.
16. Baritz L (1970) The Culture of the Twenties. Bobbs-Merrill, Indianapolis.
The official court transcript had Bryan's response as "I think it would be just as easy for the kind of God we believe in to make the earth in six days as in six years or in six million years or in six hundred million years. I do not think it important whether we believe one or the other."
17. Adams, N (2005) Timeline: Remembering the Scopes Monkey Trial. National Public Radio. 5 Jul 2005.
18. Hunter, GW (1914) A Civic Biology Presented in Problems. American Book Company. p. 195-196 and 261-265.
19. Unites States Supreme Court (1927) Buck v. Bell, 274 U.S. 200.
20. Stern, AM (nd) Sterilization Operations Performed in California Mental Hospitals and Institutions for Mental Defectives, to June 30, 1941. Department of Institutions of the State of California, Sacramento, CA. Table 29. In 2003, California Attorney General Bill Lockyer apologized for these programs, writing "I am moved to offer an apology for the injustice done to California men and women when the state permitted and promoted involuntary sterilization."
21. Harrison, SL (1994) The Scopes 'Monkey Trial' Revisited: Mencken and the Editorial Art of Edmund Duffy. J. Of American Culture 17:55-63.
22. Larson, EJ (2006) Summer for the Gods: The Scopes Trial and America's Continuing Debate over Science And Religion. Basic Books, NY.
23. Kuhl, S (1994) The Nazi Connection: Eugenics, American Racism and German National Socialism. OUP. p. 103. Kuhl is a professor of Sociology at the University of Bielefeld, Germany, and a leading expert of the Nazi eugenics programs.
24. Muller, H (1961) Science, 134:643-649.
25. Dunn, L C. and T Dobzhansky (1946) Heredity, Race, and Society. New American Library, NY.
26. Dobzhansky, T (1968) Introduction in F. Osborn (1968) The Future of Human Heredity: An Introduction to Eugenics in Modern Society. Weybright and Talley, NY. p v.
27. Gannett, L (2001) Racism and Human Genome Diversity Research: The Ethical Limits of "Population Thinking. Philosophy of Science 68:S479.
28. Lewontin, RC (2003) Public Consciousness of Science. Conversations with History; Institute of International Studies, University of California, Berkeley. 20 Nov 2003.
29. Dobzhansky, T (1964) Heredity and the Nature of Man. Harcourt, Brace and World, NY.
30. Yudell, M (nd)A Short History of the Concept of Race. Yudell is at the Drexel University School of Public Health.
31. UNESCO (1951) The Race Concept: Results of an Inquiry. UNESCO. It includes Fisher's dissenting commentary, "The Race Concept: Results of an Inquiry."
32. Affonço, D (2008) To the End of Hell: One Woman's Struggle to Survive Cambodia's Khmer Rouge. Reportage Press, London.
33. Diamond, J (2005) Guns, Germs, and Steel: The Fates of Human Societies. Norton, NY. p 19-21.
34. Lewontin, RC (2005) The Wars Over Evolution. NYRB 20 Oct 2005.
35. Wellcome Trust (2003) Francis Crick's controversial archive on first public display in major exhibition to celebrate 50th anniversary of discovery of structure of DNA. 31 March 2003.
36. Browne, J (2008) Birthdays to remember. Nature 456:324-325.
37. Huxley J (1964) Essays of a Humanist.
38. Dawkins, R (1997) Is Science a Religion? The Humanist Jan/Feb 1997.
39. Trillin, C (2006) Personal History, Alice, Off the Page. The New Yorker, 27 Mar 2006. p 44.
40. Appleyard, B (2009) For God's sake, have Charles Darwin's theories made any difference to our lives? The Sunday Times (UK), 11 Jan 2009.
41. Giubilini, A and F Minerva (2011) After-birth abortion: why should the baby live? J Med Ethics, 1 Mar 2011.
42. Adams, S (2012) Killing babies no different from abortion, experts say. The Telegraph 29 Feb 2012.

Chapter 11 Making Genes Magical

1. Rose, MR and TH Oakley (2007) The New Biology: Beyond the Modern Synthesis. Biology Direct, 2:30. Much about biology was still unknown in the 1940s. This left those promoting the "modern synthesis free to imagine and made it possible to use simple models to explain important problems.... some of these theories were repeated so often...they found their way into numerous biology texts."
2. Bohr, N (1933) Light and Life. Nature 131:421-423.
3. Pauling, L (1939) The Nature of the Chemical Bond. Ithaca. Cornell University Press, NY.
4. Schrödinger, E (1944) What is Life? CUP. He described the molecule an "aperiodic crystal."
5. In his autobiography, What Mad Pursuit, Crick says that Schrödinger's idea of how genetic information was stored influenced his insight into the structure of DNA.
6. Florkin, M (1949) Biochemical Evolution. Academic Press, NY.
7. Zuckerkandl E and L Pauling (1965) Evolutionary divergence and convergence in proteins. In Evolving Genes and Proteins, Bryson V, HJ Vogel (eds). Academic Press, NY. pp. 97–166.
8. Zuckerkandl E and L Pauling (1962) Molecular Disease, Evolution, and Genic Heterogeneity. In Horizons in Biochemistry. M Kash, E Zuckerkandl, L Pauling, B. Pullman (eds). Academic, NY. p 189–225.above, n. 3, 200–204.
9. Dietrich, MR (1998) Paradox and Persuasion. op cite.
10. Mayr, E (1959) Where Are We? Cold Springs. Harbor Symposiam on Quantitative Biology, pg 13. Mayr, E (1963) The New versus the Classical in Science. Science, 141:763.
11. Beatty, J (1994) The Proximate/Ultimate Distinction, in The Multiple Careers of Ernst Mayr. Biol.

Phil., 9:333–356.
Mayr, R (1959) Darwin and the Evolutionary Theory in Biology. In B.J. Meggers (ed.), Evolution and Anthropology: A Centennial Appraisal. Theo Gaus' Sons. Brooklyn, NY.
12. Hamburger, V (1980) "Embryology and the Modern Synthesis in Evolutionary Theory", in E Mayr and W Provine (eds.) The Evolutionary Synthesis: Perspectives on the Unification of Biology. HUP, pp. 97-112.
13. Riedl, R. (1983) The Role of Morphology In the Theory of Evolution. In Dimensions of Darwinism: Themes and Counter themes in Twentieth Century Evolutionary Theory, M Grene (ed). CUP. p 205-238. Mayr dismissed opposition from developmental biologists as "nothing but German idealistic philosophy."
14. Mayr, E (1959) Darwin and the Evolutionary Theory in Biology. In Evolution and Anthropology: A Centennial Appraisal, BJ Meggers (ed). Theo Gaus' Sons, NY. "Any author who uses findings from the ontogeny of an individual to prove one or another evolutionary theory proves thereby that he completely misunderstands the working of evolution."
15. Kirschner, MW and JC Gerhart (2005) The Plausibility of Life: Resolving Darwin's Dilemma. YUP. p 31. "The Modern Synthesis made the concept of adaptation paramount in evolutionary theory. The organism was like modeling clay, and remolding of the clay meant that each of the billions of little grains was free to move a little bit in any direction to generate a new form. This was close to saying that not only was the input of genotypic variation random but the output of phenotypic variation was random as well, or at least constrained very little. With this approach, the problem of how the processes of embryonic development and cellular function create the phenotype could be largely dismissed as interesting but not informative for evolutionary change, further segregating evolutionary biology from its peer disciplines."
16. Niles Eldredge, 1985, Unfinished Synthesis: Biological Hierarchies and Modern Evolutionary Thought. OUP. p 6.
17. Shapere, D (1980) The Structure of Scientific Revolution. In G Gutting (ed.) Paradigm and Revolution. University of Notre Dame Press. South Bend, IN. pp. 27-38.
18. Dobzhansky, T (1961) Taxonomy, Molecular Biology, and the Peck Order. Evolution, 15:263.
19. Mayr, E (1959) Where Are We? Cold Springs. Harbor Symposiam on Quantitative Biology, pg 13.
20. Ochiai K, T Yamanaka, M Kimura and O Sawada (1959) Inheritance of drug resistance (and its transfer) between Shigella strains and Between Shigella and E.coli strains. Hihon Iji Shimpor 1861: 34.
Kimura, Motoo (1968) Evolutionary rate at the molecular level, Nature 217:624–6.
Kimura, M (1969) The rate of molecular evolution considered from the standpoint of population genetics. PNAS 63:1182–1188. Kimura call his discovery the "Neutral Theory" because genetic change was "neutral" or unaffected by natural selection.
21. Sueoka N(1988) Directional mutation pressure and neutral molecular evolution. PNAS:2653-7.
22. Crow, J (1997) Motoo Kimura, 13 November 1924 - 13 November 1994. Biographical Memoirs of Fellows of the Royal Society 43:254–265.
23. King, JL and TH Jukes (1969) Non-Darwinian Evolution. Science 164:788–798.
24. Dobzhansky, T (1937 [1982]) Genetics and the Origin of Species, Columbia University Press. p 22.
25. Ayala, FJ (2009) Evolution. Encyclopædia Britannica.
26. Williams, GC (1966) Adaptation and Natural Selection: A Critique of some Current Evolutionary Thought. PUP.
27. Science Daily (2006) Unraveling Where Chimp And Human Brains Diverge. 17 Nov 2006.
28. Fisher, RA (1930 [2000]) The Genetical Theory of Natural Selection, JH Bennett (ed). OUP. p84. Fisher said evolution occurs from the effects on gene distribution of the "random survival" and "random extinction." of creatures.
In Darwinian terms, Fisher said "variation"(how a gene changes) is irrelevant in evolution. Only "selection"(which genes survive) matters, and "selection" is random. The New Darwinists reversed this saying only "variation" is random, not "selection." "Natural selection is decidedly nonrandom." (Charlesworth, B (2009) Summary. The 74th Cold Spring Harbor Laboratory Symposium on Quantitative Biology on Evolution. Cold Spring Harbor Laboratory Press, Cold Spring Harbor, NY.) Dawkins and other New Darwinists have repeatedly said the same thing.
29. Dawkins, R (1976) The Selfish Gene. OUP. p 11.
30. Mayr, E (2001) What Evolution Is. op cite.
31. Gould, S J (1980) Caring Groups and Selfish Genes. In The Panda's Thumb. Norton, NY. p 89-90.

Chapter 12 Making Evolution "Everybody's Toy and Plaything"

1. Antonovics, J (1987) The Evolutionary Dys-Synthesis. op cite.
2. Lewontin, RC, S Rose and L Kamin (1984) Not in Our Genes. Penguin, NY.
3. E.g. Migliano, AB, L Vinicius and MM Lahr (2007) Life history trade-offs explain the evolution of human pygmies. PNAS, 104:20216–20219. "Explanations for the evolution of human pygmies continue to be a matter of controversy... Traditional hypotheses assume that the small body size of human pygmies is an adaptation to special challenges, such as thermoregulation, locomotion in dense forests, or endurance against starvation...However...none of these explanations account for the worldwide distribution of human pygmies—some... are found outside forests, and many live in cool and dry areas; furthermore, long-standing poor nutrition does not necessarily lead to pygmy size, as shown by groups who, like certain pygmies, experience frequent food shortages and yet are among the tallest populations in the world."
4. Kettlewell, HBD (1955) Selection experiments on industrial melanism in the Lepidoptera.

Heredity, 9:323–342.
5. Wright S (1978 [1984]) Evolution and the Genetics of Populations, volume 4. UCP. p 186.
6. Coyne, JA (2002) Evolution under pressure. Nature, 418:19 – 20.
7. Sargent, TD, CD Millar, and DM Lambert (1998) The "Classical" Explanation of Industrial Melanism: Assessing the Evidence. In Evolutionary Biology, MK Hecht, RJ MacIntyre and MT Clegg. Springer, NY. p 318.
8. Coyne, JA (1998) Not black and white. Nature, 396: 35-36.
In 2002, Coyne reviewed the peppered moth issue again and still found no experimental evidence for natural selection. [Coyne, JA (2002) Evolution under pressure. Nature, 418:19 – 20.]
In 2010 Coyne apparently claimed that more careful experiments demonstrate color does in fact protect dark winged moths from predators. But as Richard Lewontin pointed out "the caterpillars of the dark-winged forms also have a slightly higher survival rate than those of the speckled-wing form, even though they are not black, so something more is going on." [Lewontin, RC (2010) Not So Natural Selection. NYRB 27 May 2010; Lewontin, RC (2010) What Darwin Got Wrong, Colin Wells, reply by Richard C. Lewontin. NYRB 30 Sept 2010.]
Despite the claim, I am not aware of any published experimental proof showing natural selection is the cause of the change in the peppered moth groups.
9. Mark Ridley (2003) Evolution, 3 edition. Wiley-Blackwell, NY. p114.
10. Dart RA (1953) The predatory transition from man to ape. Internat. Anthropological Linguistic Review, 1:201–218.
11. Brain, C K (2009) Were our early ancestors murderers and headhunters? Quest: Science for South Africa, 5:15-19. Brain is the Emeritus Curator of the Transvaal Museum, South Africa.
12. Sussman, R (2003) The Myth of Man the Hunter, Man the Killer and the Evolution of Human Morality. Zygon, 34:453 – 471.
13. Johanson, D (1997) In Search of Human Origins Part Two. PBS, Airdate: 10 June 1997.
14. Sussman, R (2003) The Myth of Man the Hunter, Man the Killer. op cite.
15. Dawkins, R (1976) The Selfish Gene. OUP. p 19.
16. Dawkins, R (2006) Preface to The Selfish Gene, thirtieth anniversary edition. OUP. p xvii.
17. Gould, S J (1980) Caring Groups and Selfish Genes, in The Panda's Thumb. Norton, NY. p 89-90.
18. Dawkins, R (1999) The Extended Phenotype: The Long Reach of the Gene. OUP. p 4. Individual genes "are not... selected directly, but by proxy; they are judged by their phenotypic effects."
19. Ridley, M (2001) The Cooperative Gene: How Mendel's Demon Explains the Evolution of Complex Beings. Free Press, NY.
20. Lewontin, RC (1990) Response to John Maynard Smith "Are We Robots?" NYRB, 37:9
21. Wilson, DS and E O Wilson (2008) Evolution for the Good of the Group. American Scientist, Sept-Oct 2008.
Wilson EO and B Hölldobler (2005) Eusociality: origin and consequences. PNAS- USA, 102:16119.
22. Riley, M A and JE Wertz (2002) Annu. Rev. Microbiol., 56:117–137.
23. Nowak, MA (2008) Generosity: A winner's advice. Nature, 456, 579.
24. Krieger, MJB and KG Ross (2002) Science, 295:328–332
25. Nowak, MA, C Tarnita and E Wilson (2010) The evolution of eusociality, Nature,466:1057-1062.
26. Thomas, L (1979) The Medusa and the Snail: More Notes of a Biology Watcher. Viking, NY.
27. Schreiber, H and DA Rowley (2008) Cancer: Quo Vadis, Specificity? Science,319: 164-165.
28. Savage, PA et a31(2008) Recognition of a ubiquitous self-antigen by prostate cancer-infiltrating CD8+ T lymphocytes. Science, 319:215-220.
29. Lewontin, RC (2003) Public Consciousness of Science. Conversations with History; Institute of International Studies, UC Berkeley 20 Nov 2003.
30. Lewontin, RC (1983) The Corpse in the Elevator. NYRB, 20 Jan 1983. "As Richard Dawkins puts it in The Selfish Gene, we are "lumbering robots" controlled "body and mind" by our genes. So politics becomes a branch of molecular biology and our social and political institutions are immutable as the chemicals of which we are made."
Lewontin, RC (2003) Public Consciousness of Science. op cite. "My wife and I went yesterday to the Berkeley Art Museum, and there we saw an exhibit called "Gene Genesis." The poster for that exhibit, which is all over, has a quote from Jim Watson. And what does the quote say? It says, "We used to think our fate was in the stars, but now we know it's in our genes." Now, I mean, that's rubbish, you know, but that's what's pushed; and if you're a Nobel prizewinner like Jim, you get all the more credit for it."
31. Thornhill, R and CT Palmer (2000) A Natural History of Rape: Biological Bases of Sexual Coercion. The MIT Press.
32. Sewell, D (2009) Charles Darwin and the Children of the Evolution. The Sunday Times (UK), 8 Nov 2009.

CHAPTER 13 THE NEW WAR AGAINST THE NEW EVIDENCE

1. Hertela, LA, CJ Bayneb and ES Lokera (2002) The symbiont Capsaspora owczarzaki, nov. gen. nov. sp., isolated from three strains of the pulmonate snail Biomphalaria glabrata is related to members of the Mesomycetozoea. International J. for Parasitology, 32:1183-91.
2. Sebe-Pedros A, A de Mendoza, F Lang, B Degnan, I Ruiz-Trillo (2011) Unexpected repertoire of metazoan transcription factors in the unicellular holozoan Capsaspora owczarzaki. Molecular Biology and Evolution 28(3): 1241-54.
Zmasek CS and A Godzik (2013) Evolution of the Animal Apoptosis Network. Cold Spring Harb Perspect Biol. 2013 Mar 1;5(3):a00864.

3. Green, R, et al. (2010) A Draft Sequence of the Neandertal Genome. Science, 328:710-722.
4. Morelle, R (2006) Gene reveals mammoth coat colour. BBC News, 6 July 2006.
5. Li, Q, et al. (2010) Plumage Color Patterns of an Extinct Dinosaur. Science 327, 1369-1372.
6. Mirsky, S (2005) How Did We Get Here? BioInteractive. Howard Hughes Medical Institute, Chevy Chase, MD.
7. Carroll, SB (2006) Endless Forms Most Beautiful: The New Science of Evo Devo and the Making of the Animal Kingdom. Norton, NY. p 72.
8. Discovery Channel (2008) Dinosaurs: Return to Life. Broadcast 14 June 2008. "As far as we know, we were really built with a common vertebrate kit The same one that built guppies and built dinosaurs and built elephants and built mice, that's what's built us It's not an overstatement to say it blew biologists' minds."
9. Mirsky, S (2005) How Did We Get Here? BioInteractive. Howard Hughes Medical Institute, Chevy Chase, MD.
10. Kirschner, MW and JC Gerhart (2005) The Plausibility of Life: Resolving Darwin's Dilemma. YUP. p 35-6.
11. Kirschner, MW and JC Gerhart (2005) The Plausibility of Life: Resolving Darwin's Dilemma. YUP. p 34.
12. Discovery Channel (2008) Dinosaurs: Return to Life. Broadcast 14 June 2008.
13. Sagan, L (1967) On the origin of mitosing cells. Journal of Theoretical Biology 14:225–193. (Sagan is Margulis).
14. Margulis, L and D Sagan (2002) Acquiring Genomes: A Theory of the Origins of Species, Basic Books, NY. p 11-12.
15. Sapp, J (1994) Evolution by association: a history of symbiosis. OUP.
Sapp, J (2005) The Prokaryote-Eukaryote Dichotomy: Meanings and Mythology. Microbiology and Molecular Biology Reviews, 69:292-305.
16. Lawton, G (2009) Why Darwin was wrong about the tree of life. New Scientist 21 Jan 2009.
17. Goldscheider, E. (2009) Evolution Revolution. Wisconsin Alumni Magazine, Fall 2009.
18. Woese C and G Fox (1977) Phylogenetic structure of the prokaryotic domain: the primary kingdoms. PNAS, 74:5088–90.
Woese C, L Magrum and G Fox (1978). Archaebacteria. J Mol Evol 11: 245–51.
19. Moeller, R (2002) New Theory of Cell Evolution Rejects Single-Ancestor Doctrine. Sci Am, 19 June 2002.
20. Wolfe, R (2006) The Archaea: A Personal Overview of the Formative Years. Prokaryotes, 3:3-9.
21. Morell, V (1997)Microbiology's Scarred Revolutionary. Science 276: 699-702.
22. Woese, RC and N Goldenfeld (2009) How the Microbial World Saved Evolution from the Scylla of Molecular Biology and the Charybdis of the Modern Synthesis. op. cite.
23. University Of Illinois At Urbana-Champaign (2002). New Cellular Evolution Theory Rejects Darwinian Assumptions. 18 June 2002. ScienceDaily.
24. Sample, I (2009) Evolution: Charles Darwin was wrong about the tree of life. The Guardian, 21 Jan 2009.
25. Editors (2009) Charles Darwin's tree of life is 'wrong and misleading', claim scientists. Telegraph (UK). 22 Jan 2009.
26. Highfield, R (2009) Uprooting Darwin's Tree. New Scientist, 21 Jan 2009.
27. Lawton, G (2009) Why Darwin was wrong about the tree of life. New Scientists, 2692:34-39.
28. Mann, C (1991) Lynn Margulis: Science's Unruly Earth Mother. Science, 252:379-381.
29. Zimmer, C (2009) Crunching the Data for the Tree of Life. NYT, 9 Feb 2009.
30. In 2008 Dagan and colleagues examined a half a million genes from 181 prokaryotes and found that 80% showed signs of horizontal transfer (PNAS, vol 105, p 10039).
31. Margulis, L (1981) Symbiosis in cell evolution. W. H. Freeman, San Francisco, CA.
32. Meyers, P (2009) New Scientist says Darwin was wrong. Evolution, development, and random biological ejaculations from a godless liberal. Posted 28 Jan 2009. http://scienceblogs.com/pharyngula/2009/01/new_scientist_says_darwin_was.php
33. Dennett, D, J Coyne, R Dawkins and P Myers (2009) Letter: Darwin was right. New Scientist, 18 Feb 2009.
34. Mazur, S (2008) Will the Real Theory of Evolution Please Stand Up? (Emphasis added.)
35. Antonovics, J (1987) The Evolutionary Dys-Synthesis. op cite.
36. Editors (2009) 15 Evolutionary Gems. Nature, 457:8.
37. Reid, RGB (2000) Biological Emergences, MIT Press, p2. Reid is Emeritus Professor of Biology at the University of Victoria, British Columbia.
38. Michael B. Berkman and Eric Plutzer (2011) Defeating Creationism in the Courtroom, But Not in the Classroom, Science, 331:404-405.
39. Nagel, T (2012) Mind and Cosmos: Why the Materialist Neo-Darwinian Conception of Nature Is Almost Certainly False. OUP. p 9-10. "Even though writers like Michael Behe and Stephen Meyer are motivated at least in part by their religious beliefs, the empirical arguments they offer... are of great interest in themselves. Another skeptic, David Berlinski, has brought out these problems vividly without reference to the design inference."
40. Nagel, T (2012) Mind and Cosmos. op cite. p 6.
41. Nagel, T (2012) Mind and Cosmos. op cite. p 7, p 9.
42. Mann, C (1991) Lynn Margulis: Science's Unruly Earth Mother. Science, 252:379-381.
43. Hoenigsberg, H (2002) The future of selection: individuality, the twin legacies of Lamarck & Darwin. Genet. Mol. Res. 1:39-50
44. Shapiro, J (2012) What Is the Best Way to Deal With Supernaturalists in Science and Evolution? Huffington Post 16 Mar 2012.
45. Mazur, S (2008) Will the Real Theory of Evolution Please Stand Up?

46. Young, E (2008) Rewriting Darwin: The new non-genetic inheritance. New Scientist, 09 July 2008.
47. Kelly, K (1995) Post Darwinism, in Out of Control. Basic Books, NY.
48. Hall, B (2001) Commentary. Integrative and Comparative Biology, 31:1049-1051.
49. Eldredge, N (2000) The Pattern of Evolution. Freeman, NY. p 6.
50. Eldredge, N (2000) The Pattern of Evolution. Freeman, NY. p152, 174.
51. Ruse, M (20111) Evolution and Catholic Theology. The Huffington Post, 11 May 2011.
52. Mazur, S (2008) Will the Real Theory of Evolution Please Stand Up?
53. Erwin, D H (2007) Darwin Still Rules, but Some Biologists Dream of a Paradigm Shift. NYT , 26 June 2007.
54. Woese, RC and N Goldenfeld (2009) How the Microbial World Saved Evolution from the Scylla of Molecular Biology and the Charybdis of the Modern Synthesis. op. cite.
55. Woese, C (2004) A New Biology for a New Century. Microbiology and Molecular Biology Reviews, June 2004.
Shapiro, JA (2005) A 21st century view of evolution: genome system architecture, repetitive DNA, and natural genetic engineering. Gene,17:91-100.
56. Zimmer, C (2009) The Ever Evolving Theories of Darwin,. Time 12 Feb 2009.

CHAPTER 14 THE HIDDEN ORDER IN NATURE

1. Lorenz, EN (1963) Deterministic Nonperiodic Flow. J. of the Atmospheric Sciences, 20:130–141.
2. Whitesides, GM and B Grzybowski (2002) Self-Assembly at All Scales. Science, 29: 2418-2421.
3. Lawrence Berkeley National Laboratory staff (1999) Complex Systems: Science for the 21st Century. op cite.
4. BBC Radio 4 (2008) The Six O'clock News, 18 April 2008. "Along with Relativity Theory and Quantum Mechanics, Chaos theory, was one of three fundamental theories which revolutionized science" of the 20th century."
5. Pearce, N and F Merletti (2006) Complexity, simplicity, and epidemiology. Internat. J of Epidemiology, 35:515-519.
6. Goldenfeld N and LP Kadanoff (1999) Simple lessons from complexity. Science, 284:87–89.
7. Whitesides, GM and RF Ismagilov (1999) Complexity in chemistry. Science, 284:89–92.
8. Chaisson, EJ (2001) Cosmic Evolution: The Rise of Complexity in Nature. HUP.
9. Werner, BT (1999) Complexity in natural landform patterns. Science, 284:102–04.
10. Rind, D (1999) Complexity and climate. Science,284:105–07.
11. Oltvai, ZN and A-L Barabasi (2002) Life's complexity pyramid. Science, 298:763–64.Weng GZ, US Bhalla and R Iyengar (1999) Complexity in biological signaling systems. Science, 284:92–96.
12. Parrish JK and L Edelstein-Keshet (1999) Complexity, pattern, and evolutionary trade-offs in animal aggregation. Science, 284:99–101.
13. Medicine, Wilson T and T Holt (2001) Complexity and clinical care. BMJ , 323:685–88.
14. Koch, C and G Laurent (1999) Complexity and the nervous system. Science, 284:96–98.
15. Stacey, RD (2003) Complexity and Group Processes: A Radically Social Understanding of Individuals. Routledge, London.
16. Lansing, JS (2003) Complex Adaptive Systems. Annual Review of Anthropology, 32:183-204.
17. Arthur WB (1999) Complexity and the economy. Science, 284:107–09.
18. Olson, EC and GH Eoyang (2001) Facilitating Organisation Change: Lessons From Complexity Science. Jossey-Bass/Pfeiffer, San Francisco, CA.
Stacey, R D (2001) Complex Responsive Processes in Organisations: Learning and Knowledge Creation. Routledge, London.
19. Misteli, T (2009) Self-organization in the genome. PNAS 106: 6885-6886 .
"One of the most important recent discoveries in the field of genome biology has been the demonstration that genomes are nonrandomly organized...The Holy Grail in this field is to elucidate the mechanisms that determine where a gene... localizes within the cell nucleus... [Recent] findings support the provocative idea that genomes are self-organizing entities."
20. Chaisson, EJ (2001) Cosmic Evolution: The Rise of Complexity in Nature. HUP.
21. Hoyle, F (1954) Astrophys. J. Suppl., 1:121.; Clayton, D (2007) Hoyle's Equation. Science Vol. 318 p 1876 - 1877.
22. Whitesides, GM and M Boncheva (2002) Beyond molecules: Self-assembly of mesoscopic and macroscopic components, Proc Natl Acad Sci U SA 99:4769–4774.
Also see T. Misteli, (2001).The concept of self-organization in cellular architecture J. Cell Biol. 155:181. "The dynamic properties of several cellular structures are consistent with a role for self-organization in their formation, maintenance, and function; therefore, self-organization might be a general principle in cellular organization."
23. Grantcharova V, EJ Alm, D Baker and AL Horwich (2001) Curr Opin Struct Biol.11:70–82, and Neidle, S. (1999) Oxford Handbook of Nucleic Acid Structure. OUP, as cited in Whitesides, GM and M Boncheva (2002) Beyond molecules: Self-assembly of mesoscopic and macroscopic components, Proc Natl Acad Sci U SA, 99:4769–4774.
Also see, for example, Chung, I, H Leonhardt and K Rippe (2011) De novo assembly of a PML nuclear subcompartment occurs through multiple pathways and induces telomere elongation. J. Cell Sci., 124: 3603-3618. "We show... proteins were able to initiate the formation of bona fide APBs with high efficiency in a self-organizing and self-propagating manner."
24. Kaiser, TE, RV Intine, M Dundr (2008) De Novo Formation of a Subnuclear Body. Science, 5908:1713-1717. "The mammalian cell nucleus contains structurally stable functional compartments. We show here that one of them, the Cajal body (CB)...does not follow a hierarchical assembly pathway and exhibits hallmarks of a self-organizing structure."
Also see Misteli, T (2009) Self-organization in the genome. PNAS 106: 6885-6886,

Dundr, M and T Misteli (2010) Biogenesis of Nuclear Bodies. Cold Sp. Harbor Pers. in Bio., 10 Nov 2010, and
Volkova, EG, SY Kurchashova, VY Polyakov and EV Sheval (2011) Self-organization of cellular structures induced by the overexpression of nuclear envelope proteins: a correlative light and electron microscopy study. J Electron Microsc., 65 57-71. This study concluded that nuclear envelope formation is achieved by the self-organization mechanism. "This study concluded that nuclear envelope formation is achieved by the self-organization mechanism."
25. Gell-Man, M (1995) The Quark and the Jaguar: Adventures in the Simple and Complex. St. Martins, NY.
26. Water properties such as fluidity, viscosity, and transparency are not simply the aggregate properties of oxygen and hydrogen. Through their interactive effects they are transformed into new phenomena. Bunge (1977).
27. Disney, MJ, et al. (2008) Nature 455:1082–1084. All galaxies form the same way, based on the same single factor - the internal pull of gravity - leading to the same types of galaxies everywhere in the universe.
28. Holland, J (1995) Hidden Order: How Adaptation Builds Complexity. Perseus, NY.
29. O'Hear, A (1996) Karl Popper: Philosophy and Problems. CUP. p145.
30. Carey, S (2009) The Origin of Concepts. OUP. p 538.
31. Strohman, R (2000) Genetic Determinism as a Failing Paradigm in Biology and Medicine. In Promoting Human Wellness: Frontiers for Research, Practice and Policy, MS Jamner and D Stokols (eds). Univ of California Press. p 100.
32. Grobstein, P (1994) Variability in Brain Function and Behavior. The Encyclopedia of Human Behavior, Volume 4, VS Ramachandran (ed). Academic Press, NY. pp. 447-458.
33. Laughlin, RB (2005) A Different Universe: Reinventing Physics from the Bottom Down. Basic Books, NY. p 31.
34. Laughlin, RB and D Pines (2000) The Theory of Everything. PNAS, 97:28-31.
35. Schrödinger, E (1944) What Is Life? The Physical Aspects of the Living Cell. Emphasis added.

Chapter 15 The Hidden Plan of Life

1. Wills, C (2009) Evolution: The next 200 years. New Scientist, 28 Jan 2009.
2. Jacob, F and J Monod (1961) Genetic Regulatory Mechanisms in the Synthesis of Proteins. J. of Mol. Bio., 3: 318-356.
3. Comfort, NC (1999) The real point is control: The reception of Barbara McClintock's controlling elements. J History of Biology, 32:133–136.
4. Dimmock, NJ and SB Primrose (1994) Introduction to Modern Virology, fourth ed. Blackwell Pub., London. p 80.
5. Portin, P (1993) The Concept of the Gene: Short History and Present Status. Quarterly Rev of Biology, 68:173-223.
6. Dawkins, R (1986) The Blind Watchmaker. Norton, NY. p 296.
7. Prescott, DM (1984) Molecular genetics of a ciliate. Carlsberg Research Communications 49: 341-350.; Marande, M and G Burger (2007) Mitochondrial DNA as a Genomic Jigsaw Puzzle. Science 318:415.
8. Angier, N (2008) Pursuing Synthetic Life, Dazzled by Reality. NYT 5 Feb 2008.
9. Landweber, L (2007) Why Genomes in Pieces? Science, 318:405 - 407.
10. Strohman, R (2000) Genetic Determinism as a Failing Paradigm in Biology. Op cite.
11. Keller, EF (2008) Decoding the Genetic Program Or, Some Circular Logic in the Logic of Circularity. In The Concept of the Gene in Development and Evolution, PJ Beurton, R Falk and H-J Rheinberger (eds) CUP. p 159.
12. Turing, AM (1952) The chemical basis of morphogenesis. Phil. Trans. R. Soc. London B 237: 37-72; in The Collected Works of A. M. Turing, PT Saunders (ed). North-Holland, Amsterdam.
13. As one example see the work on modeling biomolecular systems by Mikko Karttunen and his group at Helsinki University's Biophysics & Statistical Mechanics Group.
Leppänen, T, M Karttunen, K Kaski, and RA Barrio (2004) Turing systems as models of complex pattern formation. Braz. J. Phys, 34:368-372.
14. Kaufman, S (1996) At Home in the Universe. OUP. p 71.
15. Levy, ED, EB Erba, CV Robinson and SA Teichmann (2008) Assembly reflects evolution of protein complexes. Nature, 453:1262-1265. "Self-assembly allows the reliable prediction of the evolution...of a complex solely from its crystal structure." "We have shown that the evolutionary pathway of a homomer [the elements that form proteins] can be inferred from its atomic structure morphology. This allowed us to predict the... assembly pathway of homomers."
"Our model of self-assembly allows reliable prediction of evolution and assembly of a complex solely from its crystal structure."
"we have shown that the evolutionary pathway of a homomer can be inferred from its atomic structure morphology."
"This result implies that the evolutionary route of a homomer can be predicted solely from its interface sizes."
"These results—together with previous studies where reassembly was found to be strongly dependent on factors such as ionic strength, temperature and concentration of denaturant—indicate that disassembly is the reverse of assembly under the appropriate conditions."
Also see: Luke K and P Wittung-Stafshede (2006) Folding and assembly pathways of co-chaperonin proteins 10: Origin of bacterial thermostability. Arch. Biochem. Biophys. 2006;456:8–18.
Cheesman C, LW Ruddock and RB Freedman (2004) The refolding and reassembly of

Escherichia coli heat-labile enterotoxin B-subunit: analysis of reassembly-competent and reassembly-incompetent unfolded states. Biochemistry, 43:1609–1617.
16. Grueninger, D, et al (2008) Designer Protein-Protein Association. Science, 319:206.
17. Locke, M (1990) Is there somatic inheritance of intracellular patterns? Journal of Cell Science 96:563-567. "In considerations of cell architecture we accept the idea that the innate properties of molecules in particular environments lead to their self-assembly into stable, oriented, three-dimensional objects."
18. Zimmer, C (2009) From Developing Limbs, Insights That May Explain Much Else. NYT, 6 Apr 2009.
19. Camazine, S, J-L Deneubourg, N Franks, J Sneyd, G Theraulaz and E Bonabeau (2003) Self-Organization in Biological Systems (Princeton Studies in Complexity). PUP. p 3.
20. Cherniak, C, Z Mokhtarzada and U Nodelman (2002) Optimal-Wiring Models of Neuroanatomy. In Computational Neuroanatomy: Principles and Methods, G Ascoli (ed). Humana, p 71-82.
21. A similar idea was advanced in 1990 by University of Georgia biologists Maria Jerka-Dziadosz, in a remarkable summary that could have explained the entire idea of "context based development." Writing about the development of the cortex of ciliates Jerka-Dziadosz said "Genetic analysis demonstrates that pattern formation [of development] relies both on local constraints imposed by the pre-existing organization, and on the differential interpretation of inductive signals by different cell territories whose individual developmental properties are reset at each division."
Jbrka-Dziadosz, M. and J Beisson (1990) Genetic approaches to ciliate pattern formation: from self-assembly to morphogenesis. Trends Genet. 6:41-45.
22. Eiraku, M. et al. (2008) Self-organized formation of polarized cortical tissues from ESCs and its active manipulation by extrinsic signals. Cell Stem Cell,3:519–532.
"Here, we demonstrate self-organized formation of apico-basally polarized cortical tissues from ESCs...The generated cortical neurons are functional, transplantable, and capable of forming proper long-range connections in vivo and in vitro...In addition, the in vivo-mimicking birth order of distinct cortical neurons permits the selective generation of particular layer-specific neurons... Importantly, cortical tissues generated from mouse and human ESCs form a self-organized structure that includes four distinct zones (ventricular, early and late cortical-plate, and Cajal-Retzius cell zones) along the apico-basal direction."
23. Gaspard, N. et al. (2008) An intrinsic mechanism of corticogenesis from embryonic stem cells. Nature 455:351–357.
"...embryonic stem cells... in the presence of a sonic hedgehog inhibitor, recapitulate in vitro the major milestones of cortical development, leading to the sequential generation of a diverse... neurons that display most salient features of genuine cortical pyramidal neurons.... The discovery of *intrinsic* corticogenesis sheds new light on the mechanisms of neuronal specification" Emphasis added.
24. Brüstle, O (2013) Developmental neuroscience: Miniature human brains. Nature, 501:319–320.
25. Eiraku, M. et al. (2011) Self-organizing optic-cup morphogenesis in three-dimensional culture. Nature, 472:51–56.
"Here we report the dynamic, autonomous formation of the optic cup (retinal primordium) structure from... mouse embryonic stem cell aggregates... We demonstrate that optic-cup morphogenesis... depends on an *intrinsic self-organizing program* involving stepwise and domain-specific regulation of local epithelial properties." Emphasis added.
26. Brüstle, O (2013) Developmental neuroscience: Miniature human brains. Nature, 501:319–320.
Lancaster, M. A. et al. (2013) Cerebral organoids model human brain development and microcephaly. Nature, 501:373–379.
"we have developed a human... stem cell-derived... system, termed cerebral organoids, that develop various discrete, although interdependent, brain regions. These... are shown to recapitulate... human cortical development, namely characteristic progenitor zone organization with abundant outer radial glial stem cells...these data show that three-dimensional organoids can recapitulate development... even in this most complex human tissue."
27. Camazine, S, J-L Deneubourg, N Franks, J Sneyd, G Theraulaz and E Bonabeau (2003) Self-Organization in Biological Systems (Princeton Studies in Complexity). PUP. p 3
28. Jensen, J (2004) Gene Regulatory Factors in Pancreatic Development. Developmental Dynamics, 229:176-200.
29. Marcus, G (2003) The Birth Of The Mind: How A Tiny Number Of Genes Creates The Complexities Of Human Thought. Basic Books, NY.. p 147-148. Marcus is Director NYU Center for Child Language.
30. Macy, MM and R Willer (2002) From Factors to Actors: Computational Sociology & Agent-Based Modeling. Annual Rev of Sociology, 28:143-166.
31. Kroo, I (2002) Aeronautical Applications of Collaborative Optimization and Collective Design. Collective Cognition Workshop: Mathematical Foundations of Distributed Intelligence. Santa Fe Institute, Jan 2002.
32. A discovery called the "fractional quantum Hall effect," which won Robert Laughlin the 1998 Nobel Prize in Physics.
33. Goldenfeld, N and LP Kadanoff (1999) Simple Lessons from Complexity. Science, 284:87-89.
34. Gordon, D (2002) The Organization of Work in Social Insect Colonies. Complexity, Sept-Oct 2002.
35. Hemelrijk, C (2005) Self-organisation and Evolution of Biological and Social Systems. CUP. p ix.
36. Kull, K (2000) Organisms can be proud to have been their own designers. Cybernetics and Human Knowing 7:45–55.
37. Angier, N (2007) Small, Yes, but Mighty: The Molecule Called Water. NYT 10 July 2007.

38. Lehn, J.-M (1990)Supramolecular Chemistry-Scope and Perspectives: Molecules, Supermolecules, and Molecular Devices (Nobel Lecture). Angew. Chem. Int. Ed. Engl. 29: 1304-1319.

Chapter 16 The Hidden Code of Evolution

1. Rosenfield, I and E Ziff (2006) Evolving Evolution. NYRB, 11May 2006.
2. Dobzhansky, T (1950). The genetic basis of evolution. Sci. Am. 182:32-41.
3. Wright, BE (2000) A Biochemical Mechanism for Nonrandom Mutations and Evolution. Journal of Bacteriology 182:2993-3001.
4. Goldscheider, E (2009) Evolution Revolution, Wisconsin Alumni Magazine, Fall 2009.
5. Margulis, L and D Sagan (2002) Acquiring Genomes: A Theory of the Origins of Species, Basic Books, NY. p 11-12.
6. Freeland, SF, RD Knight, LF Landweber and LD Hurst (2000) Early Fixation of an Optimal Genetic Code. Mol. Biol. Evol., 17:511–518.
   "The canonical code is at or very close to a global optimum for error minimization."
   Kunkel TA (2009) Evolving Views of DNA Replication (In)Fidelity. The 74th Cold Spring Harbor Laboratory Symposium on Quantitative Biology on Evolution. Cold Spring Harbor Laboratory Press, Cold Spring Harbor, NY. Elaborate cellular machinery ensures the high accuracy of DNA replication and correspondingly low rate of random mutation.
7. Friedberg, EC (2003) DNA damage and repair. Nature 421, 436-440.
8. Janion C (2001) Some aspects of the SOS response system-a critical survey. Acta Biochim Pol. 48:599–610.
9. Friedberg, EC (2003) DNA damage and repair. Nature 421, 436-440.
   Friedberg, EC (1997) Correcting the Blueprint of Life: An Historical Account of the Discovery of DNA Repair Mechanisms. Cold Spring Harbor Laboratory Press. Cold Spring, NY.
10. Fry RC, TJ Begley and LD Samson (2004) Genome-wide responses to DNA-damaging agents. Annu Rev Microbiol.59:357–7.
11. Hopfield, JJ (1974) Kinetic proofreading: A new mechanism for reducing errors in biosynthetic processes requiring high specificity. PNAS, 71:4135–4139.
12. Cromie GA, JC Connelly and DR Leach (2001) Recombination at double-strand breaks and DNA ends: conserved mechanisms from phage to humans. Mol Cell. 8:1163-74.
   Wood, RD, M Mitchell, J Sgouros, T Lindahl, (2001) Human DNA Repair Genes. Science, 291:1284–1289.
13. Knight, RD (2001) The Origin and Evolution of the Genetic Code: Statistical and Experimental Investigations. PhD thesis, Princeton University.
14. Campbell, BG. (1966 [1985]) Human Evolution: An Introduction to Man's Adaptations. Aldine, NY.
15. Bernhard, R (1967) Heresy in the halls of biology: mathematicians question Darwinism. Sci. Res.2:59-66.
16. Kumar, A (2009) An Overview of Nested Genes in Eukaryotic Genomes. Eukaryotic Cell, 8:1321-1329.
17. McGary, KL, TJ Park, JO Woods, HJ Cha, JB Wallingford, and EM Marcotte (2010) Systematic discovery of nonobvious human disease models through orthologous phenotypes. PNAS, 107: 6544-6549.
18. Hinegardner, RT and J Engelberg (1963) Rationale for a Universal Genetic Code. Science,142:1083–1085.
19. Kirschner, MW and JC Gerhart (2005) The Plausibility of Life: Resolving Darwin's Dilemma. YUP. p 35.
20. Ohno S (1970) Evolution by Gene Duplication. Springer-Verlag, NY.
21. Ohno S (1972) An argument for the genetic simplicity of man and other animals. J Hum Evol, 1:651–662.
22. Garcia-Bellido, A. and EB Lewis (1976) Autonomous cellular differentiation of homoeotic bithorax mutants of Drosophila melanogaster. Develop. Biol.48: 400-410.
   Lewis, EB (1978) A gene complex controlling segmentation in Drosophila. Nature, 276: 565-570.
   Lewis, EB (1982) Control of body segment differentiation in Drosophila by the bithorax gene complex. In Embryonic Development, Part A: Genetic Aspects, MM Burger and R Weber (eds). Alan R. Liss, NY. p 269-288.
23. Belting HG, CS Shashikant and FH Ruddle (1998) Modification of expression and cis-regulation of Hoxc8 in the evolution of diverged axial morphology. PNAS, 95:2355–2360.
   Wittkopp PJ, K Vaccaro, SB Carroll (2002) Evolution of yellow gene regulation and pigmentation in Drosophila. Curr Biol 12:1547–1556.
24. Britten RJ, Davidson EH (1969) Gene regulation for higher cells: A theory. Science 165:349–357.
   Britten RJ, Davidson EH (1971) Repetitive and non-repetitive DNA sequences and a speculation on the origins of evolutionary novelty. Q Rev Biol. 46:111–138.
   Davidson, EH (2006) Genomic Regulatory Systems: In Development and Evolution. Academic Press, NY.
   Shen-Orr, SS, R Milo, S Mangan, U Alon (2002) Network motifs in the transcriptional regulation network of Escherichia coli. Nature Genet. 31:64-68
25. Davidson EH (2013) Personal communication.
   Arendt, D (2006) Evolution and Development: A Kernel Bears Fruit" Science, 314:1085-1086. Arendt, a biologist at the European Molecular Biology Laboratory in Heidelberg, Germany calls Davidson "the world's leading expert on the gene regulatory networks that control animal development.

26. Mattick, JS (2007) A new paradigm for developmental biology. J. of Experimental Bio., 210:1526-1547.
27. Jacob F (1977) Evolution and tinkering. Science 196:1161–1166.
28. Lewontin, RC (1998) Higher Superstition: An Exchange. NYRB, 3 Dec 1998.
"All biologists know that DNA is not "self-replicating," but is manufactured by a complex protein machinery in the cell, yet when they describe the results of molecular biology to students in lectures and textbooks and to the public in trade books, newspapers, or television, they almost always speak of "the master molecule that is self-replicating.""
29. Prasanth, KV and DL Spector (2007) Eukaryotic regulatory RNAs: an answer to the 'genome complexity' conundrum. Genes and Development, 21:11-42.
30. Park, A (2012) Junk DNA - Not So Useless After All. Time, 6 Sept 2012.
Susumu Ohno may have been the first to use the term "junk genes," in the 1970s. But he did not use the term the way Darwinists did in the 2000s.
31. Innan, H and F Kondrashov (2010) The evolution of gene duplications: classifying and distinguishing between models. Nature Reviews Genetics 11:97-108.
32. Kosinski, R. A.; M. Zaremba (2007). "Dynamics of the Model of the Caenorhabditis elegans Neural Network". Acta Physica Polonica B 38 (6): 2202. Retrieved 2009-07-31.
33. Kass, JH (2009) Evolutionary Neuroscience. Academic Press, San Diego, CA p 109.
34. Cherniak, C, Z Mokhtarzada, R Rodriguez-Esteban and K Changizi (2004) Global optimization of cerebral cortex layout PNAS,101:1081-1086.
35. Cherniak, C (2005) Innateness and Brain-Wiring Optimization: Non-Genomic Nativism. In Evolution, Rationality and Cognition, A Zilhao (ed). Routledge, London. p 103-112.
36. Levy, ED, EB Erba, CV Robinson and SA Teichmann (2008) Assembly reflects evolution of protein complexes. Nature, 453:1262-1265.
"we show...the...assembly pathway mimics the evolutionary pathway. These data represent a molecular analogy to Haeckel's evolutionary paradigm of embryonic development, where an intermediate in the assembly of a complex represents a form that appeared in its own evolutionary history... Results revealed that the... assembly pathway, which takes place on a protein-folding timescale (seconds), mimics the evolutionary pathway that has taken place over a considerably longer timescale (millions of years)."
37. Locke, M (1990) Is there somatic inheritance of intracellular patterns? Journal of Cell Science 96:563-567.

CHAPTER 17 STRESS ALTERS THE BODY AND BEHAVIOR

1. Medawar, PB (1967) The Art of the Soluble. Methuen. London.
2. Biologist Walter Cannon originally developed the idea of homeostasis in the 1930s. Biologists Eugene Yates has updated this what he calls homeodynamics, the idea physiology adjusts to maintain balance in the face of internal or external stress, creating adaptive evolution.
Yates, FE (1994) Order and Complexity in Dynamical Systems: Homeodynamics as a Generalized Mechanics for Biology. Mathematical and Computer Modelling, 19:49–74.
Yates, FE (2008) Homeokinetics/Homeodynamics: A Physical Heuristic for Life and Complexity. Ecological Psychology, 20:148–179.
3. Sultan SE (2000) Phenotypic plasticity for plant development, function and life history. Trends Plant Sci. 5:537–542.
Miura T (2005) Developmental regulation of caste-specific characters in social-insect polyphenism. Evol. Dev,7: 122–129. Insect colonies depend on new members to become a specific type (worker, guard) based on current need, not predetermined genetics. Thus, actual development is "not inherited but they arise during development" based on how eggs treated by the workers who alter factors like embryonic diet and incubation temperature.
4. Schlichting C.D and H Smith H.(2002) Phenotypic plasticity: linking mechanisms with evolutionary outcomes. Evol. Ecol. 16:189-211.
5. Laughlin, SB and TJ Sejnowski (2003) Communication in Neuronal Networks. Science, 301:18701874. "neuronal networks can adapt to circumstances, self-assemble, auto calibrate, and store information by changing their properties according to experience."
6. West-Eberhard, MJ(2003). Developmental Plasticity and Evolution. OUP. p vii. "The universal environmental responsiveness of organisms...influences individual development and organic evolution, and this realization compels us to reexamine the major themes of evolutionary biology in a new light".
7. Waterland, RA and RL Jirtle (2003) Transposable elements: targets for early nutritional effects on epigenetic gene regulation. Mol. Cell. Biol. 23:5293-300.
Waterland, RA and RL Jirtle (2004) Early nutrition, epigenetic changes at transposons and imprinted genes, and enhanced susceptibility to adult chronic diseases. Nutrition 20: 63-68.
8. Watters, E (2006) DNA is Not Destiny. Discover, Nov 2006.
9. Burkhardt, RW (1980) Lamarckism in Britain and the United States. op cite. p 346.
10. Griffith, F. (1928) The Significance of Pneumococcal Types. J of Hygiene.27:113-159.
11. Guyer, MF (1924) Further Studies on Inheritance of Eye Defects Induced in Rabbits. J of Exp Zoology, 38:449-475.
12. Crick, F (1970) Central Dogma of Molecular Biology. Nature 2227:461-563. See note 8, p 561.
13. Gould, SJ (1980) The Panda's Thumb. Norton, NY. p.181.
14. Temin, H and S Mizutani (1970) RNA-dependent DNA Polymerase in Virions of Rous Sarcoma Virus. Nature, 226: 1211-1213.
Baltimore, D (1970) Viral RNA-dependent DNA Polymerase. Nature, 226: 1209-1211.
15. Editors (1970) Central Dogma Reversed. Nature 226:1198-1199.

16. A few examples among thousands include:
Cropley JE, el al. (2006) Germ-line epigenetic modification of the murine Avy allele by nutritional supplementation. PNAS,103:17308–17312.
Liu Y (2007) Like father like son: A fresh review of the inheritance of acquired characteristics. EMBO Rep 8:798-803. "endogenous retroviral vectors would capture RNA from somatic cells and transduce them into germline cells."
Skinner MK, et al (2012) Epigenetic transgenerational inheritance of somatic transcriptomes and epigenetic control regions. Genome Biol, 13:R91. "Combined observations demonstrate that all tissues derived from the epigenetically altered germ line develop transgenerational transcriptomes unique to the tissue...This systems biology approach provides insight into the molecular mechanisms involved in the epigenetic transgenerational inheritance of a variety of adult onset disease phenotypes."
17. E.g. Frésard, L, F Pitel, et al (2013) Epigenetics and phenotypic variability: some interesting insights from birds. Genetics Selection Evolution, 5:16. "A first wave of DNA demethylation takes place in male or female germ cells... this occurs throughout the genome."
18. Bukovsky, A, et al (2008) Origin of Germ Cells and Formation of New Primary Follicles in Adult Human and Rat Ovaries. Methods in Molecular Biology, 450:233-265.
White, YAR, et al (2012) Oocyte formation by mitotically active germ cells purified from ovaries of reproductive-age women. Nature Medicine, 26 Feb 2012.
19. Halfmann, R and S Lindquist (2010) Epigenetics in the Extreme: Prions and the Inheritance of Environmentally Acquired Traits, Science, 330:629-632.
20. BBC Horizon (2005) The Ghost in Your Genes.
21. Bird, A (2007) Perceptions of epigenetics. Nature 447:396-398.
22. Nanney, DL (1958) PNAS,44:712–717.
23. Marris, E (2008) Language: Disputed definitions. Nature 455:1023-1028.
24. Jablonka, E and M Lamb (1995) Epigenetic Inheritance and Evolution: The Lamarckian Dimension. OUP. "In recent years, molecular biology has shown that the genome is far more fluid and responsive to the environment than previously supposed. It has also shown that information can be transmitted to descendants in ways other than through the base seqeunce of DNA."
Lamm, E and E Jablonka (2008) The Nurture of Nature: Hereditary Plasticity in Evolution" Philosophical Psych. 21:305-319. "The epigenetic mechanisms involved in repatterning can be activated by both environmental and genomic stress, and lead to phylogenetic as well as ontogenetic changes... The effects and the mechanisms of plasticity directly contribute to evolvability"
25. Hoenigsberg, H (2002) The future of selection: individuality, the twin legacies of Lamarck & Darwin. Genet. Mol. Res. 1: 39-50.
Henderson, IR and SE Jacobsen (2007) Epigenetic inheritance in plants. Nature 447, 418-424.
Cubas, P, C Vincent and E Coen (1999) An epigenetic mutation responsible for natural variation in floral symmetry. Nature, 401:157–161.
26. Rassoulzadegan, M, et al (2006) RNA-mediated non-mendelian inheritance of an epigenetic change in the mouse. Nature ,441:469-474.
Chong, S and E Whitelaw (2004) Epigenetic germline inheritance. Curr. Opin. Genet. Dev.14, 692–696..
Herman. H, et al (2003) Trans allele methylation and paramutation-like effects in mice. Nature Genetics, 34:199-202.
Fire, A. et al. (1998) Potent and specific genetic interference by double-stranded RNA in Caenorhabditis elegans. Nature 391:806–811
27. Weaver, IC, et al. (2004). Epigenetic programming by maternal behavior. Nature Neurosci, 7:847–854.
Krawetz, SA (2005) Paternal contribution: new insights and future challenges. Nature Rev. Genet. 6:633–642.
Young, RL and AV Badyaev (2007) Evolution of ontogeny: linking epigenetic remodeling and genetic adaptation in skeletal structures. Integrative and Comparative Biology, 47:234-244.
28. Chandler, VL (2007) Paramutation: From Maize to Mice. Cell, 128: 641-645.
29. Keim, B (2005) Whew! Your DNA Isn't Your Destiny. Wired 16 Aug 2005.
30. Cairns, J, J Overbaugh and S Miller (1988) The origin of mutants. Nature 335, 142–145. Emphasis added.
31. Hall, BG (1990) Spontaneous point mutations that occur more often when they are advantageous than when they are neutral. Genetics 126: 5-16.
Hall, BG (1991) Increased Rates of Advantageous Mutations in Response to Environmental Challenges. ASM News 57: 82-86.
Hall BG (1997) On the Specificity of Adaptive Mutations. Genetics,145:39-44.
"Adaptive mutations... appear to be specific to the challenge... in that the only mutations that arise are those that provide a growth advantage to the cell. The issue of the specificity has been controversial because it violates... assumptions about the randomness of mutations... Here I ... provide evidence... mutation occurs at that site but not at an alternative and equally mutable site within the same gene."
32. Hall, BG (1998) Adaptive mutagenesis: a process that generates almost exclusively beneficial mutations. Genetica,102-103:109-125.
33. Long M, et al (2003) The origin of new genes: glimpses from the young and old. Nat Rev Genetics 4:865-875. "An obvious feature that is associated with the formation of new genes by these mechanisms is the rapid evolution of new sequences, structures and expression. Most of these evolutionary changes are probably adaptive."

Galloway LF and JD Etterson (2007) Transgenerational plasticity is adaptive in the wild. Science 318:1134-6. In plants, the maternal transfer of response to light to the next generation. "Transgenerational plasticity evolved in response to natural variation in light and provides a flexible mechanism by which sedentary organisms cope with heterogeneous [i.e. different] environments."

34. Shapiro, JA (2007) Bacteria are small but not stupid: cognition, natural genetic engineering and socio-bacteriology. Studies in the Hist. and Phil. of Biology and Biomed Science. 38:807-819.
Shapiro JA (2009) Revisiting the central dogma in the 21st Century. Annal. NY Acad. Sci. 1178: 6-28.
Pennisi, E (1998) Molecular Evolution: How the Genome Readies Itself for Evolution. Science, 28:1131-1134.
35. Bhanoo, SA (2010) Plants Near Chernobyl Appear to Grow a Shield. NYT, 20 Sept. 2010.
36. Klubicov, K, et al (2010) Proteomics Analysis of Flax Grown in Chernobyl Area Suggests Limited Effect of Contaminated Environment on Seed Proteome. Environmental Science and Technology, 44:6940–6946.
37. Price TD, A Qvarnström and DE Irwin (2003) The role of phenotypic plasticity in driving genetic evolution. Proc. Biol. Sci. 270:1433–40. "Models of population divergence and speciation are often based on the assumption that differences between populations are due to genetic factors, and that phenotypic change is due to natural selection. It is equally plausible that some of the differences among populations are due to phenotypic plasticity."
Massimo Pigliucci (2008): "Phenotypic plasticity as a causal factor in evolution" Conference: Towards an Extended Evolutionary Synthesis, Konrad Lorenz Institute for Evolution and Cognition Research, Altenberg, Austria. 12 July 2008.
38. Brunetta L and CL Craig (2010) Spider Silk: Evolution and 400 Million Years of Spinning, Waiting, Snagging, and Mating. Yale University Press, New Haven, CT.
39. Belko, Mark (2006) The Thinkers: Pitt anthropologist thinks Darwin's theory needs to evolve on some points. Pittsburgh Post-Gazette, 29 May 2006.
40. Hunt, T, et al (2007) A Comprehensive Phylogeny of Beetles Reveals the Evolutionary Origins of a Superradiation. Science, 318:1913-19161. "The extreme diversity of beetles reflects the Jurassic origin of numerous modern lineages, high lineage survival, and the diversification into a wide range of niches, including the *utilization of all parts of plants*. These switches into new niches occur repeatedly as, for example, the multiple shifts from terrestrial to aquatic habits in the evolutionary history of beetles, which occurred at least 10 times." Emphasis added.
41. Grant, PR (1986) Ecology and Evolution of Darwin's Finches. PUP.
42. Ernst Mayer, in Schwartz, J (1999) Sudden Origins: Fossils, Genes, and the Emergence of Species. Wiley, NY. p 314.
43. Wu, P et al (2004) Molecular Shaping of the Beak. Science, 305:1465-1466.
44. E.g. Pennisi, E (2004) Bonemaking Protein Shapes Beaks of Darwin's Finches. Science, 305:1383.
45. E.g. Grant, PR and BR Grant (2002) Unpredictable Evolution in a 30-Year Study of Darwin's Finches. Science 296:707-711.
46. Mayell, H (2003) When Did "Modern" Behavior Emerge in Humans? National Geographic News 20 Feb 2003, reporting on the symposium Revolution and Evolution in Modern Human Origins: When, Where, and Why? Denver, Colorado. 20 Feb 2003.
47. Miller, B, B Wood, A Balansky, J Mercader, M Panger (2006). Anthropology. Allyn and Bacon, Boston. p 232. Miller, is director of the Culture in Global Affairs Research Program, and Wood is Director of the Center for the Advanced Study of Hominid Paleobiology, both at George Washington University, Washington, DC.
48. Zimmer, C, (2013) Vitamins' Old, Old Edge. NYT 9 Dec 2013.
49. Adaptive changes in the body precede changes in genes, a pattern some geneticists call "phenotype precedes genotype.," e.g. Palmer, AR (2004) Science, 306:828-833.
50. West-Eberhard, MJ (2005) Developmental plasticity and the origin of species differences. PNAS 102:6543-6549.
51. Schmalhausen, II (1986 [1949]) Factors of Evolution: The Theory of Stabilizing Selection. Translated by I Dordick and edited by T Dobzhansky. UCP. p 2.
52. Rosenberg, SM, C Shee, RL Frisch and PJ Hastings (2012) Stress-induced mutation via DNA breaks in Escherichia coli: A molecular mechanism with implications for evolution and medicine. Bioessays 34: 885–892.
"Evolutionary theory assumed that mutations occur constantly, gradually, and randomly over time. This formulation from the "modern synthesis" of the 1930s was embraced decades before molecular understanding of genes or mutations. Since then, our labs and others have elucidated mutation mechanisms activated by stress responses. Stress-induced mutation mechanisms produce mutations, potentially accelerating evolution, specifically when cells are maladapted to their environment, that is, when they are stressed."
53. Yamanaka, K., L Fang and M Inouye (1998) The CspA family of Escherichia coli: multiple gene duplication for stress adaptation. Molec. Microbio. 27:247-256.
54. Lin, R, L Ding, C Casola, DR Ripoll, C Feschotte and H Wang (2007) Transposase-Derived Transcription Factors Regulate Light Signaling in Arabidopsis. Science, 23:1302-1305.
55. Purnell, BA (2007) Domesticated Mobile Elements Modulate Light Responses. Science STKE,414: tw433.
56. Kim, B (2009) Early Life Didn't Just Divide, It United. Wired, 19 Aug 2009.
57. Butterfield, NJ (2011) Animals and the invention of the Phanerozoic Earth system. Trends Ecol Evo, 26:81-87.
58. Eichler, EE (1998) Masquerading Repeats: Paralogous Pitfalls of the Human Genome. Genome Res. 8:758-762.

59. Agrawal, A, QM Eastman and DG Schatz (1998) Transposition mediated by RAG1 and RAG2 and its implications for the evolution of the immune system. Nature 394:744-7511.
60. Editors, (2005) "Jumping Gene" Helps Explain Immune System's Abilities. Science Daily, 7 Jan 2005
61. Lemke, H, A Coutinho and H Lange (2004) Lamarckian inheritance by somatically acquired maternal IgG phenotypes. Trends in Immunology, 25:180-186. The appearance "of maternal IgG in offspring appears phenotypically as a non-genetic inheritance, which thus reveals a Lamarckian dimension of the immune system"
Boulinier, T and V Staszewski (2008) Maternal transfer of antibodies: raising immuno-ecology issues. Trends in Ecology & Evolution, 25:282-288.
62. Science Daily (1998) Ancient "Jumping DNA" May Have Evolved Into Key Component Of Human Immune System. 21Aug 1998.
63. Capy, P (1998) Evolutionary biology: A plastic genome. Nature 396:522-523.
64. Brant, J, et al (2005) Transposable elements as a source of genetic innovation: expression and evolution of a family of retrotransposon-derived neogenes in mammals. Gene, 345:101-111.
65. Pennisi, E (1998) Molecular Evolution: How the Genome Readies Itself for Evolution. Science 28:1131-1134.

CHAPTER 18 NEW RELATIONSHIPS CREATES NEW FORMS

1. Lewin, R (1983) Origin of species in stressed environments. Science 222:1112-1124.
2. Valentine, JW (1992) The macroevolution of phyla. Topics in Geobiology,10:525-553.
3. Whittington, HB (1985) The Burgess Shale. YUP.
4. Gordon MS (1999) The Concept of Monophyly: A Speculative Essay. Biology and Philosophy,14: 331-348. "Recent research make it seem improbable that there could have been single...forms for many...categories of evolution... common descent apparently does not apply to kingdoms, it probably does not apply to many, if not all, phyla, and possibly... many classes."
5. Browne, J (2008) Birthdays to remember. Nature, 456: 324-325.
6. Heimberg, AM, et al (2010) microRNAs reveal the interrelationships of hagfish, lampreys, and gnathostomes and the nature of the ancestral vertebrate. PNAS, 107:19379-19383.
7. Philippe, H (2011) Acoelomorph Flatworms Are Deuterostomes Related to Xenoturbella. Nature, 470:255-258.
8. Keim, B (2011) Animal Kingdom Gains Phylum But Loses Link. Wired, 09 Feb 2011.
9. Gould, SJ (1982) Is a new and general theory of evolution emerging? in Evolution Now: A Century After Darwin, JM Smith (ed).Freeman, NY p 140.
10. Keim, B (2010) Hagfish Analysis Opens Major Gap in Tree of Life. Wired, 19 Oct 2010.
11. Mayr, E (1982) The Growth of Biological Thought: Diversity, Evolution, and Inheritance. HUP. p 524.
12. Raup, DM (1981) Evolution and the Fossil Record. Science 213:289.
13. Eldredge, N and I Tattersall (1982) The Myths of Human Evolution, Columbia University Press. p 163.
14. Eldredge, N and I Tattersall (1982) The Myths of Human Evolution. op site. p 163.Emphasis added.
15. Goldschmidt, RB (1952) Evolution, As Viewed By One Geneticist. American Scientist, 40:84-94; Gould, SJ (1977) Evolution's Erratic Pace. Natural History, vol.86, no 5.
16. Eldredge, N and SJ Gould (1972) Punctuated equilibria: an alternative to phyletic gradualism. In Models in Paleobiology, TJM Schopf (ed). Freeman Cooper, San Francisco. p 82-115.
17. Smith, JM (1995) Genes, Memes, & Minds. NYRB, 30 Nov 1995.
18. Ridley M (1981)Who doubts evolution? New Scientist, 90:831.
19. Gould, SJ (1997) Darwinian Fundamentalism. NYRB, Vol 44, N 10.
20. Davidsion, K (2002) Theory still rocks scientists' equilibrium. San Francisco Chronicle, 27 May 2002. Jablonski is head of the Committee on Evolutionary Biology, University of Chicago.
21. Horner, JR, DJ Varricchio and MB Goodwin (1992) Marine transgressions and the evolution of Cretaceous dinosaurs. Nature, 358:59-61.
22. Kemp, TS (2005) The Origin and Evolution of Mammals. OUP. p 83.
TS Kemp (1982) The Reptiles that Became Mammals. New Scientist, 93:581-584
23. Barabási, A-L (2005) The origin of bursts and heavy tails in human dynamics, Nature 207:435.
24. Schwartz, J (1999) Sudden Origins: Fossils, Genes, and the Emergence of Species. Wiley, NY. p 3.
25. Erwin DH. (2000) Macroevolution is more than repeated rounds of microevolution. Evol Dev 2:78–84.
26. Cuvier, G, (1816) Le règne animal distribué d'après son organisation, pour servir de base à l'histoire naturelle des animaux et d'introduction à l'anatomie comparée. "Natural history...has a rational principle...The different parts of each being must be coordinated in such a way as to make possible the total being, and the analysis of these conditions often leads to general laws."
27. Rupke, N (1994) Richard Owen: Victorian Naturalist, YUP. p 172 ff.
28. Bak, P., C Tang, and K Wiesenfeld (1987) Self-organized criticality: an explanation of 1 / f noise. Physical Review Letters, 59: 381–384.
29. Jensen, MH (2002) Obituary: Per Bak (1947–2002). Nature 420: 284.
30. Pfister, C. (2001) History of Climate. IESBS p 2000-03, Pfister at the University of Bern, Switzerland.
31. Sole, RV and SC Manrubia (1995) Are rainforests self- organized in a critical state? J Theoretical Biology, 173:31-40.
32. Sole, RV and O Miramontes (1995) Information at the edge of chaos in fluid neural networks. Physica D 80:171- 180.

33. Bak, P and M Paczuski (1995) Complexity, contingency and criticality. PNAS, 92:6689–6696.
34. Bak, P (1996) How Nature Works: The Science of Self-Organized Criticality. Copernicus, NY.
35. The idea that all elements must be balanced is a central concept of two of biology's most important modern developments: systems biology and "Evo-Devo." Systems biology, now widely adopted, says all biological processes must be view within a whole that includes all other processes (consistent with "systems theory" in the other physical sciences). This means all traits are functionally linked, and thus must evolve in coordination with each other. Evo-Devo (evolutionary developmental biology) says something similar, only focusing on the connections between evolutionary change and the development processes. For example, Evo-Devo pioneer Wallace Arthur, in *The Origin of Animal Body Plans: A Study in Evolutionary Developmental Biology* (1997), says that increased complexity (with more parts to coordinated), creates more constraints on change and less chance those changes can be incorporated into the development process.
Similar ideas are emerging throughout biology. For example, in a remarkable analysis Tom Kemp, Curator of Zoology at Oxford University, describes "the systems-related concept of correlated progression, in which all the traits are functionally linked." Kemp believes this is the "basis of evolutionary origin of new taxa [what we commonly call new species]." Kemp, TS (2007) The concept of correlated progression as the basis of a model for the evolutionary origin of major new taxa. Proceedings of the Royal Society B274:1667-1673. [Kemp believes these correlated changes must occur in tiny increments, which may be true of some adaptive changes, but not the emergence of new species where speed is essential, otherwise biological chaos and death quickly ensue.]
36. Salazar-Ciudad, I and J Jernvall (2005) Graduality and innovation in the evolution of complex phenotypes: insights from development. J of Experimental Zoology Part B: Molecular and Developmental Evolution, 304B:619 -631.
37. Camazine, S, J-L Deneubourg, N Franks, J Sneyd, G Theraula, E Bonabeau (2003) Self-Organization in Biological Systems (Princeton Studies in Complexity), PUP. p 7.
38. Conway Morris, S (2003) Chance and Design. In Our Time, BBC Radio 4.
39. El Albani, A, S Bengtson, et al (2010) Large colonial organisms with coordinated growth in oxygenated environments 2.1 Byr ago. Nature, 466:100-104.
40. Valentine, JW, SM Awramik, PW Signor and PM Sadler (1991) The biological explosion at the Precambrian-Cambrian boundary. Evolutionary Biology, 25:279-356;
Valentine, JW (1992) The macroevolution of phyla. Topics in Geobiology,10:525-553.
41. Wignall, PB, et al. (1996) The timing of palaeo-environmental changes at the Permo-Triassic (P/Tr) boundary using conodont biostratigraphy. Hist. Biol. 12:39-62.
Erwin, DH (1993) The Great Paleozoic Crisis: Life and Death in the Permian. Columbia University Press.
42. Wagner, GP, C-H Chiu, and M Laubichler (2000) Developmental Evolution as a Mechanistic Science: The Inference from Developmental Mechanisms to Evolutionary Processes. American Zoologist, 40:819-831.
43. Koonin, EV (2007) The Biological Big Bang model for the major transitions in evolution. Biology Direct 2:21 http://www.biology-direct.com/content/2/1/21
44. Stanley, S (1998) Macroevolution: Pattern and Process. TJUP.
Stanley, S (1984) The New Evolutionary Timetable: Fossils, Genes, and the Origin of Species. Basic Books, NY.
45. Kirchner, JW and A Weil (2005) Fossils make waves. Nature 434:147-148.
46. Takhtajan, A (1991) Evolutionary Trends in Flowering Plants. Columbia University Press. pg. 7.
Cui, L, CW dePamphilis, et al (2006) Widespread genome duplications throughout the history of flowering plants. Genome Res.16: 738-749.
47. Keim, B (2010) Hagfish Analysis Opens Major Gap in Tree of Life. Wired, 19 Oct 2010. "There are more microRNAs acquired at the origin of vertebrates than at any other time in animal evolutionary history."
48. Valentine, JW, D Jablonski and DH Erwin (1999) Fossils, molecules and embryos: new perspectives on the Cambrian explosion. Development,126:851-859.
49. Biémont, C and C Vieira (2006) Junk DNA as an Evolutionary Force. Nature 443:521.
50. Pennisi, E (1998) Molecular Evolution: How the Genome Readies Itself for Evolution. Science 28:1131-1134.
51. Frost, LS, R Leplae, AO Summers and A Toussaint (2005) Mobile genetic elements: the agents of open source evolution. Nature Reviews Microbiology 3:722-732.
52. Lin, R, L Ding, C Casola, DR Ripoll, C Feschotte and H Wang (2007) Transposase-Derived Transcription Factors Regulate Light Signaling in Arabidopsis. Science, 23:1302-1305.
53. Solé, RV, P Fernández and SA Kauffman (2003) Adaptive walks in a gene network model of morphogenesis: insights into the Cambrian explosion. Int. J. Dev. Biol 47:685–693.

CHAPTER 19 DARWINISTS RESPOND TO THESE DISCOVERIES

1. Highfield, R (2009) Uprooting Darwin's Tree. New Scientist, 21 Jan 2009.
2. Smith, JM (1995) Life at the Edge of Chaos? NYRB 2 Mar 1995.
3. Dawkins, R (2000) Unweaving the Rainbow: Science, Delusion and the Appetite for Wonder. Houghton Mifflin, NY. p 187-188. "Recently my incoming mail has registered a sharp rise in the normal load of 'chaos theory', 'complexity theory', 'non-linear criticality' and similar phrases. Now I'm not saying that these correspondents lack the faintest, foggiest clue what they are talking about. But I will say it's hard to discover whether they do. New Age cults of all kinds are swimming in bogus scientific language, regurgitated, half-understood (no, less than

half) jargon."
4. Padian, K (2008) Darwin's enduring legacy. Nature 451:632-634.
5. Eldredge, N (1995) Reinventing Darwin: The Great Debate at the High Table of Evolutionary Theory. Wiley, NY.
6. Darwin, C (1859 [1998]) On the Origin of Species By Means of Natural Selection. Kindle Ed. Loc. 3914-3915.
Darwin, C (1872 [1988]) Origin of the Species, 6th ed. New York University Press. p 154.
7. Darwin, C (1859 [1998]) On the Origin of Species By Means of Natural Selection. Kindle Ed. Loc. 3477-3480.
8. Zahavi, A and Zahavi, A (1997) The handicap principle: a missing piece of Darwin's puzzle. OUP.
9. Williams GC (1966) Adaptation and Natural Selection: A Critique of Some Current Evolutionary Thought. PUP. p139. Emphasis added.
10. Dawkins, R (2006) The Selfish Gene, 30th anniversary edition. OUP. p 2.
11. Dawkins, R (2006) The Selfish Gene, 30th anniversary edition. OUP. p 7.
12. Dawkins, R (1984) The Extended Phenotype: The Gene as a Unit of Selection. OUP. p 164-165.
13. Dawkins, R (1986) The Blind Watchmaker. Norton, NY. p 300. "The Darwinians maintains that the only reason even a minority of acquired characters are improvements is that there is an underpinning of past Darwinian selection. In other words, the Lamarckian theory can explain adaptive improvement in evolution only by.. riding on the back of the Darwinian theory."
14. Dawkins, R. (2007) Review of Michael Behe's The Edge of Evolution. NYT, 1 July 2007.
15. Fisher, RA (1930 [2000]) The Genetical Theory of Natural Selection. OUP , p. 37.
"[Framing] the principle of Natural Selection in the form of a theorem... puts us in a position to judge of the validity of the objection which has been made, that the principle of Natural Selection depends on a succession of favourable chances... It is easy... to perceive the difference between a succession of favourable deviations from the laws of chance, and on the other hand, the continuous and *cumulative action* of these laws. It is on the latter that the principle of Natural Selection relies." [emphasis added] I.e., natural selection depends on the *cumulative effect* of a series of random events, each of which aids (is "favorable" to) survival. That, of course does not change the Darwinist claim that the changes themselves occur randomly.
16. Dobzhansky, T (1937 [1982]) Genetics and the Origin of Species, Columbia University Press. p 22
17. Dobzhansky, T (1950) The genetic basis of evolution. Sci. Am.182:32-41.
18. Rando, O and K Verstrepen (2007) Timescales of genetic and epigenetic inheritance. Cell 128: 655
19. Dawkins, R (1985) What Was All The Fuss About? Nature, 316:683-684.
20. Orr, HA (2007) A Mission to Convert. NYRB, 11 Jan 2007.
21. Wade, N (2009) Evolution All Around. NYT 8 Oct 2009.
22. Smith, JM and E Szathmary (1998) The Major Transitions and Evolution. OUP. p 145. Emphasis added.
23. Crick, F and LE Orgel (1973) Directed Panspermia. Icarus, 19:341-346.
24. Crick, F and LE Orgel (1973) Directed Panspermia. Icarus, 19:341-346. Crick writes that "several... explanations of the universality of the code can be suggested, but none is generally accepted to be completely convincing." That, of course is only true if you think in Darwinian terms. Self-assembly predicts that naturally forming relationships will push evolution along a very narrow path, and a tiny number of possibilities.
25. Mann, C (1991) Lynn Margulis: Science's Unruly Earth Mother. Science, 252:379-381.
26. Levy, ED, EB Erba, CV Robinson and SA Teichmann (2008) Assembly reflects evolution of protein complexes. Nature, 453:1262-1265. Explains how the evolution of proteins can be predicted solely from their atomic structure.

CHAPTER 20. A POWERFUL NEW MODEL OF EVOLUTION: ENERGY, EYES AND EVOLUTIONARY CLOCKS

1. Pennisi, E (2008) Modernizing the Modern Synthesis. Science, 321:196-197.
2. There are a few deep sea creatures which gain their energy through another process. But they have the same remarkable ability to transform energy into matter, and then back again into energy.
3. Dunlap, JC (2006) Running a Clock Requires Quality Time Together. Science 311:184-186.
4. Hansford, D (2007) Moonlight Triggers Mass Coral "Romance. Nat Geo. News, 22 Oct 2007.
5. Levy, O, L Appelbaum, W Leggat, Y Gothlif, DC Hayward, DJ Miller, and O Hoegh-Guldberg (2007) Light-Responsive Cryptochromes from a Simple Multicellular Animal, the Coral Acropora millepora. Science, 318:467-470.
6. Young, MW (2000) Circadian Rhythms: Making Time for a Kingdom. Science, 288:451-453.
7. Carroll, SB (2006) Endless Forms Most Beautiful. op cite. p 72.
8. Plachetzki,DC, CR Fong and TH Oakley (2013) Cnidocyte discharge is regulated by light and opsin-mediated phototransduction. BMC Biology (in press).
9. Arendt, D (2003) Evolution of eyes and photoreceptor cell types. Int. J. Dev. Biol. 47:563-571.
10. Shubin, N (2009) Your Inner Fish: A Journey into3.5 Billion-Year History of the Human Body. Vintage, NY.
11. Gallessich, G (2009) Scientists Find That Squid's Bioluminescence Comes From Eye-Related Genes Public Affairs Department, Univ of California Santa Barbara, 11 June 2009.
12. Gallessich, G (2012) UC Santa Barbara Researchers Discover Genetic Link Between Visual Pathways of Hydras and Humans. Univ of California, Santa Barbara, 5 March 2012.
13. Borowsky, R (2008) Restoring sight in blind cavefish. Current Biology 18:R23-R24.
14. Rose, MR and TH Oakley (2007) The New Biology: Beyond the Modern Synthesis. Biology Direct, 2:30.

15. Burian, R (2000) General Introduction to the Symposium on Evolutionary Developmental biology: Paradigms, Problems, and Prospects. American Zoologist, 40:711-717.
16. Mogilner, A and G Oster (2003):Enhanced: Shrinking Gels Pull Cells. Science, 302:1340-1341.
17. Pourquie, O (2003) Vertebrate somitogenesis: a novel paradigm for animal segmentation? Int. J. Dev. Biol. 47:597.
18. Vavylonis D, J-Q Wu, S Hao, B O'Shaughnessy, TD Pollard (2007) Assembly Mechanism of the Contractile Ring for Cytokinesis by Fission Yeast. Science, published online 13 Dec 2007.
19. Shubin, N (2009) Your Inner Fish: A Journey into3.5 Billion-Year History of the Human Body. Vintage, NY.
20. Hogeweg, P (2002) Interlocking of self-organization and evolution. Theoretical Biology and Bioinformatics Group, University of Utrecht.
21. Nowak, MA (2008) Generosity: A winner's advice. Nature, 456, 579.
22. Conway Morris, S (2000) The Crucible of Creation: The Burgess Shale and the Rise of Animals. OUP. p 14.

CHAPTER 21 A POWERFUL NEW MODEL OF EVOLUTION: EXPLAINING, INTEGRATING AND PREDICTING

1. Shapiro, JA (2007) Bacteria are small but not stupid: cognition, natural genetic engineering and socio-bacteriology, Studies in Hist. and Phil. of Biology and Biomedical Science, 38:807-819.
2. Pearlman, S (2012) Oldest living thing on earth discovered. Telegraph, 07 Feb 2012.
3. Finkel, T, M Serrano and MA Blasco (2007) The common biology of cancer and ageing. Nature, 448:767-774.
4. Antonovics, J (1987) The Evolutionary Dys-Synthesis. op site.
5. Darwin, C (1862) On the various contrivances by which British and foreign orchids are fertilised by insects.
6. Minkel, JR (2006) Most Desirable Mates May Not Sire Prolific Offspring. Sci Am, 24 Oct 2006.
7. Pischedda A, AK Chippindale (2006) Intralocus Sexual Conflict Diminishes the Benefits of Sexual Selection. PLoS Biol 4(11): e356.
8. Bernstein, H (1983) Recombinational repair may be an important function of sexual reproduction. Bioscience, 33:326-331.
Bernstein H, RE Michod, et al (1985) Genetic damage, mutation, and the evolution of sex. Science, 229:1277-81.
Bernstein H, FA Hopf, RE Michod (1987) The molecular basis of the evolution of sex. Adv Genet. 24:323-70.
9. Chasteen, S (2003) Gender scientists explore a revolution in evolution. Stanford Univ. News Service, 19 Feb 2003.
10. Levy, D (2003) Sex and gender scientists explore a revolution in evolution. Stanford Univ. News Service, 16 Feb 2003.
11. Chasteen, S (2003) Gender scientists explore a revolution in evolution. Stanford Univ. News Service, 19 Feb 2003.
12. Harrison, CJ, et al (2005) Independent recruitment of a conserved developmental mechanism during leaf evolution. Nature, 434:509-514.
13. Boyce CK and AH Knoll (2002) Evolution of developmental potential and the multiple independent origins of leaves in Paleozoic vascular plants. Paleobiology, 28:70-100.
14. Bak, P (1996) How Nature Works: The Science of Self-Organized Criticality. Copernicus, NY.
Smalley Jr, RF, DL Turcotte and SA Solla (1985) A renormalization group approach to the stick-slip behavior of faults. J. of Geophysical Research, 90: 1894.
Turcotte, DL (1997) Fractals and Chaos in Geology and Geophysics. CUP.
Adami, C (1995) Self-organized criticality in living systems. Physics Letters A, 203: 29–32.
Poil, SS, A van Ooyen, K Linkenkaer-Hansen (Jul 2008) Avalanche dynamics of human brain oscillations: relation to critical branching processes and temporal correlations. Human Brain Mapping, 29 :770–7.
Kitzbichler, MG, ML Smith, SR Christensen and E Bullmore (2009). Broadband Criticality of Human Brain Network Synchronization. PLoS Comput Biol 5 (3): e1000314.
de Arcangelis, L., C Perrone-Capano and HJ Herrmann (2006) Self-organized criticality model for brain plasticity. Phys. Rev. Lett, 96.
Maslov, S (2000) Simple Model of a limit order-driven market. Physica A, 278:571-578.
Kron, T and T Grund (2009) Society as a Self-organized Critical System. Cybernetics and Human Knowing 16: 65–82.
Maslov, S (2000) Simple Model of a limit order-driven market. Physica A, 278:571-578.
Bak, P and K Sneppen (1993) Punctuated equilibrium and criticality in a simple model of evolution. Physical Review Letters, 71:4083–4086.
Krink, T and R Tomsen (2001) Self-Organized Criticality and Mass Extinction in Evolutionary Algorithm. Proc. IEEE Int Conf on Evolutionary Computing, 2001:1155-1161
15. Strohman, R (2000) Genetic Determinism as a Failing Paradigm in Biology and Medicine. In Promoting Human Wellness: Frontiers for Research, Practice and Policy, MS Jamner and D Stokols (eds). University of California Press.
16. Gould, SJ (1977) Ontogeny and Phylogeny. HUP.
17. Raff, RA (2000) Evo-Devo: the evolution of a new discipline. Nature Reviews Genetics 1:74-79.
18. Orr, HA (2005) Turned On: A revolution in the field of evolution? The New Yorker, 24 Oct 2005.
19. Owen, R (1848) The Archetype and Homologies of the Vertebrate Skeleton. John van Voorst, London. p. 172 "specific organizing principle or force would...mold specific forms."

20. Owen, R (1849) On the Nature of Limbs. John van Voorst, London. p 2. The "essential character of a part belongs to it [and establishes] it's relation to a predetermined pattern," and that "archetype or primal pattern is the basis supporting all [future] modifications."
Also, Rupke, N (1994) Richard Owen: Victorian Naturalist. YUP. p 172 ff.
21. Miller, SP, M Lunzer, AM Dean (2006) Direct Demonstration of an Adaptive Constraint. Science, 314:458-461.
22. Conant, GC and A Wagner (2003) Convergent evolution of gene circuits. Nat Genet, 34:264-266.
Paul, S, et al (2008) Molecular signature of hypersaline adaptation: insights from genome and proteome composition of halophilic prokaryotes. Genome Biology 2008, 9:R70.
23. Lopez, D, MA Fischbach, F Chu, R Losick, R Kolter (2009) Structurally diverse natural products that cause potassium leakage trigger multicellularity in Bacillus subtilis. PNAS, 106:280–285.
24. Gough, NR (2008) Complexity in the Bacterial Community. Sci. Signal. 1 (14), ec125.
25. Vlamakis, H, C Aguilar, R Losick, R Kolter (2008) Control of cell fate by the formation of an architecturally complex bacterial community. Genes Dev. 22: 945-953.
"Bacteria form architecturally complex communities known as biofilms in which cells are held together by an extracellular matrix...Here we... present evidence that formation of the biofilm governs differentiation. We show that motile, matrix-producing, and sporulating cells localize to distinct regions within the biofilm... Importantly, mutants that do not produce extracellular matrix... are deficient in sporulation. We propose that sporulation is a culminating feature of biofilm formation, and that spore formation is coupled to the formation of an architecturally complex community of cells."
Newman SA, HL Frisch (1979) Dynamics of skeletal pattern formation in developing chick limb. Science 205:662–668.
Newman SA, Frenz DA, Tomasek JJ, Rabuzzi DD (1985). "Matrix-driven translocation of cells and nonliving particles". Science 228 (4701): 885–889.
Newman SA, Forgacs G, Müller GB (2006). "Before programs: the physical origination of multicellular forms". Int J Dev Biol 50 (2-3): 289–299.
Newman SA, R Bhat (2009) Dynamical patterning modules: a 'pattern language' for development and evolution of multicellular form. Int J Dev Biol 53 693–675..
Zhu J, Zhang YT, Alber MS, Newman SA (2010). "Bare bones pattern formation: a core regulatory network in varying geometries reproduces major features of vertebrate limb development and evolution". PLoS ONE 5 (5): e:10892.
Moss, ML (1997) The functional matrix hypothesis revisited. 2. The role of an osseous connected cellular network. Am J Orthod Dentofacial Orthop.112:221-226.
26. Chuong C-M (2009). "Limb pattern, physical mechanisms and morphological evolution - an interview with Stuart A. Newman". Int J Dev Biol 53 (5-6): 663–671
27. Shapiro, MD, et al (2006) Parallel genetic origins of pelvic reduction in vertebrates, PNAS, 103:13753.
28. Cebra-Thomas, J, SF Gilbert, et al (2005) How the turtle forms its shell: a paracrine hypothesis of carapace formation. J. of Experimental Zoology Part B: Molecular and Developmental Evolution, 304B:558-569.
29. Shubin, N (2009) Your Inner Fish: A Journey into3.5 Billion-Year History of the Human Body. Vintage, NY.
30. Harrison, CJ, et al (2005) Independent recruitment of a conserved developmental mechanism during leaf evolution. Nature, 434:509-514
31. Boyce CK and AH Knoll (2002) Evolution of developmental potential and the multiple independent origins of leaves in Paleozoic vascular plants. Paleobiology, 28:70-100.
32. Sherwood, J (2002) Darwin's Time Machine: Scientists Begin Predicting Evolution's Next Step. Newsroom, University of Rochester, 20 Mar 2002

Part 5: How the New Evolution Changes Everything

1. Lawrence Berkeley National Laboratory staff (1999) Complex Systems: Science for the 21st Century. op cite.

Chapter 22 Healing, Health and Wellness

1. Barabási, A-L and ZN Oltvai (2004) Network biology: understanding the cell's functional organization. Nature Reviews Genetics 5:101-113.
2. Xu, X, Heimberg H, et al (2008) Cells Can Be Generated from Endogenous Progenitors in Injured Adult Mouse Pancreas. Cell 132:197-207.
3. Macchiarini, P, et al (2008) Clinical Transplantation of a Tissue-Engineered Airway. The Lancet, 372:2023–2030.
4. Rose, D (2008) Claudia Castillo gets windpipe tailor-made from her own stem cells. The Times (UK) 19 Nov 2008.
5. Fountain, H (2012) Doctors transplant synthetic windpipe in Maryland man. NYT. 12 Mar 2012.
6. Winslow, R (2012) To Fix a Heart, Doctors Train Girl's Body to Grow New Part. WSJ. 20 Mar 2012.
7. Kolata, G (2010) Glimpsing a Scientific Future as Fields Heat Up. NYT. 8 Nov 2010.
8. Editors (2012) Complex Body Parts Could Soon Be Lab-Grown. Beta News, 13 Dec 2012.
9. Swaminathan, N (2007) Fetal Neurons Still Operate in Adult Brain. Sci Am. 13 Sept 2007.
10. Cho, Y, R Shi, and RB Borgens, (2010) Chitosan produces potent neuro-protection and physiological recovery following traumatic spinal cord injury. J. Exp. Biol, 213:1513-1520.

11. Sargent, M (2008) Preparing for pandemics. Nature, 456:574.
12. Kadner, RJ. (2008) Bacteria. Encyclopædia Britannica.
13. Croucher, NJ, SD Bentley, et al (2011) Rapid Pneumococcal Evolution in Response to Clinical Interventions. Science, 331:430-434.
14. Enright, MC and BG Spratt (2011) The Genomic View of Bacterial Diversification. Science, 331:407-409.
15. Sargent, M (2008) Preparing for pandemics. Nature,456:574.
16. Wenner, M (2008) Virology: The Battle Within. Nature, 451:388-389.
17. Xu, X, Heimberg H, et al (2008) Cells Can Be Generated from Endogenous Progenitors in Injured Adult Mouse Pancreas. Cell 132:197-207.
18. Keim, B (2005) Whew! Your DNA Isn't Your Destiny. Wired 16 Aug 2005.
19. Lamm, E and E Jablonka (2008) The Nurture of Nature: Hereditary Plasticity in Evolution. Phil. Psych. 21:305-319.
    Lemke, H, A Coutinho and H Lange (2004) Lamarckian inheritance by somatically acquired maternal IgG phenotypes. Trends in Immunology, 25:180-186.
20. Lemke, H, A Coutinho and H Lange (2004) Lamarckian inheritance by somatically acquired maternal. op cite.
21. Riggins, GJ, et al (2005) Sequence survey of receptor tyrosine kinases reveals mutations in glioblastomas. PNAS, 102:14344-14349.
22. MacLeod, AR and MJ Szyf (1995) Expression of Antisense to DNA Methyltransferase mRNA Induces DNA Demethylation and Inhibits Tumorigenesis. Biol. Chem. 270:8037-8043.
23. Gangadharan, S, L Mularoni, J Fain-Thornton, SJ Wheelan, N. Craig (2010) DNA transposon Hermes inserts into DNA in nucleosome-free regions in vivo. PNAS 107:21966-21972.
    Evertts, A.G., C. Plymire, N.L. Craig, H.L. Levin (2007) The Hermes transposon of Musca domestica is an efficient tool for the mutagenesis of Schizosaccharomyces pombe. Genetics 177:2519-2523.
24. Editors (2005) "Jumping Gene" Helps Explain Immune System's Abilities. Science Daily, 7 Jan 2005.
25. Hayday, A, J Strid, M Girardi (2008) Nature Immunology. Identified *Rae-1* that activates natural anti-cancer response.
26. http://www.epigenomics.com/en/Press_Releases/2010/datednews/100407_Synlab.html
27. (2013) Predictive Biosciences launches prostate cancer test based on Epigenomics' DNA methylation marker. 29 Jan 2013.
28. Ornish, D, PR Carroll, et al (2008) Changes in prostate gene expression in men undergoing an intensive nutrition and lifestyle intervention" PNAS, 105:8369-8374.
29. Greer, EL, et al (2011) Transgenerational epigenetic inheritance of longevity in Caenorhabditis elegans. International Weekly Journal of Science,17 Nov 2011.
30. Varner, AE, WW Busse, RF Lemanske Jr (1998) Hypothesis: Decreased Use of Pediatric Aspirin Has Contributed to the Increasing Prevalence of Childhood Asthma. Annals of Allergy, Asthma & Immunology 81: 347-35.
31. Beasley, R, et al, (2008) Association between paracetamol use in infancy and childhood, and risk of asthma, rhinoconjunctivitis, and eczema in children aged 6-7 years: analysis from Phase Three of the ISAAC programme. The Lancet,372: 1039-1048.
32. Etminan M, M Sadatsafavi, S Jafari, M Doyle-Waters, K Aminzadeh, JM Fitzgerald. (2009) Acetaminophen use and the risk of asthma in children and adults: a systematic review and metaanalysis. Chest.136:1316-23.
33. BBC Horizon (2005) The Ghost in Your Genes.
34. Ness, AR (2004) The Avon Longitudinal Study of Parents and Children (ALSPAC)–a resource for the study of the environmental determinants of childhood obesity. European Journal of Endocrinology, 151, Suppl 3, U141-U149.
35. Jurisicova A, et al (2007) Maternal exposure to polycyclic aromatic hydrocarbons diminishes murine ovarian reserve via induction of Harakiri. J Clin Invest 17:3971-8.
    Klonisch, T, PA Fowler and S Hombach-Klonisch (2004) Molecular and genetic regulation of testis descent and external genitalia development. Developmental Biology, 270:1-18.
36. Pembrey, ME, LO Bygren, et al (2006) Sex-specific, male-line transgenerational responses in humans. European Journal of Human Genetics,14:59–166.
37. Roth, TL, FD Lubin, AJ Fun, and JD Sweatt (2009) Lasting Epigenetic Influence of Early-Life Adversity on the BDNF Gene. Biol Psychiatry, 65:760-9.
38. Shalev, I, T Moffitt, et al. (2012) Exposure to Violence During Childhood is Associated with Telomere Erosion from 5 to 10 Years of Age: A Longitudinal Study. Molecular Psychiatry, 24 Apr 2012.
39. Murgatroyd, C, D Spengler, et al (2009) Dynamic DNA methylation programs persistent adverse effects of early-life stress. Nature Neurosci.12:1559-1566.
40. McGowan, PO, M Szyf, MJ Meaney, et al (2009) Epigenetic regulation of the glucocorticoid receptor in human brain associates with childhood abuse. Nature Neuroscience,12:342-348.
41. Roth, TL. and JD Sweatt, (2010) Epigenetic marking of the BDNF gene by early-life adverse experiences, Horm. Behav. 59:315-20.
42. Gregory SG, JJ Connelly, et al (2009) Genomic and epigenetic evidence for oxytocin receptor deficiency in autism BMC Medicine, 7:62.
43. Arai J, S Li, DM, Hartley, LA Feig (2009) Transgenerational rescue of a genetic defect in long-term potentiation and memory formation by juvenile enrichment. J Neuroscience. 29:1496-1502
44. Alleyne, R (2009) Women who enjoy stimulating childhoods "have brighter children." Telegraph (London) 03 Feb.
45. Weaver, ICG, M Szyf, MJ Meaney, et al (2004) Epigenetic programming by maternal behavior,

Nature Neurosci.7:847-854.
46. Buchen, L (2011) Neuroscience: In their nurture, Nature, 467:146-148.
47. Henrich, J, S Heine, A Norenzayan (2010) The weirdest people in the world? Behavioral and Brain Sci., 33:61-135.
48. Hawks, J, et al (2007) Recent acceleration of human adaptive evolution. PNAS 104: 20753-20758.
49. Watters, E (2006) DNA is Not Destiny. Discover, Nov 2006.

CHAPTER 23 FOUR EMPOWERING IDEAS

1. Lawrence Berkeley National Laboratory staff (1999) Complex Systems: Science for the 21st Century. Lawrence Berkeley National Laboratory.
2. Laughlin, R, and D Pines (2000) The Theory of Everything. PNAS, 97:28-31.
3. Emery, C (2010) Plastic electronics could slash the cost of solar panels. News at Princeton, 30 March 2010.
4. MacPherson, K (2010) Princeton scientists find unusual electrons that go with the flow. News at Princeton, 14 July 2010.
5. Pease, R (2011) "Flying carpet" of conductive plastic takes flight. BBC News, 30 Sept 2011.
6. Princeton University Complex Materials Theory Group (2013) Optimal Multifunctional Materials and Minimal Surfaces.
7. Editors (2012) Inspiring interfaces. Nature Materials, 11:827.
8. Sober, E (1998) Three Differences between Deliberation and Evolution. In Modeling Rationality, Morality, and Evolution (Vancouver Studies in Cognitive Science), Peter Danielson (ed). OUP. p 408.
9. Jones, S (1997) Go Milk a Fruit Bat!, NYRB,17 July 1997.
10. Lewontin, RC (1983) The Corpse in the Elevator. NYRB, 20 Jan 1983.
11. Wynne-Jones, J. (2009) Poll reveals public doubts over Charles Darwin's theory of evolution. Telegraph, 31 Jan 2009.
12. Roughgarden, J (2006) Evolution and Christian Faith: Reflections of an Evolutionary Biologist. Island Press.
13. Lemaître, G (1950) The Primeval Atom - an Essay on Cosmogony, Van Nostrand, NY.
14. Mitton, S (2011) Fred Hoyle a life in science. CUP. p. 127.
15. Gallessich, G (2010) Genomic Sequencing of Marine Sponge Published; Revealed Early Genetic Complexity and Roots of Cancer. Office of Public Affairs and Communication, University of California, Santa Barbara. 4 Aug 2010.
16. Mischel, W, Y Shoda, and MI Rodriguez (1989) Delay of gratification in children. Science, 244: 933-938.

DEFINING AND PROVING DARWINIAN EVOLUTION

1. Ernst Mayr, in Schwartz, J (1999) Sudden Origins: Fossils, Genes, and the Emergence of Species. Wiley, NY. p 314.
"The development of the heavy [finch] bill was a slow process, probably involving dozens of small mutational steps, each one surviving only if proving its usefulness in the actual test of selection. It must be admitted, however, that it is a considerable strain on one's credulity to assume that finely balanced systems such as certain sense organs (the eye of vertebrates, or the bird's feather) could be improved by random mutations.... However, the objectors to random mutations have so far been unable to advance any alternative explanation that was supported by substantial evidence."
2. Pennisi, E (2004) Bonemaking Protein Shapes Beaks of Darwin's Finches. Science, 305:1383.
"Two research teams have discovered that a protein normally associated with the development of the skull and other bones is one of the molecules that tailors the shapes of beaks. Different shapes arise depending on where and when this signaling molecule, called bone morphogenic protein 4 (BMP4), is turned on during development, says Cheng-Ming Chuong, a evolutionary developmental biologist at the University of Southern California (USC) in Los Angeles."
3. Wu, P et al (2004) Molecular Shaping of the Beak. Science, 305:1465-1466.
Beaks are composed of multiple elements, e.g. the maxillary prominences, lateral nasal prominences, and mandibular prominence. "During development, these prominences are proportionally coordinated to compose a unique beak." The "proportional coordination" is achieved via complex molecular signaling networks [emphasis added]. E.g. BMP4 (bone morphogenic protein) expression level controls beak breadth, while CaM (calmodulin mediating calcium signaling) controls beak length.
4. Grant, PR and BR Grant (2002) Unpredictable Evolution in a 30-Year Study of Darwin's Finches. 296:707-711. "From 1972 to 2001, *Geospiza fortis* (medium ground finch) and *Geospiza scandens* (cactus finch) changed several times in body size and two beak traits.... Evolutionary responses of *G. fortis* to the two strongest selection episodes occurred in the following generations, as expected from the high heritabilities of the morphological traits."

THE DARWIN INDUSTRY

1. Padian, K (2008) Darwin's enduring legacy. Nature 451:632-634.
2. Palmer, D (2005) Earth Time: Exploring the Deep Past from Victorian England to the Grand Canyon. Wiley, NY. p 22, p 345 and p 402-3.
3. Darwin, C (1887) The Life and Letters of Charles Darwin: Including an Autobiographical Chapter, Volume 2, John Murray, London. p 185. (Chapter 5 has Huxley's recollections.)

4. Darwin, C (1887) The Life and Letters of Charles Darwin: Including an Autobiographical Chapter, Volume I, John Murray, London.
5. Moran, L (2012) Review: Evolution: A View from the 21st Century. Reports of the National Center for Science Education. May-June 2012 p. 27-30.
6. Shapiro, J (2012) Jerry Coyne Fails to Understand Yet Again. Huffington Post 10 Apr 2012.
7. Futuyma, DJ (2005) Evolution. Sinauer Associates, NY. p 11.
8. Wilkins, AS (2012) Review: Evolution: A View from the 21st Century by James A. Shapiro. Genome Biol Evol, 4:423-426. Emphasis added.

Index

D

H

O

P

Q

R

S

X

Y

Z

Made in the USA
Monee, IL
04 November 2026

67382842R00233